Basics of CMOS Cell Design

Basics of CMOS Cell Design

Etienne Sicard

Professor
INSA Electronic Engineering School of Toulouse, France

Sonia Delmas Bendhia

Senior Lecturer
INSA Electronic Engineering School of Toulouse, France

McGraw-Hill

New York   Chicago   San Francisco   Lisbon   London
Madrid   Mexico City   Milan   New Delhi   San Juan
Seoul   Singapore   Sydney   Toronto

Copyright © 2007 by The McGraw-Hill Companies, Inc. All rights reserved. Printed in the United States of America. Except as permitted under the United States Copyright Act of 1976, no part of this publication may be reproduced or distributed in any form or by any means, or stored in a database or retrieval system, without the prior written permission of the publisher.

1 2 3 4 5 6 7 8 9 10   FGR/FGR   0 1 3 2 1 0 9 8 7

ISBN-13: 978-0-07-148839-6
ISBN-10: 0-07-148839-1

The sponsoring editor for this book was Stephen S. Chapman and the production supervisor was Pamela A. Pelton. It was set in Times.

Printed and bound by Quebecor/Fairfield.

This book was previously published by Tata McGraw-Hill Publishing Company Limited, New Delhi, India, copyright © 2007.

McGraw-Hill books are available at special quantity discounts to use as premiums and sales promotions, or for use in corporate training programs. For more information, please write to the Director of Special Sales, McGraw-Hill Professional, Two Penn Plaza, New York, NY 10121-2298. Or contact your local bookstore.

This book is printed on acid-free paper.

In
memory of
John Uyemura

Preface

This book introduces the design and simulation of CMOS integrated circuits in an attractive way thanks to the user-friendly PC tool Microwind. The lite version of Microwind can be downloaded from *http://www.microwind.net*.

The chapters of this book have been summarized below. Chapter One describes the technology scale-down and the major improvements allowed by deep sub-micron technologies. Chapter Two is dedicated to the presentation of the single MOS device, with details on simulation at the logic and layout levels. The modelling of the MOS devices is introduced in Chapter Three. Chapter Four presents the CMOS Inverter, the 2D and 3D views, and the comparative design in micron and deep-submicron technologies. Chapter Five deals specifically with interconnects, with information on the propagation delay and several parasitic effects. Chapter Six deals with the basic logic gates (AND, OR, XOR, complex gates). Chapter Seven delineates the arithmetic functions (Adder, comparator, multiplier, ALU). The latches and counters are detailed in Chapter Eight. In Chapter Nine, analog cells are presented, including voltage references, current mirrors, and the basic architecture of operational amplifiers.

The detailed explanation of the design rules is given in Appendix A. The details of all commands are given in Appendix B for the tool Microwind, and in Appendix C for the tool Dsch. Appendix D includes a quick reference sheet for Microwind and Dsch.

A second book includes an extensive presentation of analog cells, radio-frequency analog blocks, analog-to-digital and digital-to-analog converter principles, input/output interfacing, an introduction to silicon insulator technology, and a prospective discussion about the future developments in microelectronics.

About Microwind and Dsch

The present book introduces the design and simulation of CMOS integrated circuits, and makes an extensive use of PC tools Microwind and Dsch. These tools are under the licence of ni2Design, India. The lite version 3 of the tools are available for free download at *http://www.microwind.net*.

The latest developments on MICROWIND and DSCH can be found at *http://www.microwind.org*. The commercial site for the tools is *http://www.microwind.net*.

Etienne Sicard
etienne.sicard@insa-toulouse.fr

Sonia Delmas Bendhia
sonia.bendhia@insa-toulouse.fr

Acknowledgments

We would like to thank our former colleagues, Jean-Francois Habigand, Kozo Kinoshita and Antonio Rubio, for their support throughout the development of the Microwind, Dsch tools. We would like to thank Joseph-Georges Ferrante for having faith in our ability to drive ambitious microelectronics research projects, and for having provided us continuous support over the last ten years. Productive technical discussions with Jean-Pierre Schoellkopf, Amaury Soubeyran, Thomas Steinecke, Gert Voland and Jean-Louis Noullet are also gratefully acknowledged.

Special thanks are due to the technical contributors to Dsch and Microwind software (Chen Xi, Jianwen Huang), to our colleagues at INSA who always supported this work, and to numerous professors, students and engineers who patiently debugged the technical contents of the book and the software, and gave valuable comments and suggestions. Also, we would like to thank Marie-Agnes Detourbe for having carefully reviewed the manuscript, and ni2design for the active promotion of the tools.

Finally, we would like to acknowledge our biggest debt to our parents and to our companions for their constant support.

ETIENNE SICARD

SONIA DELMAS BENDHIA

Contents

Abbreviations and Symbols

MULTIPLIERS

Value	Name	Standard Notation
10^{18}	*PETA*	P
10^{15}	*EXA*	E
10^{12}	*TERA*	T
10^9	*GIGA*	G
10^6	*MEGA*	M
10^3	*KILO*	K
10^0	–	–
10^{-3}	*MILLI*	m
10^{-6}	*MICRO*	u
10^{-9}	*NANO*	n
10^{-12}	*PICO*	p
10^{-15}	*FEMTO*	f
10^{-18}	*ATTO*	a
10^{-21}	*ZEPTO*	z

PHYSICAL CONSTANTS AND PARAMETERS

Name	Value	Description
ε_0	$8.85\,e^{-12}$ Farad/m	Vacuum dielectric constant
$\varepsilon_r\ SiO_2$	$3.9 - 4.2$	Relative dielectric constant of SiO_2
$\varepsilon_r\ Si$	11.8	Relative dielectric constant of silicon
ε_r ceramic	12	Relative dielectric constant of ceramic
k	$1.381e^{-23}$ J/°K	Boltzmann's constant
q	$1.6e^{-19}$ Coulomb	Electron charge
μ_n	600 V.cm^{-2}	Mobility of electrons in silicon
μ_p	270 V.cm^{-2}	Mobility of holes in silicon
σ_{al}	$36.5\ 10^6$ S/m	Aluminum conductivity
σ_{si}	4×10^{-4} S/m	Silicon conductivity
n_i	1.02×10^{10} cm^{-3}	Intrinsic carrier concentration in silicon at 300°K
r_{al}	0.0277 W.μm	Aluminum resistivity
g_{cu}	58×10^6 S/m	Copper conductivity
ρ_{cu}	0.0172 W.μm	Copper resistivity
$\rho_{tungstène\ (W)}$	0.0530 W.μm	Tungsten resistivity
$\rho_{or\ (Ag)}$	0.0220 W.μm	Gold resistivity
μ_0	$1.257e^{-6}$ H/m	Vacuum permeability
T	300°K (27°C)	Operating temperature

Basics of CMOS Cell Design

Introduction

The evolution of integrated circuit (IC) fabrication techniques is a unique fact in the history of modern industry. There have been steady improvements in terms of speed, density and cost for more than 30 years. In this chapter, we present some information illustrating the technology scale down.

1.1 General Trends

Inside general purpose electronics systems such as personal computers or cellular phones, we may find numerous integrated circuits (IC), placed together with discrete components on a printed circuit board (PCB), as shown in Figure 1.1. The integrated circuits appearing in this figure have various sizes and complexity. The main core consists of a microprocessor and a digital signal processor (DSP) considered as the heart of the system, that includes several millions of transistors on a single chip. The push for smaller size, reduced power supply consumption and enhancement of services, has resulted in continuous technological advances, with the possibility of ever higher integration.

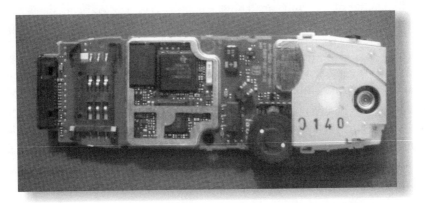

Fig. 1.1 Photograph of the internal parts of a cellular phone

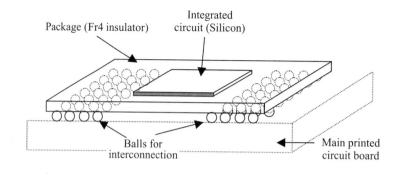

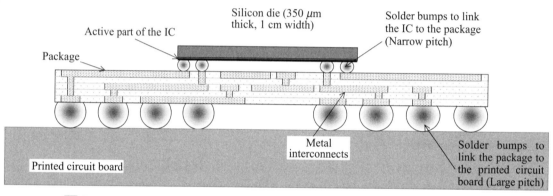

Fig. 1.2   Typical structure of an integrated circuit mounted on a Ball Grid Array (BGA)

The integrated circuit consists of a silicon die, with a size of usually around 1 cm × 1 cm in the case of microprocessors and memories. The integrated circuit is mounted on a package (Figure 1.2), which is placed on a printed circuit board. The active part of the integrated circuit is only a very thin portion of the silicon die. At the border of the chip, small solder bumps serve as electrical connections between the integrated circuit and the package. The package itself is a sandwich of metal and insulator materials, that convey the electrical signals to large solder bumps, which interface with the printed circuit board.

Around eight decades separate the user's equipment size (such as a mobile phone in Figure 1.3) and the basic electrical phenomenon, consisting in the attraction of electrons through an oxide. Inside the electronic equipment, we may see integrated circuits and passive elements sharing the same printed circuit board (1 cm scale), wire connections between the package and the die (1 mm scale), input/output structures of the integrated circuit (100 μm scale), the integrated circuit layout (10 μm), a vertical cross-section of the process, revealing a complex stack of layers and insulators (1 μm scale), the active device itself, called MOS transistor (which stands for Metal Oxide semiconductor).

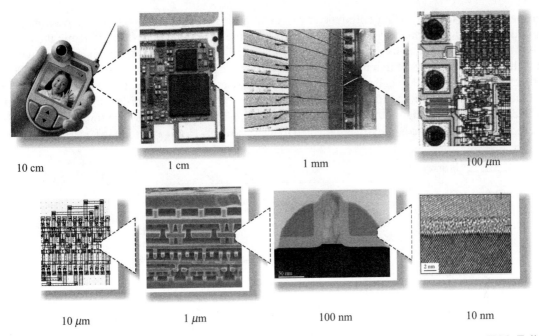

Fig. 1.3 Patterns representative of each scale decade from 10 cm to 10 nm (*Courtesy:* IBM, Fujitsu)

Figure 1.4 describes the evolution of the complexity of Intel® microprocessors in terms of number of devices on the chip [Intel]. The Pentium IV processor produced in 2003 included about 50,000,000 MOS devices integrated on a single piece of silicon not larger than 2 × 2 cm.

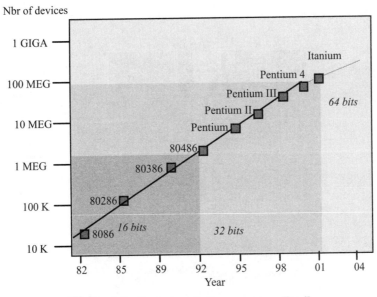

Fig. 1.4 Evolution of microprocessors [Intel]

Since the 1 Kilo-byte (Kb) memory produced by Intel in 1971, semiconductor memories have improved both in density and performances, with the production of the 256 Mega byte (Mb) dynamic memories (DRAM) in 2000, and 1Giga-byte (Gb) memories in 2004 (Figure 1.5) [Moore]. In other words, within around 30 years, the number of memory cells integrated on a single die has been multiplied by 1,000,000. Another type of memory chip called Flash memory has become very popular, due to its capabilities to retain the information without supply voltage (non-volatile memories are described in the second book "Advanced CMOS cell design"). According to the international technology roadmap for semiconductors [Itrs], the DRAM memory complexity is expected to increase up to 16 Giga-bytes (Gb) in 2008.

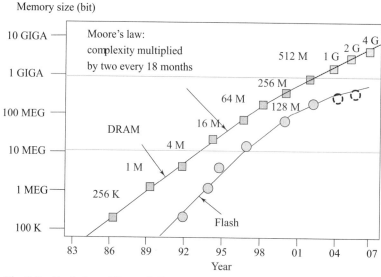

Fig. 1.5   Evolution of Dynamic RAM and Flash semiconductor memories [Itrs]

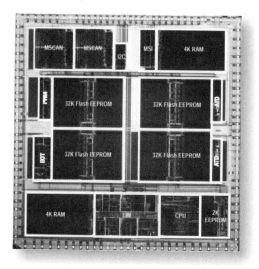

Fig. 1.6   Bird's eye view of a micro-controller die (*Courtesy:* Motorola Semiconductors)

The layout aspect of the die of an industrial micro-controller is shown in Figure 1.6 [Freescale]. This circuit is fabricated in several millions of samples for automotive applications. The micro-controller core is the central process unit (CPU), which uses several types of memory: the Electrically Erasable Read-Only Memory (EEPROM), the FLASH Memory (Rapidly Erasable Read-Only Memory) and the RAM Memory (Random Access Memory). Some controllers are also embedded in the same die: the Control Area Network (MSCAN), the debug interface (MSI), and other functional cores (ATD, ETD).

1.2   The Device Scale Down

We consider four main generations of integrated circuit technologies: micron, submicron, deep submicron and ultra deep submicron technology, as illustrated in Figure 1.7. The submicron era started in 1990 with the 0.8 µm technology. The deep submicron technology started in 1995 with the introduction of lithography thiner than 0.3 µm. Ultra deep submicron technology concerns lithography below 0.1 µm. Figure 1.7 shows that research has always kept around five years ahead of mass production. It can also be seen that the trend towards smaller dimensions has accelerated since 1996. In 2007, the lithography is expected to decrease to 65 nano-meter (nm). The lithography expressed in µm corresponds to the smallest patterns that can be implemented on the surface of the integrated circuit.

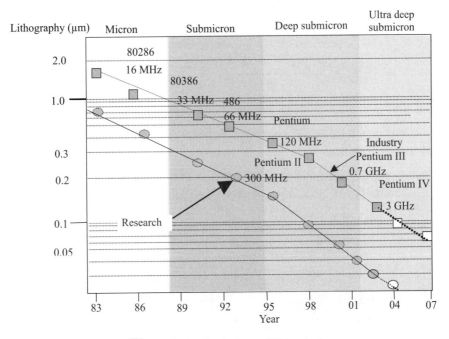

Fig. 1.7   Evolution of lithography

1.3   Frequency Improvements

Figure 1.8 illustrates the clock frequency increase for high-performance microprocessors and industrial micro-controllers with the technology scale down. The microprocessor roadmap is based on Intel processors used for personal computers [Intel], while the micro-controller roadmap is based on Freescale micro-controllers [Freescale] used for high performance automotive industry applications. The PC industry requires microprocessors running at the highest frequencies, which entails very high power consumption (30 Watts for the Pentium IV generation). The automotive industry requires embedded controllers with more and more sophisticated on-chip functionalities, larger embedded memories and interfacing protocols. The operating frequency follows a similar trend to that of PC processors, but with a significant shift.

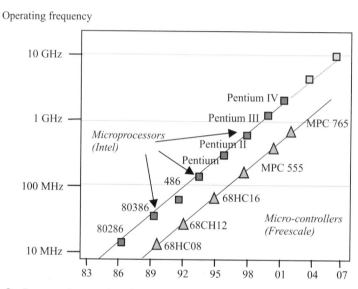

Fig. 1.8 Increased operating frequency of microprocessors and micro-controllers

1.4 Layers

Table 1.1 lists a set of key parameters, and their evolution with the technology. Special attention may be paid to the increased number of metal interconnects, the reduction of the power supply VDD and the reduction of the gate oxide down to atomic scale values. Notice also the increase in the size of the die and the increasing number of input/output pads available on a single die.

Table 1.1 Evolution of key parameters with the technology scale down [Itrs]

Lithography	Year	Metal layers	Core supply (V)	Core oxide (nm)	Chip size (mm)	Input/output pads	Microwind 2 rule file
1.2 µm	1986	2	5.0	25	5 × 5	250	Cmos12.rul
0.7 µm	1988	2	5.0	20	7 × 7	350	Cmos08.rul
0.5 µm	1992	3	3.3	12	10 × 10	600	Cmos06.rul

0.35 µm	1994	5	3.3	7	15 × 15	800	Cmos035.rul
0.25 µm	1996	6	2.5	5	17 × 17	1000	Cmos025.rul
0.18 µm	1998	6	1.8	3	20 × 20	1500	Cmos018.rul
0.12 µm	2001	6–8	1.2	2	22 × 20	1800	Cmos012.rul
90 nm	2003	6–10	1.0	1.8	25 × 20	2000	Cmos90n.rul
65 nm	2005	6–12	0.8	1.6	25 × 20	3000	Cmos65n.rul

The 1.2 µm CMOS process features n-channel and p-channel MOS devices with a minimum channel length of 0.8 µm. The Microwind tool may be configured in CMOS 1.2 µm technology using the command **File → Select Foundry**, and choosing **cmos12.rul** in the list. Metal interconnects are 2 µm wide. The MOS diffusions are around 1 µm deep. The two-dimensional aspect of this technology is shown in Figure 1.9.

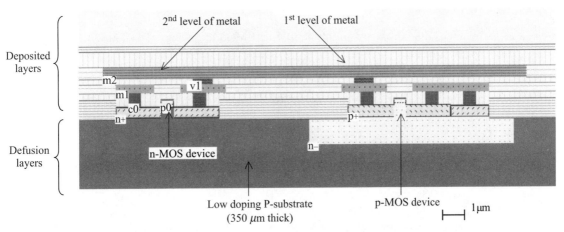

Fig. 1.9   Cross-section of the 1.2 µm CMOS technology (CMOS.MSK)

The 0.35 µm CMOS technology is a five-metal layer process with a minimal MOS device length of 0.35 µm. The MOS device includes lateral drain diffusions, with shallow trench oxide isolations. The Microwind tool may be configured in CMOS 0.35 µm technology using the command **File → Select Foundry**, and choosing "cmos035.rul" in the list. Metal interconnects are less than 1 µm wide. The MOS diffusions are less than 0.5 µm deep. The two-dimensional aspect of this technology is shown in Figure 1.10, using the layout **INV3.MSK**.

The Microwind and Dsch tools are configured by default in a CMOS 0.12 µm six-metal layer process with a minimal MOS device length of 0.12 µm. The metal interconnects are very narrow, around 0.2 µm, separated by 0.2 µm (Figure 1.11). The MOS device appears very small, below the stacked layers of metal sandwiched between oxides.

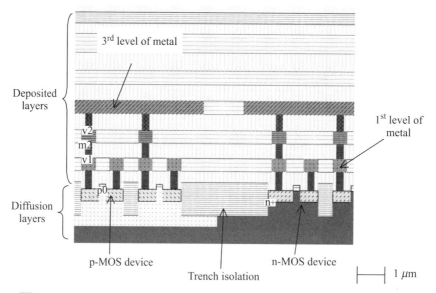

Fig. 1.10   Cross-section of the 0.35μm CMOS technology (INV3.MSK)

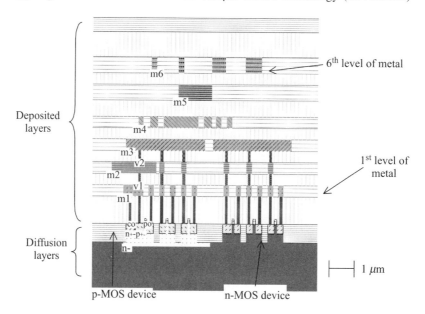

Fig. 1.11   2D view of the 0.12 μm process

1.5 Density

The main consequence of improved lithography is the ability to implement an identical function in an ever smaller silicon area. Consequently, more functions can be integrated in the same space. Moreover, the number of metal layers used for interconnects has been continuously increasing in the course of the past ten years. More layers for routing means a more efficient use of the silicon surface, as for printed circuit boards. Active areas, i.e MOS devices, can be placed closer to each other if many routing layers are provided (Figure 1.12).

The increased density provides two significant improvements. Firstly, the reduction of the silicon area goes together with a decrease in the parasitic capacitance of junctions and interconnects, thus increasing the switching speed of cells. Secondly, the shorter dimensions of the device itself speeds up the switching, which leads to further operating clock improvements.

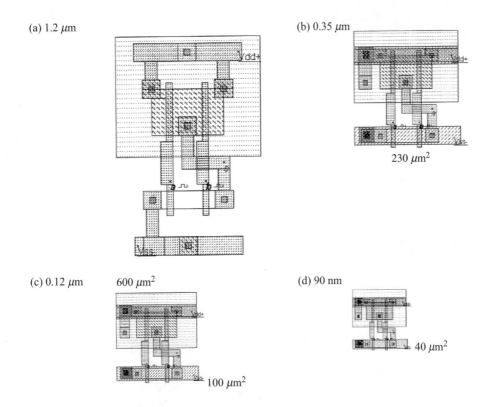

(a) 1.2 μm

(b) 0.35 μm

230 μm^2

(c) 0.12 μm   600 μm^2

(d) 90 nm

40 μm^2

100 μm^2

Fig. 1.12   Evolution of the silicon area used to implement a NAND gate, which represents 20% of logic gates used in application-specific integrated circuits

Meanwhile, the silicon wafer, on which the chips are manufactured, has constantly increased in size, with technological advances. A larger diameter means more chips fabricated at the same time, but requires ultra-high cost equipments that are able to manipulate and process these wafers with an atomic-scale precision. This trend is illustrated in Figure 1.13. The wafer diameter for 0.12 μm technology is 8 inches or 20 cm (one inch is equal to 2.54 cm). Twelve inch wafers (30 cm) have been introduced for the 90 nm technology generation. The thickness of the wafer varies from 300 to 600 μm.

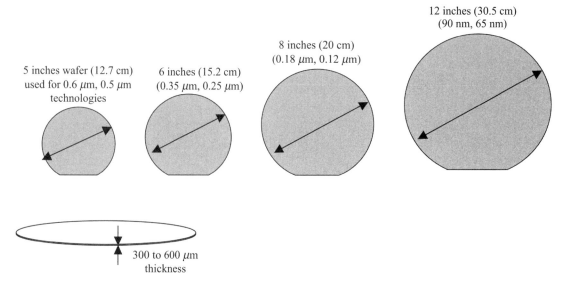

Fig. 1.13 The silicon wafer used for patterning the integrated circuits

1.6 Design Trends

Originally, integrated circuits were designed at the layout level, with the help of logic design tools, to achieve design complexities of around 10,000 transistors. The Microwind layout tool works at the lowest level of design, while Dsch operates at the logic level.

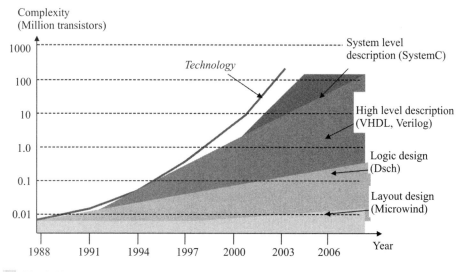

Fig. 1.14 Evolution of integrated circuit design techniques, from layout level to system level

The introduction of high level description languages such as VHDL (VHDL) and Verilog [Verilog] have made possible the design of complete systems on a chip (SoC), with complexities ranging from

1 million to 10 million transistors (Figure 1.14). Recently, languages for specifying circuit behaviour such as SystemC [SystemC] have been made available, which correspond to design complexity between 100 and 1000 million transistors. Notice that the technology has always been ahead of design capabilities, thanks to tremendous advances in process integration.

1.7 Market

Since the early days of microelectronics, the market has grown exponentially, representing more than 100 billion euros at the beginning of the twenty-first century. The average growth in a long term trend is approximately 15 per cent. Recently, two periods of negative growth have been observed: one in 1997–1999, the second one in 2002. Cycles of very high profits (1993–1995) have been followed by violent recession periods.

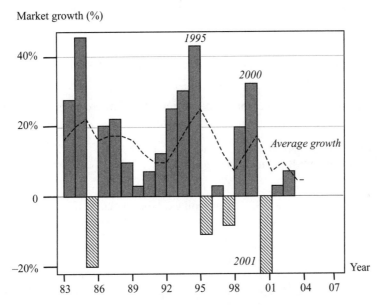

Fig. 1.15 The percentage of long term market growth over recent years is 15

1.8 Conclusion

This chapter has briefly illustrated the technology scale down, and the evolution of the microprocessor and microcontroller complexity, besides giving some general information about CMOS technology, trends and the market. The position of the Microwind layout design tool and Dsch logic design tool has been also described.

References

[Moore] G.E Moore, "VLSI: some fundamental challenges", IEEE Spectrum, N° 16 Vol 4, pp 30, 1975 [ITRS] The international roadmap for Semiconductors is regularly updated at web site *http://public.itrs.net/*

[Verilog] IEEE 1364-2001, the Verilog hardware description language (HDL) standard, also known as Verilog-2001, was approved by the IEEE as a revised standard in March 2001. See *http://standard.ieee.org/*

[VHDL] IEEE Standard 1164, Standard Multivalue Logic System for VHDL Model Interoperability (Std_logic_1164). See the IEEE Design Automation Standard Comittee at *http:// www.dasc.org/*

[SystemC] The SystemC language and its open source implementation can be downloaded at *www.systemc.org http://www.systemc.org* The IEEE and the Open SystemC Initiative (OSCI) have started to work on a standard IEEE P1666, "SystemC Language Reference Manual."

[Intel] See *http://www.intel.com/technology* for roadmaps, IC pictures, prospectives, and much more.

[Freescale] See *http://www.freescale.com* for more information on Freescale IC products

[ITRS] See SIA

EXERCISES

1. Plot the frequency improvement versus the technology for the CMOSxx technology family, using the 3-inverter ring oscillator. Can you guess the performances of the 35 nm technology?

2. Does the 3-inverter frequency performance represent the microprocessor frequency correctly? Use the data given in Figure 1.3 to build your answer.

3. From the 2D comparative aspect of 0.8 μm and 0.25 μm technologies (Figure 6), what may be the rising problems of using multiple metallization layers?

4. With the technology scale down, the silicon area decreases for the same device (see Figure 1.8), but the chip size increases (Table 1.1). Can you explain this contradiction?

The MOS Devices and Technology

This chapter presents the MOS transistors, their layout, static characteristics and dynamic characteristics. Details on the materials used to build the devices are provided. The vertical aspect of the devices and the three-dimensional sketch of the fabrication are also described.

2.1 Properties of Silicon

IA	IIA	IIIB	IVB	VB	VIB	VIIB	VII	VII	VII	IB	IIB	III	IVA	VA	VIA	VIIA	0
H 1 Hydrogen																	He 2 Helium
Li 3 Lithium	Be 4 Beryllium											B 5 Boron	C 6 Carbon	N 7 Nitrogen	O 8 Oxygen	F 9 Fluorine	Ne 10 Neon
Na 11 Sodium	Mg12 Magnesium											Al 13 Aluminum	Si 14 Silicon	P 15 Phosphorus	S 16 Sulfur	Cl 17 Chlorine	Ar 18 Argon
K 19 Potassium	Ca 20 Calcium	Sc 21 Scandium	Ti 22 Titanium	V 23 Vanadium	Cr 24 Chromium	Mn25 Manganese	Fe 26 Iron	Co 27 Cobalt	Ni 28 Nickel	Cu29 Copper	Zn30 Zinc	Ga31 Gallium	Ge32 Germanium	As33 Arsenic	Se 24 Selenium	Br 35 Bromine	Kr 36 Krypton
Rb 37	Sr 38	Y 39	Ze 40	Nb 41	Mo42	Tc 43	Ru 44	Rh 45	Pd 46	Ag 47 Silver	Cd 48 Cadmium	In 49 Indium	Sn 50 Tin	Sb 51	Te 52	I 53	Xe 54
Cs 55	Ba 56	La 57	Hf 72	Ta 73 Tantalum	W 74 Tungsten	Re 75	Os 76	Ir 77	Pt 78	Au 79 Gold	Hg 80	Tl 81	Pb 82 Lead	Bi 83	Po 84	At 85	Rn 86

Fig. 2.1 Periodic table of elements and position of silicon

The table in Figure 2.1 illustrates the table of elements. In CMOS integrated circuits, we mainly focus on silicon, situated in the column IVA, as the basic material (also called substrate) for all our designs.

The silicon atom has 14 electrons, 2 electrons situated in the first energy level, 8 in the second and 4 in the third. The four electrons in the third energy level are called valence electrons, which are shared with other atoms.

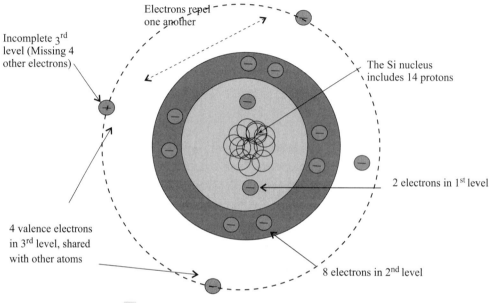

Fig. 2.2   The structure of the silicon atom

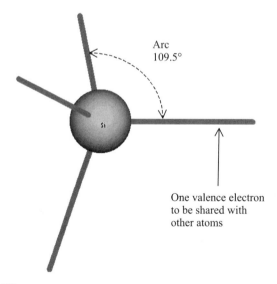

Fig. 2.3   The 3D symbol of the silicon atom

The silicon atom has four valence electrons, which tend to repel each another. The third level would be completed with eight electrons. The four missing electrons will be shared with other atoms. The position of electrons which minimizes the mutual repulsion is shown in Figure 2.3: each valence electron is represented by a line with an angle of 109.5°. In order to complete its valence shell, the silicon atom tends to share its valence electrons with four other electrons, by pairs. Each line between Si atoms in Figure 2.3 represents a pair of shared valence electrons. The distance between two Si nucleus is 0.235 nm (10^{-9} m), equivalent to 2.35 Angstrom (10^{-10} m, also represented by the letter Å).

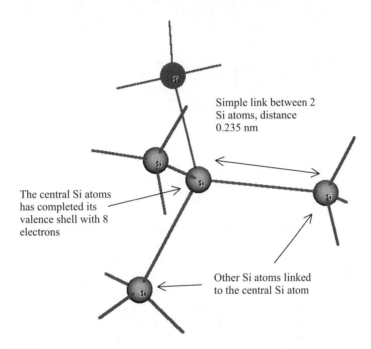

Fig. 2.4   The Si atom has four links, usually to other Si atoms

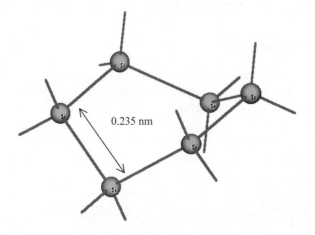

Fig. 2.5   The atom arrangement is based on a 6-atom pattern

The silicon lattice exhibits particular properties in terms of atom arrangements. The crystalline silicon is based on a 6-atom pattern shown in Figure 2.5. The structure is repeated infinitely in all directions to form the silicon substrate as used for integrated circuit design. The pure silicon crystal is mechanically very strong and hard, and electrically a very poor conductor, as all valence electrons are shared within the structure (Figure 2.6). The atomic density of a silicon crystal is about 5×10^{22} atoms per cubic centimeter (cm^{-3}).

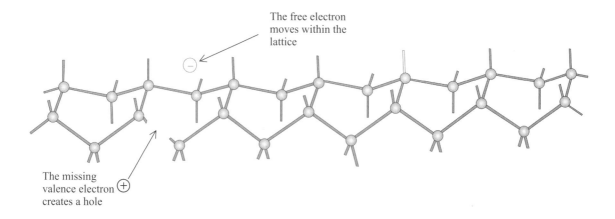

The free electron moves within the lattice

The missing valence electron $\oplus$ creates a hole

Fig. 2.6   The chain of the 6-atom pattern creates the silicon lattice

However, the random vibration of the silicon lattice due to thermal agitation may transmit enough energy to some valence electrons, for them to leave their position. The electron moves freely within the lattice, and thus participates in the conduction of electricity. The lack of electron is called a hole (Figure 2.6). This is why silicon is not an insulator, nor a good conductor. It is called a semi-conductor due to its intermediate electrical properties. The number of electrons which participate to the conduction are called intrinsic carriers. The concentration of intrinsic carriers per cubic centimeter, namely *ni*, is around 1.45×10^{10} cm^{-3}. When the temperature increases, the intrinsic carrier density also increases. The concentration of free electrons is assumed to be equal to the concentration of free holes.

2.2   N-type and P-type Silicon

In order to increase the conductivity of silicon, materials called "dopant" are introduced into the silicon lattice. In order to add more electrons in the lattice artificially, phosphorus or arsenic atoms (Group VA) are inserted in small proportions in the silicon crystal (Figure 2.7). As only four valence electrons find room in the lattice, one electron is released and participates in electrical conduction. Consequently, phosphorus and arsenic are named "electron donors", with an N-type symbol. A very high concentration of donors is coded N++ (around 1 N-type atom per 10,000 silicon atoms, corresponding to 10^{18} atoms per cm^{-3}). A high concentration of donor is coded N+ (1 N-type atom per 1,000,000 silicon atom, that is 10^{16} atoms per cm^{-3}), while a low concentration of donors is called N- (1 N-type atom per 100,000,000 silicon atom, or 10^{14} atoms per cm^{-3}).

	III	IVA	VA
	Acceptor		Donor
	Add holes		Add electrons
	P-type		N-Type

B 5 Boron	C 6 Carbon	N 7 Nitrogen
Al 13 Aluminium	Si 14 Silicon	P 15 Phosphorus
Ga 31 Gallium	Ge 32 Germanium	As 33 Arsenic

Fig. 2.7 Boron, phosphorus and arsenic are used as acceptors and donors of electrons to change the electrical properties of silicon

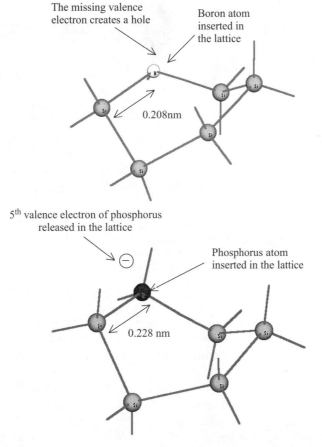

The missing valence electron creates a hole

Boron atom inserted in the lattice

0.208nm

5th valence electron of phosphorus released in the lattice

Phosphorus atom inserted in the lattice

0.228 nm

Fig. 2.8 Boron added to the lattice creates a hole (P-type property), phosphorus creates a free electron (N-type property)

In order to artificially increase the number of holes in silicon, boron is injected into the lattice, as shown in Figure 2.8. The missing valence link is due to the fact that boron only shares three valence electrons. The electron vacancy creates a hole, which gives the lattice a P-type property. A very high concentration of acceptors is coded P++ (10^{18} atoms per cm^{-3}), a high concentration of acceptors is coded P+(10^{16} atoms per cm^{-3}), a low concentration of acceptors is called P-(10^{14} atoms per cm^{-3}). The silicon substrate used to manufacture CMOS integrated circuits is lightly doped with boron, characterized by the P-symbol. The aspect of a small portion of silicon substrate is shown in 3D in Figure 2.9. It usually consists of very thick substrate (350 μm) lightly doped P–. Close to the upper surface, a buried layer saturated with P-type acceptors is usually created, to form a good conductor beneath the active region, connected to the ground voltage when the P++ layer is absent, such as in most 90 nm technologies, the substrate is highly resistive.

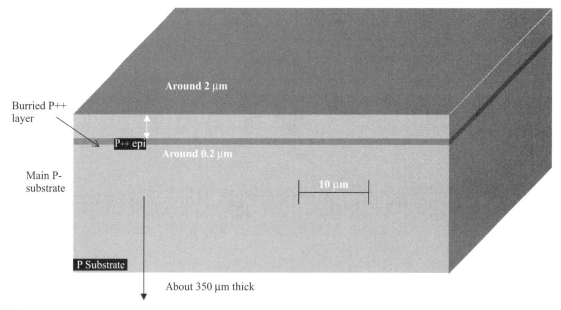

Fig. 2.9   3D aspect of a portion of silicon substrate used to manufacture CMOS integrated circuits. The substrate is based on a P-substrate with a buried P++ layer

2.3   Silicon Dioxide

The natural and most convenient insulator is silicon dioxide, noted SiO_2. Its molecular aspect is shown in Figure 2.10. Notice that the distance between Si and O atoms is smaller than for Si-Si, which leads to some interface regularity problems. Silicon dioxide is grown on the silicon lattice by high temperature contact with oxygen gas. Oxygen molecules combine not only with surface atoms, but also with underlying atoms. Silicon dioxide has an ε_r relative permittivity equal to 3.9. This number quantifies the capacitance effect of the insulator. The relative permittivity of air is equal to 1, which is the minimum value. The SiO_2 material is a very high quality insulator that is used extensively in CMOS circuits, for both devices and interconnections between devices. The "O" of CMOS is not "Oxygen" but corresponds to "oxide", as it refers to SiO_2.

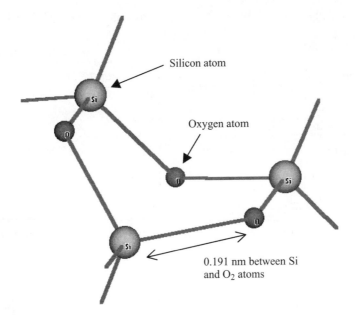

Silicon atom

Oxygen atom

0.191 nm between Si
and O_2 atoms

Fig. 2.10   Silicon is linked to oxygen to form the SiO_2 molecular structure

2.4   Metal Materials

Integrated circuits also use several metal materials to build interconnects. Aluminum (III), tungsten (IVB), gold (IB), and copper (IB) are commonly used in the manufacturing of microelectronic circuits.

Table 2.1   Metal materials used in CMOS integrated circuit manufacturing

VIB	VIIB	VII	VII	VII	IB	IIB	Al 13 Aluminum
Cr 24 Chromium	Mn 25 Manganese	Fe 26 Iron	Co 27 Cobalt	Ni 28 Nickel	Cu29 Copper	Zn 30 Zinc	Ga 31 Gallium
Mo 42	Tc 43	Ru 44	Rh 45	Pd 46	Ag 47 Silver	Cd 48 Cadmium	In 49 Indium
W 74 Tungsten	Re 75	Os 76	Ir 77	Pt 78	Au 79 Gold	Hg 80	Tl 81

Metal layers are characterized by their resistivity (σ). We notice that copper is the best conductor as its resistivity is very low, followed by gold and aluminium (Table 2.2). A highly doped silicon crystal does not exhibit a low resistivity, while the intrinsic silicon crystal is half way between a conductor and an insulator [Hastings].

Table 2.2   Conductivity of the most common materials used in CMOS integrated

Material	Symbol	Resistivity σ (Ω.cm)
Copper	Cu	1.72×10^{-6}
Gold	Au	2.4×10^{-6}
Aluminium	Al	2.7×10^{-6}
Tungsten	W	5.3×10^{-6}
Silicon, N+ doped	N+	0.25
Silicon, intrinsic	Si	2.5×10^{5}

Conductivity is sometimes used instead of resistivity. In that case, the formulation is as follows:

$$\rho = \frac{1}{\sigma}$$

(Equation 2.1)

with

ρ = conductivity $(\Omega.cm)^{-1}$

σ = resistivity $(\Omega.cm)$

2.5   The MOS Switch

The MOS transistor (MOS for Metal Oxide Semiconductor) is by far the most important basic element of the integrated circuit. The MOS transistor is the integrated version of the electrical switch [Baker]. When it is on, it allows current to flow, and when it is off, it stops current from flowing. The MOS switch is turned on and off by electricity. Two types of MOS devices exist in CMOS technology (Complementary Metal Oxide Semiconductor): the n-channel MOS device (also called nMOS) and the p-channel MOS device (also called pMOS).

2.5.1   Logic Levels

Three logic levels 0,1 and X are defined as follows:

Table 2.3   The logic levels and their corresponding symbols in Dsch and Microwind tools

Logical value	Voltage	Name	Symbol in Dsch	Symbol in Microwind
0	0.0 V	VSS	(Green in logic simulation)	(Green in analog simulation)

1	1.2 V in technology 0.12 µm	VDD	(Red in logic simulation)	(Red in analog simulation)
X	Undefined	X	(Gray in simulation)	(Gray in simulation)

2.5.2 The n-channel MOS Switch

Despite its extremely small size (less than 1 µm square), the current that the MOS transistor may switch is sufficient to turn on and off a led, for example. The MOS device consists of two electrical regions called drain and source respectively, separated by a channel. A channel of electrons may or may not exist in this channel, depending on a voltage applied to the gate. The gate is a conductor placed on the top of the channel, and electrically isolated by an ultra thin oxide. The MOS is basically a switch between the drain and source. A schematic cross-section of the MOS device is given in Figure 2.11. Theoretically, the source is the origin of channel impurities. In the case of this nMOS device, the channel impurities are the electrons. Therefore, the source is the diffusion area with the lowest voltage.

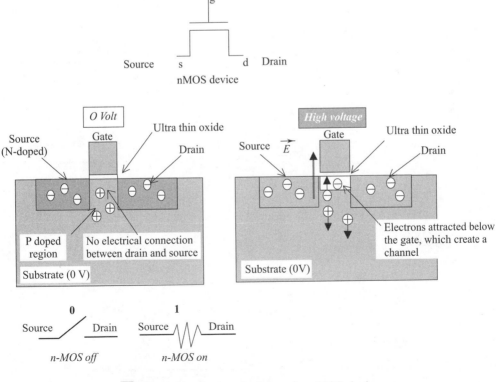

Fig. 2.11 Basic principles of an MOS device

When used in logic cell design, it can be *on* or *off*. As illustrated in the figure, the n-channel MOS device requires a high supply voltage to be on. When *on*, a current can flow between the drain and source. When the MOS device is on, the link between the source and drain is equivalent to a resistor. The resistance may vary from less than 0.1Ω to several hundred kilo-Ohm ($K\Omega$). Low resistance MOS devices are used for power application, while high resistance MOS devices are widely used in analog low power designs. In logic gate, the Ron resistance is around 1 $K\Omega$.

The 'off' resistance is considered infinite at first order, but its value is several $M\Omega$. When off, almost no current flows between the drain and source. The device is equivalent to an open switch, and the voltage of the floating node (the drain in the case of Table 2.4) is undetermined. The n-channel MOS logic table can be described as follows:

Table 2.4   The n-channel MOS switch truth-table

Gate	Source	Drain
0	0	X
0	1	X
1	0	0
1	1	1

2.5.3   The p-channel MOS Switch

In contrast, the p-channel MOS device requires a zero voltage supply to be on. The p-channel MOS symbol differs from the n-channel device with a small circle near the gate. The channel carriers for pMOS are holes.

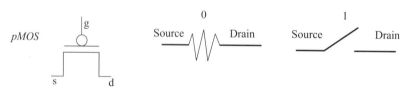

Fig. 2.12   The MOS symbol and switch

The p-channel MOS logic table can be described as follows:

Table 2.5   The p-channel MOS switch truth-table

Gate	Source	Drain
0	0	0
0	1	1
1	0	X
1	1	X

For the p-channel MOS, a high voltage disables the channel. Almost no current flows between the source and drain. A zero voltage VDD on the gate attracts holes below the gate, creates a hole channel and enables current to flow, as shown in Figure 2.13.

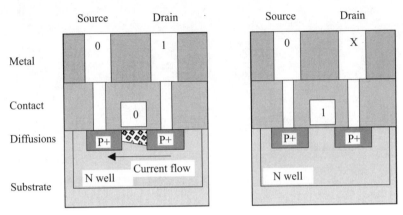

Fig. 2.13   The channel generation below the gate in a pMOS device

2.6   The MOS Aspect

A bird's eye view of the layout including one n-channel MOS device and one p-channel MOS device placed at the minimum distance is given in Figure 2.14. A two-dimensional zoom at micron scale in the active region of an integrated circuit designed in 0.12 μm technology is shown in Figure 2.15. This view corresponds to a vertical cross-section of the silicon wafer, in location X-X' in Figure 2.14. The Microwind tool has been used to build the layout of the MOS devices (the corresponding file is **allMosDevices.MSK**), and to visualize its cross-section, using the command **Simulate → 2D vertical cross-section**.

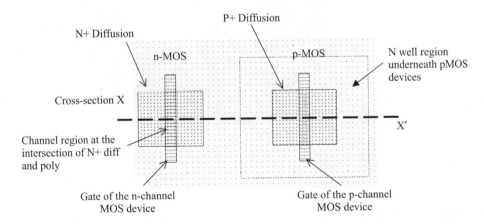

Fig. 2.14   Bird's eye view of the n-channel and p-channel MOS device layout (allMosDevices.MSK)

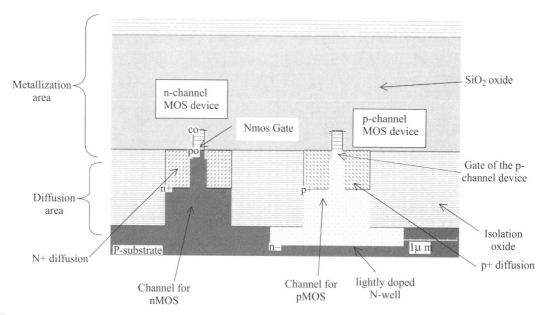

Fig. 2.15 Vertical cross-section of an n-channel and p-channel MOS devices in 0.12 μm technology (allMosDevices.MSK)

The layout of the nMOS and pMOS devices, seen from the top of the circuit, is shown in the Figure. The MOS is built using a set of layers that are summarized below. CMOS circuits are fabricated on a piece of silicon called wafer, usually lightly doped with boron, that gives a p-type property to the material.

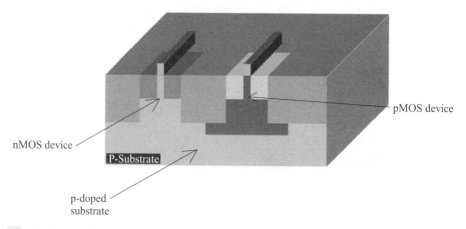

Fig. 2.16 3D view of the n-channel and p-channel MOS devices (AllMosDevices.MSK)

2.6.1 Zoom at Atomic Scale

When zooming on the gate structure, we distinguish the thin oxide beneath the gate, the low doped diffusion regions on both sides of the channel, the metal deposit on the diffusion surface, and the spacers on each side of the gate, as shown in Figure 2.17.

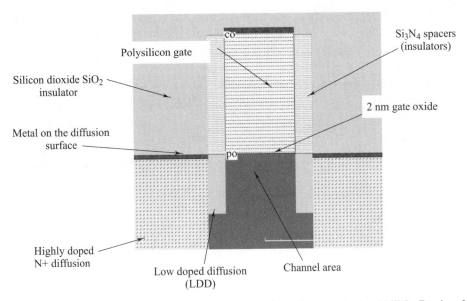

Polysilicon gate

Silicon dioxide SiO$_2$ insulator

Metal on the diffusion surface

Highly doped N+ diffusion

Low doped diffusion (LDD)

Channel area

Si$_3$N$_4$ spacers (insulators)

2 nm gate oxide

Fig. 2.17   Zoom at the gate oxide for a 0.12 μm n-channel MOS device (AllMosDevices.MSK)

When zooming at the maximum scale, we see the atomic structure of the transistor. The gate oxide accumulates between 5 and 20 atoms of silicon dioxide. In 0.12 μm, the oxide thickness is around 2 nm for the core logic, which is equivalent to 8 atoms of SiO$_2$.

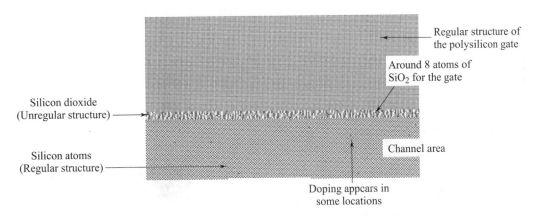

Silicon dioxide (Unregular structure)

Silicon atoms (Regular structure)

Regular structure of the polysilicon gate

Around 8 atoms of SiO$_2$ for the gate

Channel area

Doping appears in some locations

Fig. 2.18   Zoom at atomic scale near the gate oxide for a 0.12 μm n-channel MOS device (AllMosDevices.MSK)

2.7   MOS Layout

The objective of this paragraph is to draw the n-channel and p-channel MOS devices according to the design rules and usual design practices. The Microwind tool is used to draw the MOS layout and to simulate its behaviour.

The Microwind main screen shown in Figure 2.19 includes two windows: one for the main menu and the layout display, and the other for the icon menu and the layer palette. The main layout window features a grid, scaled in lambda (λ) units. The size of the grid constantly adapts to the layout. In Figure 2.19, the grid is 5 lambda. The lambda unit is fixed to half of the minimum available lithography of the technology, i.e $L_{min}/2$. For example, the default technology is a CMOS six-metal layers 0.12 µm technology, consequently lambda is 0.06 µm.

$$\lambda = \frac{L_{min}}{2}$$ (Equation 2.2)

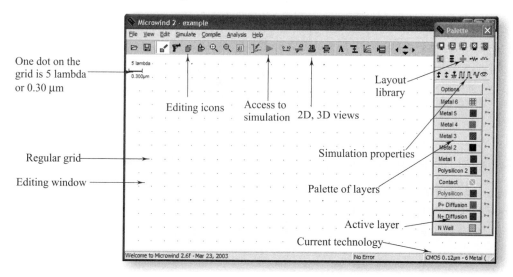

Fig. 2.19   The Microwind2 window as it appears at the initialization stage

The palette is located in the right corner of the screen. A red colour indicates the current layer. Initially the selected layer in the palette is polysilicon.

2.7.1   n-channel MOS Layout

By using the following procedure, you can create a manual design of the n-channel MOS device. The n-channel MOS device consists of a polysilicon gate and a heavily doped diffusion area. Select the "polysilicon" layer in the palette window.

1. Fix the first corner of the box with the mouse. While keeping the mouse button pressed, move the mouse to the opposite corner of the box. Release the button. This creates a narrow box in polysilicon layer as shown in Figure 2.20. The box width should not be inferior to 2 λ, which is the minimum and optimal thickness of the polysilicon gate.

2. Change the current layer into N+ diffusion by a click on the palette of the N+ diffusion button. Make sure that the red layer is now the N+ diffusion. Draw an n-diffusion box at the bottom of the drawing as in Figure 2.20. The N+ diffusion should have a minimum of 4 lambda on both sides of the polysilicon gate. The intersection between diffusion and polysilicon creates the channel of the nMOS device.

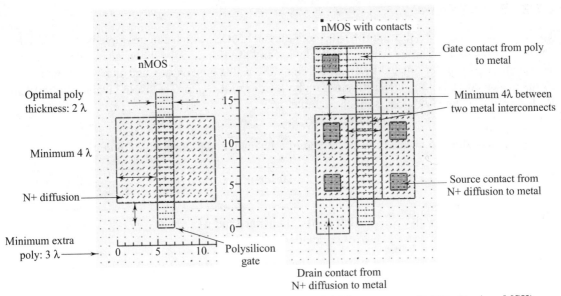

Fig. 2.20   Creating the n-channel MOS transistor and adding contacts (AllMosDevices.MSK)

Now, we add the metal contacts to enable an electrical access to the source and drain regions. In the palette, such contacts are ready to instantiate on the layout (Figure 2.21). Click on the appropriate icon, and then on the appropriate location in the left N+ diffusion. Repeat the process and add another contact on the right part of the N+ diffusion. The layout aspect should correspond to the layout shown in Figure 2.21.

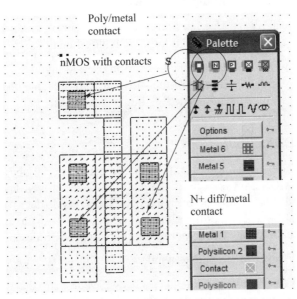

Fig. 2.21   Access to the n-diffusion/metal and poly/metal contacts

2.7.2   Basic Layers

The wafer serves as the substrate (or bulk) to n-channel MOS, which can be implemented directly on the p-type substrate. The n-channel MOS device is based on a polysilicon gate, deposited on the surface of the substrate, isolated by an ultra thin oxide (called gate oxide), and an N+ implantation that forms two electrically separated diffusions on both sides of the gate. The list of layers commonly used for the design of MOS devices is given in Table 2.6.

Table 2.6   Materials used to build n-channel and p-channel MOS devices

Layer name	Code	Description	Colour in Microwind
Polysilicon	Poly	Gate of the n-channel and p-channel MOS devices.	Red
N+ diffusion	Diffn	Delimits the active part of the n-channel device. Also used to polarize the N-well	Dark green
P+ diffusion	Diffn	Delimits the active part of the p-channel device. Also used to polarize the bulk.	Maroon
Contact	Contact	Makes the connection between diffusions and metal for routing. The contact plug is fabricated by drilling a hole in the oxide and filling the hole with metal.	White cross
First level of metal	Metal1	Used to rout devices together, in order to create the logic or analog function.	Blue
n-well	n-well	Low doped diffusion used to invert the doping of the substrate. All p-channel MOS are located within N well areas.	Dotted green

2.7.3   p-channel MOS Layout

The p-channel MOS is built by using polysilicon as the gate material and P+ diffusion to build the source and drain. The pMOS device requires the addition of the n-well layer to the polysilicon and diffusion layers. The P+ diffusion must be completely included inside the n-well layer, in order to work properly (Figure 2.22). The extension of n-well is 6 λ around the P+ diffusion. By using the following procedure, you can manually create the layout of the p-channel MOS device:

- Select the *polysilicon* layer in the palette window.

- Create a narrow polysilicon box to create the p-channel MOS gate (the minimum value 2 λ is often used). The material is the same as for the n-channel MOS.

- Change the layer into P+ *diffusion*. Draw a p-diffusion box at the bottom of the drawing as in Figure 2.22. The P+ diffusion should have a minimum of 4 λ on both sides of the polysilicon gate.

- Select the *n-well* layer. Add an n-well region that completely includes the P+ diffusion, with a border of 6 λ, as illustrated in Figure 2.22.

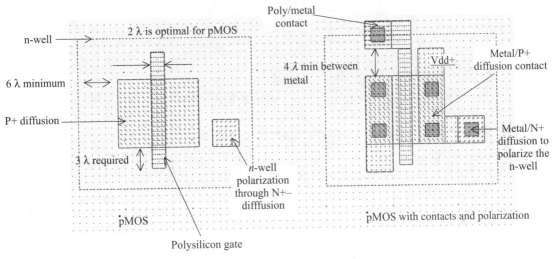

Fig. 2.22 Creating the p-channel MOS transistor (AllMosDevices.MSK)

Moreover, the n-well region cannot be kept floating. A specific contact, that can be seen on the right side of the n-well, serves as a permanent connection to high voltage. Why high voltage? Let us consider the two cross-sections in Figure 2.23. On the left side, the n-well is floating. The risk is that the n-well potential decreases enough to turn on the P+/n-well diode. This case corresponds to a parasitic PNP device. The consequence may be the generation of a direct path from the VDD supply of the drain to the ground supply of the substrate. In many cases the circuit can be damaged.

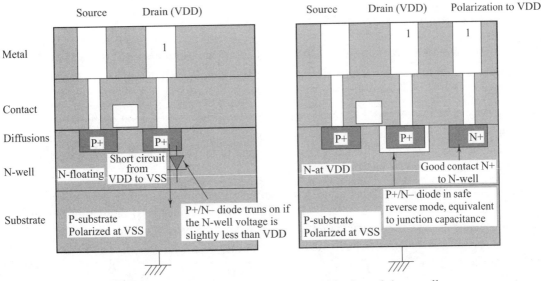

Fig. 2.23 Incorrect and correct polarization of the n-well

The correct approach is indicated in the right part of Figure 2.24. A polarization contact carries the VDD supply down to the n-well region, thanks to an N+ diffusion. A direct contact with the n-well would generate parasitic electrical effects. Consequently, the N+ region embedded in the n-well area is mandatory. There is no more fear of parasitic PNP device effect as the P+/n-well junctions are in inverted mode, and thus may be considered as junction capacitance.

2.7.4 Useful Editing Tools

Editing layout is rarely a simple task at the beginning, when cumulating the discovery of the editor user's interface and the application of new microelectronics concepts. The following commands may help you in the layout design and verification processes:

Table 2.7 A set of useful editing tools

Command	Icon/Short cut	Menu	Description
UNDO	CTRL+U	Edit menu	Cancels the last editing operation.
DELETE	CTRL+X	Edit menu	Erase some layout included in the given area or pointed by the mouse.
STRETCH		Edit menu	Changes the size of one box, or moves the layout included in the given area.
COPY	CTRL+C	Edit menu	Copy of the layout included in the given area.

2.7.5 Vertical Aspect of the MOS

Click on this icon to access *process simulation* (Command **Simulate → Process section in 2D**). The cross-section is given by a click of the mouse at the first point and the release of the mouse at the second point.

In the example shown in Figure 2.24, three nodes appear in the cross-section of the n-channel MOS device: the *gate* (red), the left diffusion called *source* (green) and the right diffusion called *drain* (green), over a substrate (gray). A thin oxide called the gate oxide isolates the gate. Various steps of oxidation have led to stacked oxides on the top of the gate.

The lateral drain diffusion (LDD) is a small region of lightly doped diffusion, at the interface between the drain/source and the channel. A light doping reduces the local electrical field at the corner of the drain/source and gate. Electrons accelerated below the gate at the maximum electrical field, such as in Figure 2.25, acquire sufficient energy to create a pair of electrons and holes in the drain region. Such electrons are called "hot electrons".

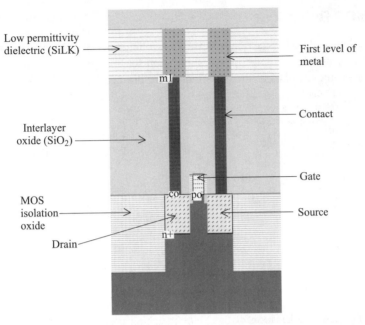

Fig. 2.24 The cross-section of the nMOS devices (AllMosDevices.MSK)

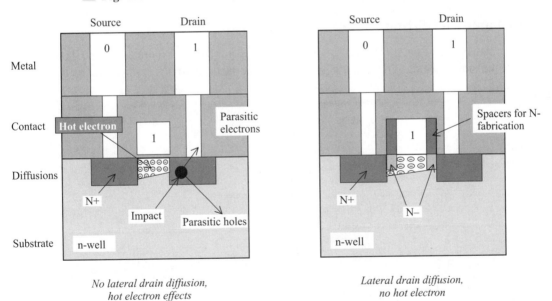

No lateral drain diffusion,
hot electron effects

Lateral drain diffusion,
no hot electron

Fig. 2.25 Lateral drain diffusion reduces the hot electron effect

Consequently, parasitic currents are generated in the drain region. One part of the current flows down to the substrate while another part is collected at the drain contact. The lateral drain diffusion efficiently reduces this parasitic effect. This technique has been introduced since the 0.5 μm process generation.

2.8   Dynamic MOS Behaviour

In this paragraph, we stimulate the MOS device with variable voltages in order to verify by analog simulation their correct behaviour as switches. The proposed simulation set-up (Figure 2.26) consists in applying 0 and 1 to the gate and the source, and in seeing the effect on the drain, as outlined in the schematic diagram below.

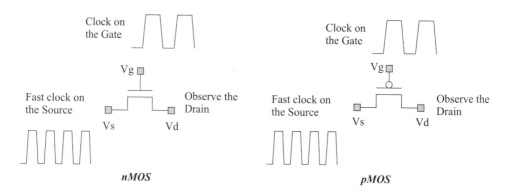

Fig. 2.26   Verification of the MOS switching properties using clocks

2.8.1   n-channel MOS Behaviour

The expected behaviour of the n-channel MOS device is summarized in Figure 2.27. The 0 on the gate should leave the drain floating. The 1 on the gate should link the drain to the source, via a resistive path.

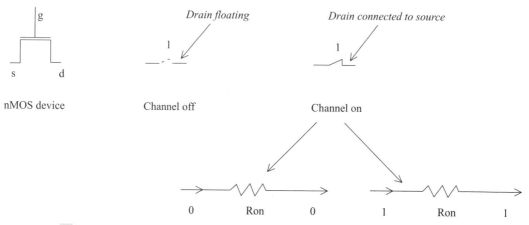

Fig. 2.27   Expected n-channel MOS switching characteristics (MosExplain.SCH)

The most convenient way to operate the MOS is to apply a clock property to the gate and another to the source, and to observe the drain. The summary of the available properties that can be applied to the layout is reported in Figure 2.28.

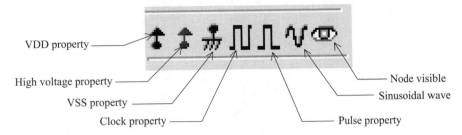

VDD property
High voltage property
VSS property
Clock property
Node visible
Sinusoidal wave
Pulse property

Fig. 2.28   Simulation properties used to conduct the simulation from layout

A clock should be applied to the source, which is situated on the green diffusion area on the left side of the gate. Click on the *Clock* icon and then click on the polysilicon gate. The clock menu appears again. Change the name into *Vdrain* and click on *OK* to apply a clock with 1ns period (0.225 ns at 0, 25 ps rise, 0.225 ns at 1, 25 ps fall). The label *Vdrain* appears at the desired location in *italics*, meaning that the waveform of this node will appear in the next simulation.

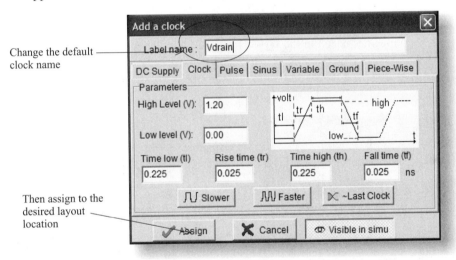

Change the default clock name

Then assign to the desired layout location

Fig. 2.29   Details on the clock parameters and clock menu

Now, to apply a clock to the gate, click again on the *Clock* icon, and click on the polysilicon gate. The clock menu appears. Notice that the clock parameters *Time low* and *Time high* have been automatically increased by 2, to create a clock with a period twice slower than previously. Change the name into *Vclock* and click on *OK*. The clock property is sent to the gate and appears on the right hand side of the label *Vgate*.

In order to see the source, click on the *Node Visible* property situated on the right of the palette menu, represented by an eye (see Figure 2.30), and click on the right diffusion. Change the label name into "Vout". Click OK. The visible property is then sent to the node. The associated text *Vout* is in *italics*, meaning that the waveform of this node will appear in the next simulation.

The layout should then appear as illustrated in Figure 2.31. The clock properties are situated on the gate and the left N+ diffusion, the "node visible" property is located on the right part of the N+ diffusion. The layout is now ready for analog simulation.

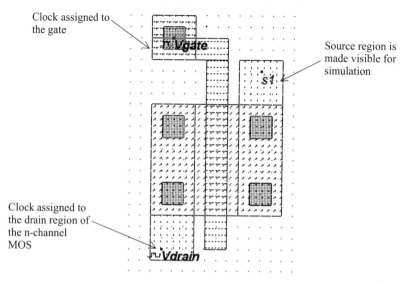

Clock assigned to the gate

Source region is made visible for simulation

Clock assigned to the drain region of the n-channel MOS

Fig. 2.30   Properties added to the layout for controlling the analog simulation (MosN.MSK)

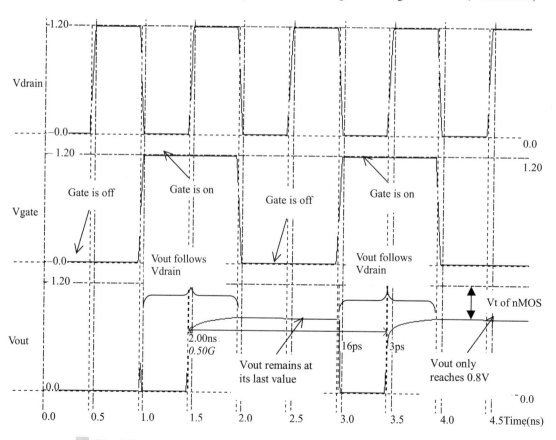

Fig. 2.31   Analog simulation of the n-channel MOS device (MosN.MSK)

Click on **Simulate → Run Simulation → Voltage vs. Time** (Or CTRL + S, or the icon *Run Simulation* in the main icon menu). The timing diagrams of the nMOS device appear, as shown in Figure 2.31. Most of the logic and analog behaviour of the MOS device are summarized in this single figure.

The upper waveform corresponds to *Vdrain*, with a clock from 0 to 1.2 V (*VDD* is 1.2 V in 0.1 2 µm), exhibiting a 1ns period. Below is the gate voltage. The n-channel MOS is off when *Vgate* is zero, and on if *Vgate* is 1.2 V. The third waveform concerns *Vout*. Its aspect is very interesting: from 0 to 1 ns, its value is zero, which is the default voltage value of all nodes in the layout. The channel is off, consequently nothing happens to the drain. When the gate is on, the channel enables *Vout* to copy the value of *Vdrain* (Time 1.0 to 2.0 ns).

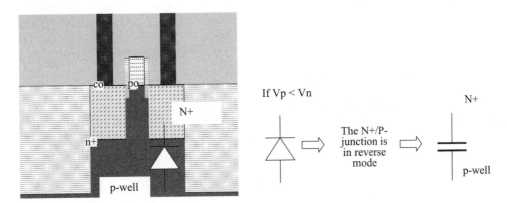

Fig. 2.32   The P/N junction in reverse mode is equivalent to a capacitor that memorizes the voltage when the channel is off (MosN.MSK)

Then, when the gate is off again (Time 2.0 to 3.0), the voltage maintains almost the same value as before. The reason for this is illustrated in Figure 2.32: the junction p-well/N+ diffusion is in reverse mode and can be considered as a capacitor. The charges are stored in this junction capacitor while the channel is off, which keeps its voltage stable, independently of the source fluctuations.

Notice a very small decrease in the voltage, due to the parasitic leakage effect between the source and drain. Click **More** in order to perform more simulations. Click on **Close** to return to the editor.

2.8.2 Threshold Voltage

You probably noticed that the voltage *Vout* never reaches 1.2 V. It saturates at around 0.8 V. The reason for this is the parasitic effect called threshold voltage, which is around 0.4 V in the 0.12 µm default technology. In summary, the n-channel MOS device behaves as a switch, but when on, it does not correctly pass the high voltages. A zero on one side leads to a good zero, a logic 1 on one side leads to a poor 1. The main reason is the threshold voltage of the MOS.

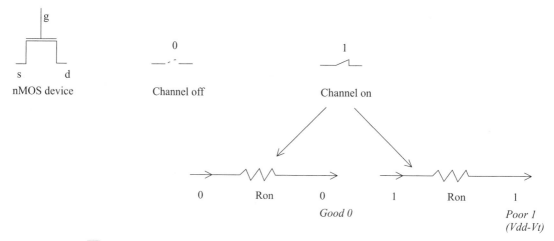

Fig. 2.33   The nMOS device behaviour summary (MosExplain.SCH)

2.8.3   p-channel MOS Behaviour

The expected behaviour of the n-channel MOS device is summarized in Figure 2.34. The 0 on the gate should link the drain to the source, via a resistive path. The 1 on the gate should leave the drain floating. In other words, the p-channel transistor simulation features the same functions as the n-channel device, but with opposite voltage control of the gate.

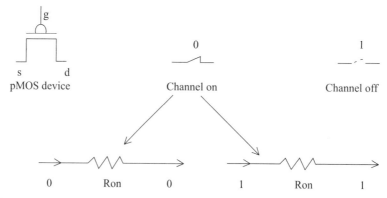

Fig. 2.34   Expected p-channel MOS switching characteristics (MosExplain.SCH)

Again, we operate the p-channel MOS device by using two clocks, one on the gate, another on the source and we observe the drain. Be sure to add the polarization contact inside the n-well region, and add a supplementary VDD property on top of the contact.

Click on **Simulate → Run Simulation → Voltage vs. Time**. The timing diagrams of the pMOS device appear as shown in Figure 2.36. The upper waveform corresponds to *Vdrain*, with a clock from 0 to 1.2 V (VDD is 1.2 V in 0.12 μm), exhibiting a 1 ns period. Below is the gate voltage. The p-channel MOS is on when *Vgate* is zero, and off when *Vgate* is 1.2 V. When the gate is on, the channel enables *Vout* to copy the value of *Vdrain* (Time 0 to 1.0 ns). Then, when the gate is off again, the voltage remains almost at its last value. The reason why *Vout* kept around 1.0 V from 1.0 to 2.0 ns is that the channel turned off synchronously with a change in the value of *Vdrain*.

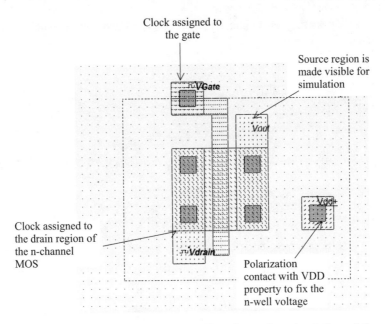

Fig. 2.35   Properties added to the layout for controlling the analog simulation of the p-channel device (Mosp.MSK)

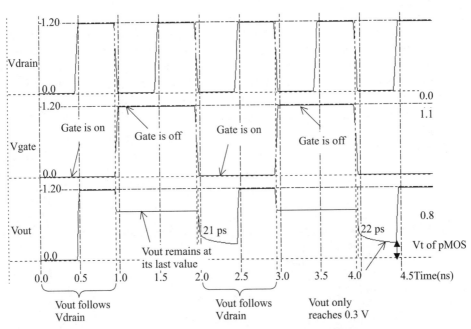

Fig. 2.36   Time domain simulation of the p-channel MOS (Mosp.MSK)

Notice that for the p-channel MOS, the voltage *Vout* never reaches 0.0 V. It saturates to around 0.3 V due to the threshold voltage of the device. In short, the p-channel MOS device behaves as a switch, but when on, it does not correctly pass the low voltages (Figure 2.37). A zero on one side leads to a poor zero, a logic 1 on one side leads to a good 1.

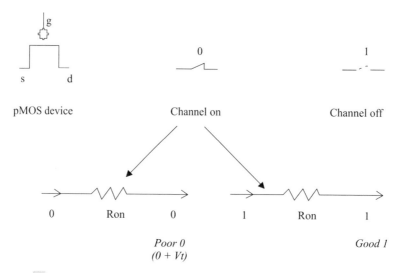

Fig. 2.37   Summary of the performances of a pMOS device

2.9   The Perfect Switch

Both nMOS and pMOS devices exhibit poor performances when transmitting one particular logic information. The nMOS degrades the logic level 1 while the pMOS degrades the logic level 0. Thus, a perfect pass gate can be constructed from the combination of nMOS and pMOS devices working in a complementary way, leading to improved switching performances [John] [Razari]. Such a circuit, presented in Figure 2.38, is called the transmission gate. In DSCH, the symbol may be found in the "advance" menu in the palette. The main drawback of the transmission gate is the need for two control signals, *Enable* and */Enable*, which is why an inverter is required.

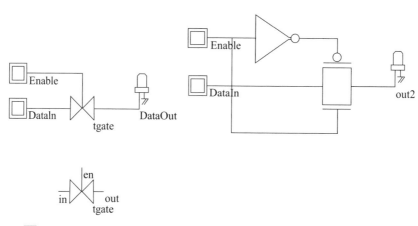

Fig. 2.38   Schematic diagram of the transmission gate (Tgate.SCH)

The transmission gate lets a signal flow if *en* = 1 and ~*en* = 0. In that case, both the n-channel and p-channel devices are on. The n-channel MOS transmits low voltage signals, while the p-channel device preferably transmits high voltage signals (Figure 2.39).

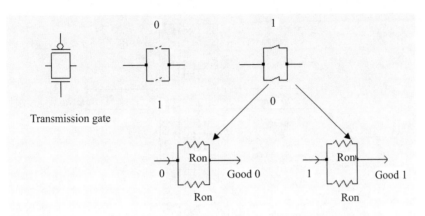

Fig. 2.39 The transmission gate used to pass logic signals (Tgate.SCH)

2.9.1 Implementation

We need to create an electrical connection between the N+ and P+ regions, in order to comply with the schematic diagram shown in Figure 2.40. The most efficient solution consists in using metal and contacts to create a bridge from the N+ region to the P+ region. Figure 2.41 shows the cross-section of the bridge.

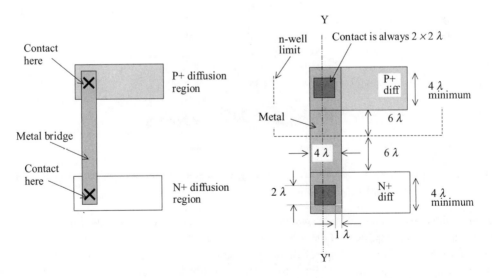

Fig. 2.40 Principles for metal bridge between N+ diffusion and P+ diffusion regions, and associated design rules

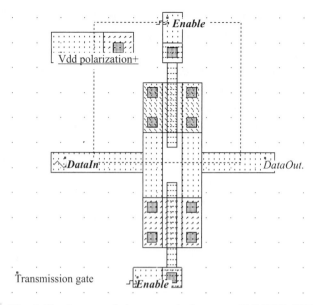

Fig. 2.41 Layout of the transmission gate (TGATE.MSK)

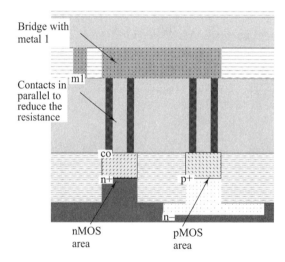

Fig. 2.42 Cross-section at location Y-Y′ showing the metal bridge and contacts with the N+ diffusion and P+ diffusion regions

The layout of the transmission gate is shown in Figure 2.41. The n-channel MOS is situated arbitrarily at the bottom, while the p-channel MOS at the top. Notice that the gate controls are not connected, as ~*Enable* is the opposite of *Enable*. The operation of the transmission gate is illustrated in Figure 2.43. A sinusoidal wave with a frequency of 2 GHz is assigned to *DataIn*. The sinusoidal property may be found in the palette of Microwind2, near the clock and pulse properties. With a zero on *Enable* (and a 1 on ~*Enable*), the switch is off, and no signal is transferred. When *Enable* is asserted, the sinusoidal wave appears nearly identical to the output.

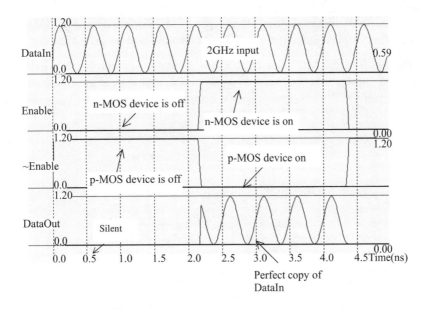

Fig. 2.43   Simulation of the transmission gate (TGATE.MSK)

2.10   Layout Considerations

2.10.1   MOS Generation

The safest way to create an MOS device is to use the MOS generator. In the palette, click the MOS generator icon. A window appears as reported below. The main parameters are the MOS type (either n-channel or p-channel), the width and the length. By default, the proposed MOS is an n-channel device, with a 0.6 μm width and 0.12 μm minimum length. The maximum current that can flow in the MOS channel is given for information (0.39 mA in that case). The units for width and length are μm by default. You may change the units and use lambda values instead.

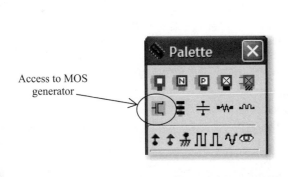

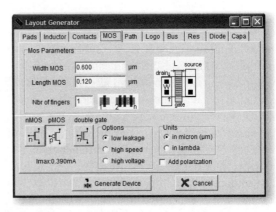

Fig. 2.44   Access to the MOS generator menu

Since the advent of the 0.18 μm technology, three types of MOS devices have been made available: low leakage, high speed and high voltage devices. These concepts will be discussed in the next chapter.

If we increase the width of the MOS device to 2 μm, the layout generated by Microwind2 takes the aspect of Figure 2.45. Notice that the length is equal to 2 λ, that is 0.12 μm in the default technology, while the width is 2 μm.

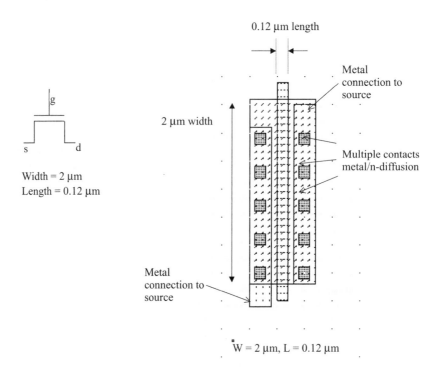

Fig. 2.45   An n-channel MOS with 2 μm width generated by Microwind2 (MosLayout.MSK)

2.10.2 Multiple Contacts

In the layout in Figure 2.45, the surprise comes from the multiple contacts on the drain and source regions. The reason for this addition of contacts is the intrinsic current limitation of each elementary contact plug, as well as the high resistance of one single contact. One single contact can stand less than 1 mA current without any reliability problem. When the current is stronger than 1 mA, the contact can be damaged (Figure 2.46). The effect is called electromigration: if too much current flows within the contact, the metal structure starts to change as atoms move inside the conductor. A very strong current such as 10 mA would destroy even one lonely contact.

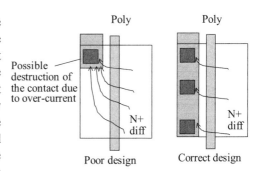

Fig. 2.46   A strong current through a single contact can damage the metal structure

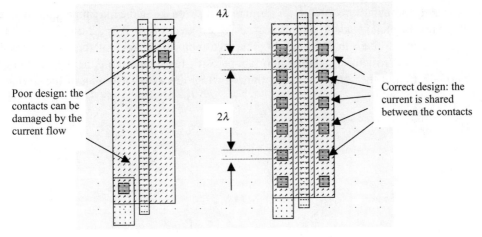

Fig. 2.47   A single contact cannot handle more that 1 mA. A series of contacts is preferred (MosLayout.MSK)

The illustration of this important limitation is given in Figure 2.48. Basic rules for contact design are also reported in the figure. The contact is $2 \times 2\ \lambda$, and the separation is $4\ \lambda$.

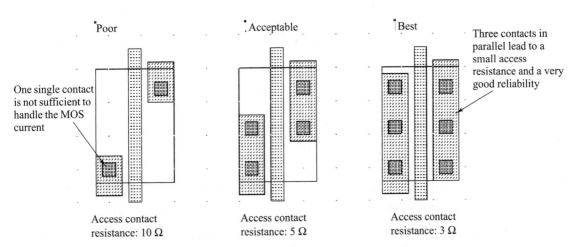

Fig. 2.48   A series of contacts also reduced the serial access resistance (MosContacts.MSK)

Adding as many contacts as the design rules permit also limits the contact resistance. The equivalent resistance of the access to the drain and source regions is reduced proportionally to the number of contacts.

2.10.3 Multiple Gates

The use of MOS devices with long width is very common, as for example in analog design and buffer design such as clocks and interface structures. Let us try to design a 5 mA MOS switch, which can be found in a standard output buffer structure. A rapid investigation using the maximum current evaluation in the MOS generator menu leads to the need of an MOS device with W = 12 μm, L = 0.12 μm. The corresponding layout is given in Figure 2.49. The two main drawbacks of this layout are the very unpractical shape of the structure, and the important parasitic resistance along the polysilicon gate, that delays the propagation of the control voltage, thus slowing down the switching of the device.

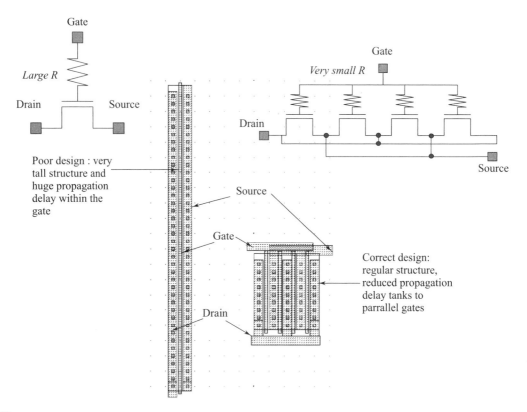

Fig. 2.49   MOS devices with large width must be designed with parallel gates to reduce the delay (MosLayout.MSK)

The most efficient solution is to connect MOS devices in parallel. Firstly, the polysilicon gate length is divided by four in the case of Figure 2.49. Secondly, the structures become regular, easing the future interconnections. Notice how drains and sources are interleaved to create an equivalent device with the same channel width and length, and consequently the same current ability.

2.11 CMOS Process

The process steps to fabricate the integrated circuit [Chang] are illustrated in this paragraph. The starting material is an extremely pure silicon substrate, entering the foundry as a thin circular wafer. The wafer diameter for most 0.12 μm technologies is 8 inches. Most CMOS processes use lightly doped P-type wafers.

2.11.1 Masks

The complete fabrication includes a series of chemical steps in order to diffuse doping materials (n-well, N+ and P+ diffusions), or to deposit materials (polysilicon, contacts and metals). The chemical attack is driven by optical masks on which the patterns are drawn. The masks are shown in Figure 2.50.

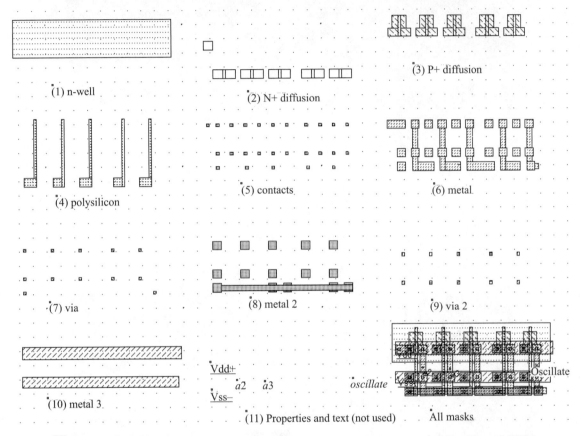

Fig. 2.50   Splitting the layers of the INV5 circuit into separate masks (Inv5Steps.MSK)

In deep submicron technologies, the price of these mask accounts for a significant percentage of the total cost of the chip fabrication. The reason is the extreme precision of each mask, which must have no defect at all, in order to succeed in fabricating the chip correctly. The cost of a complete set of masks in 0.12 μm is around 100,000 euros. This price should rise significantly in nano-scale technologies (90 nm and below), and become the primary limiting factor.

2.11.2 CMOS Process Steps

Let us first load the ring oscillator used in Chapter 1 to illustrate the increase in operating frequency with the scale down (INV5.MSK). The masks can be split into 11 layers as described in Figure 2.50. The illustration of the process steps piloted by these masks is included in Microwind2. The access command is **Simulate → Process steps in 3D**.

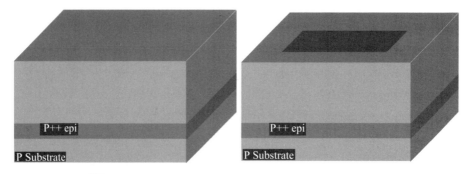

Fig. 2.51   The fabrication of the n-well (Inv5.MSK)

A portion of the substrate is shown in Figure 2.51. The substrate is a P-type wafer, with a resistivity of around 10 Ωcm. A P++ layer is situated some microns below the surface of the wafer. Due to its very low resistivity, it serves as a ground plane. Not all CMOS processes use this P++ layer, mainly for cost reasons. In the 3D process window, click *Next step* to skip to the next technological step.

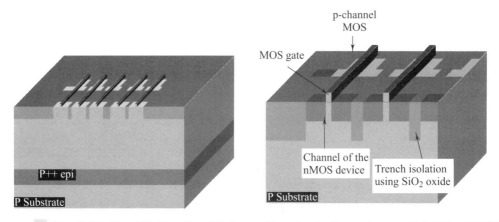

Fig. 2.52   N+ diffusion, P+ diffusion (left) and polysilicon deposit (right) (Inv5.MSK)

The n-well mask is used here to build the n-well area, into which p-channel devices will operate. n-channel devices use the native P-type substrate, without the need for a p-well. Next, a thick oxide is created to isolate MOS devices. In 0.12 μm, this step is called shallow trench isolation. Then, a crucial step consists in growing a very high quality, extremely thin oxide that isolates the gate from the channel. Extraordinary precautions are taken to create millions of atomic scale gate oxides, around 18 Å thick in 0.12 μm, that is 8 atoms of SiO_2. On the top of that ultra-thin oxide, the polysilicon gate is deposited (Figure 2.52).

Then, N-type dopant ions, usually using arsenic of phosphor, are implanted to form the drain and source regions of the n-channel MOS devices. The polysilicon gates block the ions and protect the underlying channel from any n+ dopant. The gate serves as a mask to separate the implanted area into two electrically different regions, the source and the drain. Consecutively, P-type boron ions are implanted to form the p-channel drain and sources.

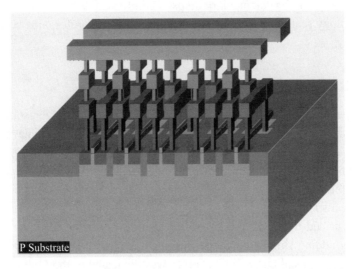

Fig. 2.53   Fabrication of the metal interconnects (oxide not shown for clarity) (Inv5.MSK)

The next steps are related to metallization. Up to six metal layers can be fabricated on top of each other in 0.12 µm. When the contact and metal steps are completed, the chip takes the aspect shown in Figure 2.53. Notice that oxides have been removed for clarity (Figure 2.53). You may see the oxide in the window showing the process aspect in 2D, accessible from the simulation menu of Microwind2.

At the end of the process comes the passivation, consisting in the growth of an oxide, usually Si_3N_4. The structure of the oxides that separate the metal layers is not homogeneous. Recent advances in lithography have generalized the use of low dielectric oxides together with the traditional native oxide SiO_2. The combinational use of SiO_2 and low dielectric oxide will be justified in Chapter 5, dedicated to interconnects.

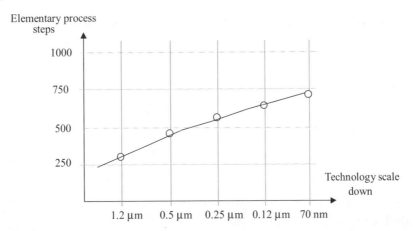

Fig. 2.54   Increase in elementary process steps with the technology scale down

Each process step appearing in the process simulation window of Microwind2 is the sum of elementary chemical, mechanical and optical steps. Consequently, a complete technological process such as 0.12 µm is made up of more than 600 elementary steps. The global trend is an increase of technological steps with the technology scale down, as shown in Figure 2.54. Therefore, the cost of fabrication also increases accordingly.

2.12 Conclusion

In this chapter, we have described the atomic structure of silicon, and given some information about P-type and N-type materials. Silicon dioxide has also been studied. Then, we have presented the MOS device from a simple functional point of view, and focused on its switching properties. We have presented the two types of MOS devices: n-channel and p-channel MOS. Through analog simulation, we have exhibited the good and poor performances of these switches, depending on the logical information. We have also proposed a good switch that takes advantage of nMOS and pMOS properties. Finally, we have described the MOS layout editing using Microwind, and showed how to conduct analog simulation by adding simulation properties such as clocks, supplies, and sinusoidal voltages directly on the layout. The final part of this chapter was dedicated to the CMOS process steps.

References

1. R.J Baker, H. W.Li, D. E. Boyce "CMOS design, layout and simulation", IEEE Press, 1998, Chapters 3 and 4, *www.ieee.org*.

2. M. John, S. Smidth, "Application-specific integrated circuits", Addison Wesley, 1997, ISBN 0-201-50022-1, Chapter 2, *www.awl.com/cseng*.

3. B. Razavi "Design of analog CMOS integrated circuits", McGraw-Hill, ISBN 0-07-238032-2, 2001, *www.mhhe.com*.

4. C. Y. Chang, S.M. Sze "ULSI technology", McGraw-Hill, 1996, ISBN 0-07-063062-3 [Hastings]

EXERCISES

1. Layout a single gate n-channel MOS with W = 1.0 µm, L = 0.12 µm. What is the maximum current? Build the equivalent MOS device with parallel gates.

2. Using the current evaluator in the MOS generator menu in Microwind2, find a simple relationship between the maximum current Imax and the width and length of the device.

3. What are the equivalent width and length of the device drawn in the figure below? What is the possible application of such a design style?

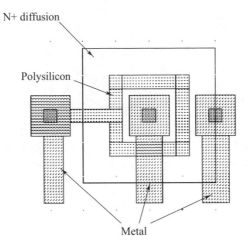

4. What is the width and length of this device? What is the main drawback of this design style?

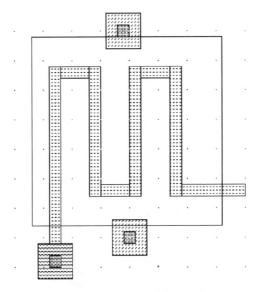

5. What is the maximum current which can flow within the interconnect without any damage?

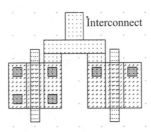

6. Add the appropriate layers to create the biggest MOS you can create with the following structure. What is the MOS size? Find the maximum current which can flow within this MOS without any damage.

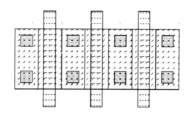

7. Consider a sample of silicon (at room temperature) doped P-type with a boron density of around 10^{16} cm^{-3}. Find the number of minority carriers in this P-type sample.

8. For each transistor nMOS and pMOS show the drain and the source.

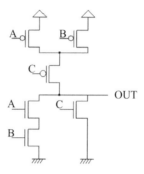

9. Considering each transistor as a perfect switch, find all the combinations that allow the connection between X and Y.

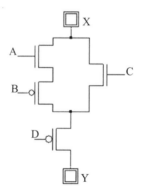

3

The MOS Modelling

This chapter introduces CMOS transistor modelling. The static characteristics of n-channel and p-channel MOS devices are shown, with details of the maximum current and its relationship with the sizing, the threshold voltage and various second order effects. Three generations of MOS device models are introduced. Firstly, the original MOS model 1 is presented, as it was proposed in the early versions of the SPICE simulator developed by the University of Berkeley, California. We demonstrate the inaccuracies of this model. Secondly, we introduce the semi-empirical model 3, which is still in use for MOS device simulation with a channel length greater than 1 μm. Thirdly, we present a simplified version of the BSMI4 models, developed by the University of Berkeley for MOS devices with channel lengths down to 90 nm.

Details of model parameters are provided for all models. The effects of temperature on the MOS performances are then presented. Finally, the three different MOS that may be found in 0.12 μm are introduced: low threshold voltage, high speed and high voltage.

3.1 Introduction to Modelling

Modelling the MOS device consists in writing a set of equations that link voltages and currents, in order to simulate and predict the behaviour of the single device [Shockley] and consequently the behaviour of a complete circuit. A considerable research and development effort has been dedicated in the past years to modelling MOS devices in an accurate way. Many books have been published over the years about semiconductor physics and semiconductor device modelling. The most common references are [Tsividis], [Sze], [Lee] and recently [Liu]. For MOS devices, one of the key objectives of the model is to evaluate the current Ids which flows between the drain and the source, depending on the supply voltages Vd, Vg, Vs and Vb.

From the equation 3.1, we may represent the variation of the current Ids versus voltages in three different ways, as illustrated in Figure 3.1. The graphs are usually called Id/Vd, Id/Vg, and $Id(log)/Vg$. For the sake of simplicity's, we consider that the voltage Vs is grounded.

In the Id/Vd curve, the current Ids is plotted for varying gate voltage Vgs, from 0 to VDD. The parameter Ion gives the maximum available current, corresponding to maximum voltages Vds and Vgs. Ion is a very important parameter for signal switching, for example in logic gates.

In the *Id/Vg* curve, we clearly see the threshold voltage. In the previous chapter, we observed the parasitic effects due to this threshold. Analog design is greatly concerned with an accurate prediction of the threshold voltage. Then, the curve *Id(log)/Vg* is convenient for illustrating the current *Ids* for small values of the gate control. One of the most important parameters is the *Ioff* current, when *Vg* = 0, that has a direct impact on standby power consumption.

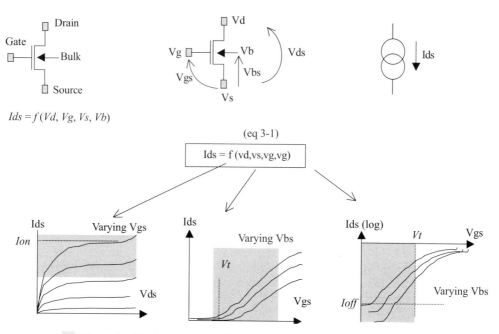

Fig. 3.1   Useful representations of the MOS device characteristics

A second objective of MOS models is to estimate the value of parasitic capacitances, mainly *Cgs*, *Cgd* and *Cgb* (Figure 3.2). Those capacitances vary with the voltages *Vs*,*Vd*,*Vg* and *Vb*. Although not considered in the static simulations *Id/Vd* and *Id/Vg*, the variation of the capacitance must be computed at each iteration of the analog simulation, to facilitate prediction of the switching delay.

$$Cgs = f_1 \ (Vd, \ Vg, \ Vs, \ Vb) \qquad \text{(Equation 3.2)}$$

$$Cgd = f_2 \ (Vd, \ Vg, \ Vs, \ Vb) \qquad \text{(Equation 3.3)}$$

$$Cgb = f_3 \ (Vd, \ Vg, \ Vs, \ Vb) \qquad \text{(Equation 3.4)}$$

A long list of MOS models has been developed for analog simulators. We choose to implement in MICROWIND three of these: the model 1, the model 3 and the model BSIM4. Details of these three models and their physical basis are provided in the next paragraphs.

The complete set of parameters for a given technology is called the model card. The procedure to build an accurate MOS model is quite complex, as it is based on a large set of measurements and sophisticated optimization procedures. The experimental data concerning an MOS device with large width and large length is used first, to fix basic parameters. Then the MOS model is tuned for small channel device measurements, and then for several sizes.

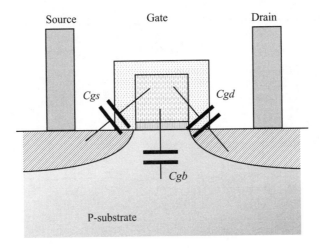

Fig. 3.2   Capacitance between the gate and the source, drain, or substrate

3.2   MOS Model 1

3.2.1   Equations

Historically, the MOS model 1 was the first proposed in 1952 [Shockley]. The equations of the MOS level 1 are provided in the next paragraphs. The evaluation of the current *Ids* between the drain and the source as a function of *Vd*, *Vg* and *Vs* is summarized in equations 3.5, 3.6 and 3.7. The model parameters appearing in the user interface of MICROWIND2 are written using the COURRIER font. The device operation is divided into three regions: cut-off, linear and saturated.

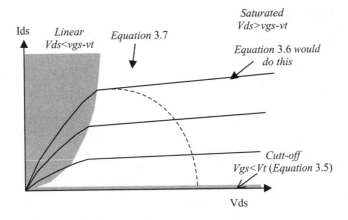

Fig. 3.3   Two main domains are considered in the model: the linear area and the saturated area

IF $V_{gs} < 0$, the device is in cut-off mode.

$$Ids = 0 \qquad \text{(Equation 3.5)}$$

IF $V_{ds} < V_{gs} - \text{VTO}$, the device is in linear mode.

$$Ids = \text{UO}\, \frac{\varepsilon_0 \varepsilon_r}{\text{TOX}} \cdot \frac{W}{L} \left(\left(V_{gs} - vt \right) . V_{ds} - \frac{(V_{ds})^2}{2} \right) \qquad \text{(Equation 3.6)}$$

IF $V_{ds} > V_{gs} - \text{VTO}$, the device is in saturated mode.

$$Ids = \text{UO}\, \frac{\varepsilon_0 \varepsilon_r}{\text{TOX}} \cdot \frac{W}{L} (V_{gs} - vt)^2 \qquad \text{(Equation 3.7)}$$

With:

$$vt = \text{VTO} + \text{GAMMA} \left(\sqrt{(\text{PHI}- vbs)} - \sqrt{\text{PHI}} \right) \qquad \text{(Equation 3.8)}$$

$\varepsilon_0 = 8.85\ 10^{-12}$ F/m is the absolute permittivity

ε_r = relative permittivity, equal to 3.9 in the case of SiO_2 (no unit)

Table 3.1   Parameters of MOS level 1 implemented into Microwind2

MOS Model 1 parameters			
Parameter	Definition	Typical Value 0.12 µm	
		n-MOS	p-MOS
VTO	Threshold voltage	0.4 V	− 0.4 V
UO	Carrier mobility	0.06 m^2/ V-s	0.02 m^2/V-s
TOX	Gate oxide thickness	2 nm	2 nm
PHI	Surface potential at strong inversion	0.3 V	0.3 V
GAMMA	Bulk threshold parameter	0.4 V$^{0.5}$	0.4 V$^{0.5}$
W	MOS channel width	1 µm	1 µm
L	MOS channel length	0.12 µm	0.12 µm

3.2.2 Implementation in Microwind

The static characteristics of the MOS model 1 may be obtained by using the command **Simulate** → **MOS characteristics** available in the main menu of Microwind.

In the top right part of the window, select the item "Level 1". The variation of *Ids* versus the voltage *Vds*, for varying gate voltage *Vgs*, is shown by default. The device width is 10 µm by default while the channel length is 0.12 µm. The parameters VTO,U0,TOX, PHI and GAMMA are listed in the right part of the window.

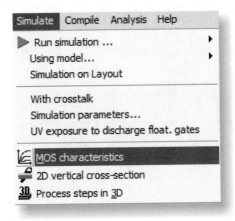

Fig. 3.4 Access to the static MOS characteristics

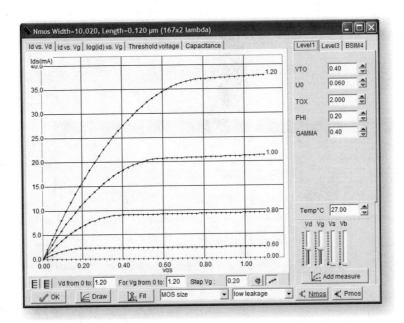

Fig. 3.5 The screen used to simulate the static characteristics of the MOS with model 1 within Microwind2

3.2.3 Mismatch between Simulation and Measurements

These old equations (1968, in [Shichman]) are not acceptable in 0.12 µm. If we consider MOS devices with very long lengths (L > 10 µm), the mismatch between the simulation and the measurement is of the order of a factor of five. Let us compare the simulation and the measurement, for a device with a width W = 10 µm, and a long channel length L = 10 µm, fabricated in 0.12 µm CMOS technology, as presented in Figure 3.6. The measurement *Ne10x10.MES* was downloaded using the button **Load Measurement**. This measurement corresponds to an n-channel MOS device with a 10 µm channel width and 10 µm length, fabricated in CMOS 0.12 µm from ST-microelectronics.

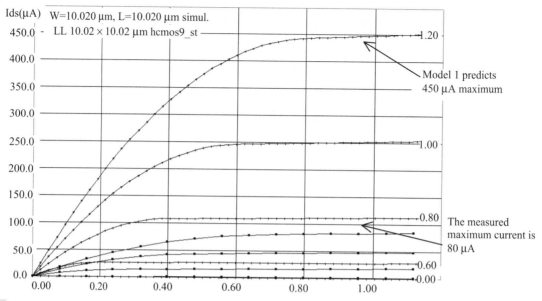

Fig. 3.6   The model 1 predicts a current five times higher than the measurement in the case of a large channel MOS device (L = 10 μm)

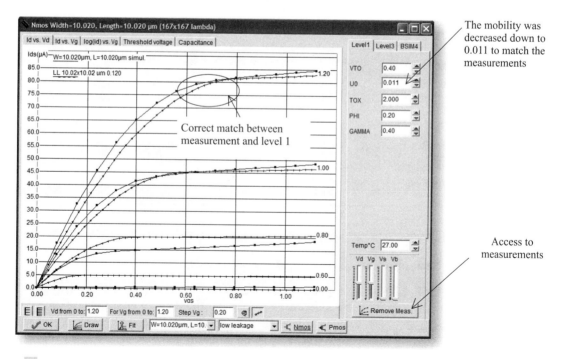

Fig. 3.7   Comparing measured Id/Vd and level 1 simulations for a 10 × 10 μm device result in a surprising similarity (Ne10x10.MES)

Initially, the simulation and measurement do not correspond at all. The mobility U0 needs to be decreased from its initial value of 0.06 down to 0.01. The curves are fitted at the price of an unrealistic change in the mobility parameter.

When dealing with submicron technology, the current predicted by model "level 1" is several times higher than the real-case measurements. This means that several parasitic effects appeared with the technology scale down, with most of them tending to reduce the effective current as compared to the early modelling equations of model 1.

3.3   MOS Model 3

For the evaluation of the current *Ids* as a function of *Vd*, *Vg* and *Vs* between the drain and source, we commonly use the following equations, close to SPICE model 3 formulations [Weste]. The formulations are derived from model 1 and take into account a set of physical limitations in a semi-empirical way.

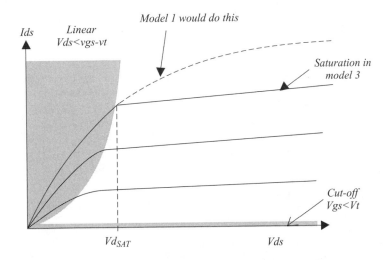

Fig. 3.8   Introduction of the saturation voltage Vd_{SAT} which truncates the equations issued from model 1

One of the most important changes is the introduction of Vd_{SAT}, a saturation voltage from which the current saturates and does not rise as the level1 model would do. This saturation effect is significant for small channel length. The main Level 3 equations are listed below.

CUT-OFF MODE. Vgs<0

$Ids = 0$ (Equation 3.9)

NORMAL MODE. Vgs>Von

$$Ids = Keff \, \frac{W}{Leff}(1 + KAPPA.Vds)Vde\left((Vgs - Vth) - \frac{Vde}{2}\right)$$ (Equation 3.10)

with

$$Von = 1.2Vth$$

$$Vth = VTO + GAMMA(\sqrt{PHI - Vbs} - \sqrt{PHI})$$

$$Vde = min(Vds, Vdsat)$$

$$Vdsat = Vc + Vsat - \sqrt{Vc^2 + Vsat^2}$$

$$Vdsat = Vgs - Vth$$

$$Vc = VMAX\frac{Leff}{0.06}$$

$$Leff = L - 2LD$$

The formulation of the effective factor *Keff* (Equation 3.11) includes a mobility degradation factor THETA, which tends to reduce the mobility at high *Vgs*. The consequence is a reduction of the current *Ids* as compared to level1.

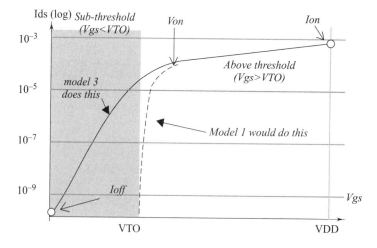

Fig. 3.9   Introduction of an exponential law to model the sub-threshold behavior of the current

$$Keff = \frac{\varepsilon_0\varepsilon_r}{TOX}\frac{UO}{(1 + THETA(vgs - vth))} \qquad \text{(Equation 3.11)}$$

In sub-threshold mode, that is for a gate voltage which is less than the threshold voltage, *Vds* is replaced by *Von* in the above equations. An exponential dependence of the current with *Vgs* is introduced by using the Equation 3.12. Notice the temperature effect introduced in the denominator *nkT*.

Without any voltage applied to the gate, the current is no longer equal to zero. The current of *Ids* for *Vgs* = 0 is called the *Ioff* current (Figure 3.9). Its value in 0.1 2 µm is around 10^{-10} A. In contrast, for *Vgs* = VDD, the maximum current *Ion* is of the order of several mA (10^{-3} A).

$$Ids = Ids(Von, Vds) \exp\left(\frac{q(Vgs - Von)}{nkT}\right)$$ (Equation 3.12)

3.3.1 Temperature Effects

The MOS device is sensitive to temperature. Three main parameters are concerned: the threshold voltage VTO, the mobility U0 and the slope in sub-threshold mode dependent on *kT/q*. Both VTO and U0 decrease when the temperature increases. The physical background is the degradation of the mobility of electrons and holes when the temperature increases due to a higher atomic volume of the crystal underneath the gate, and consequently less space for the current carriers. The modelling of the temperature effect is as follows:

$$U0 = U0_{(T = 27)}\left(\frac{T + 273}{300}\right)^{-1.5}$$ (Equation 3.13)

$$VT = VT0_{(T = 27)} - 0.002(T - 300)$$ (Equation 3.14)

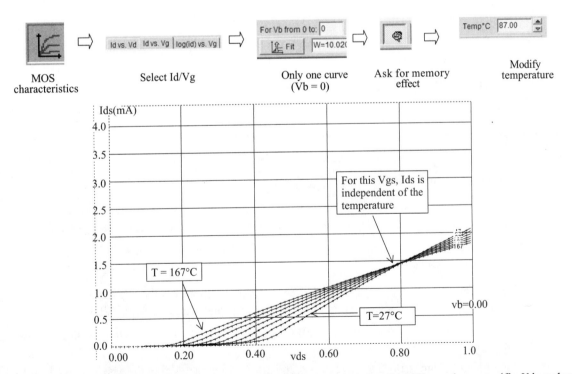

Fig. 3.10   The effect of temperature on the MOS characteristics. In *Id/Vg* mode, a specific *Vds* makes the current independent of the temperature

In order to obtain the curve of Figure 3.10, click the icon MOS characteristics, select the curve *Id/Vg*, and enter the value "0" for the upper limit of *Vb*, so as to draw only one single curve. Enable the screen memory mode by a click on the icon **Enable Memory**. When you change the temperature, the change in the slope and the temperature-independent point appears, as shown in Figure 3.10.

Table 3.2   List of parameters used in the implementation of the level3 model in Microwind

MOS Model 3 parameters			
Parameter	**Definition**	**Typical Value 0.12 μm**	
		NMOS	**pMOS**
VTO	Theshold voltage of a long channel device, at zero Vbs.	0.4 V	−0.4 V
U0	Carrier mobility	0.06 m^2/V.s	0.025 m^2/V.s
TOX	Gate oxide thickness	3 nm	3 nm
PHI	Surface potential at strong inversion	0.3 V	0.3 V
LD	Lateral diffusion into channel	0.01 μm	0.01 μm
GAMMA	Bulk threshold parameter	0.4 V$^{0.5}$	0.4 V$^{0.5}$
KAPPA	Saturation field factor	0.01 V^{-1}	0.01 V^{-1}
VMAX	Maximum drift velocity	150 km/s	100 km/s
THETA	Mobility degradation factor	0.3 V^{-1}	0.3 V^{-1}
NSS	Substhreshold factor	0.07 V^{-1}	0.07 V^{-1}
W	MOS channel width	0.5–20 μm	0.5-40 μm
L	MOS channel length	0.12 μm	0.12 μm

3.3.2   Microwind User's Interface

You may understand the action of each parameter by using the screen reported in Figure 3.11. Each parameter may be changed interactively using cursors, or by entering the appropriate value using the keyboard.

Several screens may be proposed:

- *Id vs. Vd*, for varying *Vg*. This is the default screen. Its main interest is the characterization of the Ion current, the maximum current available in the device, for *Vd* and *Vg* set to VDD.
- *Id vs. Vg*, for varying *Vb*. In this screen, the threshold voltage *Vt* (VTO) as well as its dependence on the bulk polarization are characterized.
- *Id vs. Vg*, in logarithmic scale. This screen is mandatory for characterizing the MOS device in sub-threshold mode, that is for *Vgs* < *Vt*. Two of the important parameters are the slope of the current

vs. *Vgs*, and the *Ioff* current. The *Ioff* current is the standby current appearing between the drain and source for *Vgs* = 0.

- *Threshold voltage Vt vs. Length*. This screen has been added to illustrate advances in the modelling of deep-submicron effects. With level 3, *Vt* is constant for varying length, but is impacted by the bulk voltage.

- *Capacitance vs. Vds*. This screen illustrates the variation of *Cgs* and *Cgd* versus the drain-source voltage.

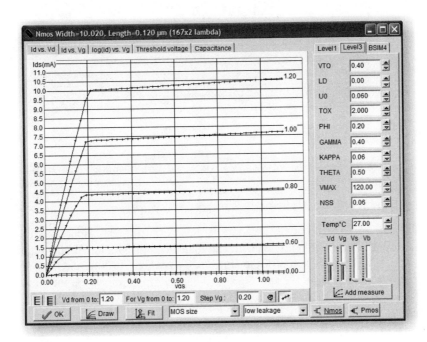

Fig. 3.11 The user interface to investigate the effect of each parameter on the current *Ids* (*W* =10 μm, *L* = 0.12 μm)

3.3.3 Current versus Drain-source Voltage

Using the display mode **Id vs. Vd**, you may see the effect of parameters U0, TOX, KAPPA and VMAX. Basically, the carrier mobility U0 moves the whole curve, as it impacts the current *Ids* in an almost linear way. As U0 is nearly a physical constant, a significant change of mobility has no physical meaning. The oxide thickness TOX does the same but in an opposite way.

A TOX increase leads to a less efficient device, with less current. KAPPA changes the slope of the current when *Vds* is high, corresponding to the saturation region. Finally, VMAX truncates the curves for low values of *Vds*, to fit the transition point between the linear and the saturated region (Figure 3.12).

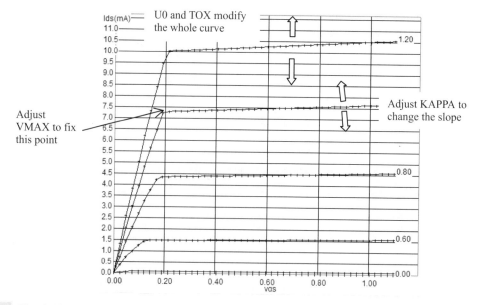

Fig. 3.12   Demonstration of the role of U0, KAPPA and VMAX in Id./Vd ($W = 10 \ \mu m$, $L = 0.12 \ \mu m$)

3.3.4   Current versus Gate Voltage

The role of VTO and GAMMA can be observed in Figure 3.13, using the display mode **Id vs. Vg**. If we use a long channel device, that is a length much greater than the minimum length, the second order effects are minimized. Act on VTO cursors in order to shift the curves right or left, and on GAMMA to fit the spacing between curves. Parameters U0 and TOX also have a direct impact on the slope for high *Vgs*.

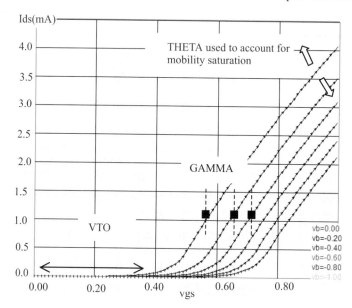

Fig. 3.13   The effects of VTO and GAMMA are illustrated in Id/Vg mode voltage ($W = 10 \ \mu m$, $L = 0.12 \ \mu m$)

Now we focus on a short channel MOS device, for example W = 2 μm, L = 0.12 μm. Using the same display mode **Id vs. Vg**, we obtain similar curves as for a long-channel device. We observe that the shape of the current is bent. This modification is due to short channel parasitic effects. The parameter THETA is used to bend the current curves at high *Vgs*. The MOS model 3 does not provide parameters to account for the VTO dependence with length.

3.3.4   Current versus Vg in Logarithmic Scale

We finally illustrate the role of NSS in the display mode **Id(log)/Vg** (Figure 3.14). The parameter NSS has a direct impact on the slope in sub-threshold mode, that is for *Vgs* < *Vt*.

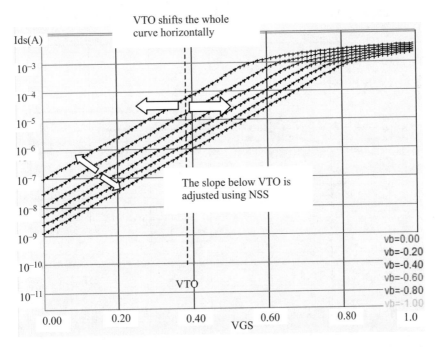

Fig. 3.14   In sub-threshold region, the *Id* dependence on *Vgs* is exponential. The slope is tuned by parameter NSS. The whole curve is shifted using VTO voltage (W = 10 μm, L = 0.12 μm)

3.3.5   Capacitance versus Vds

The five main capacitors considered in our implementation of MOS model 3 are the gate to bulk capacitance *Cgb*, the gate to source capacitance *Cgs*, the gate-to-drain capacitance *Cgd*, the junction capacitance between source and bulk *Csb*, and the junction capacitance between drain and bulk *Cdb*.

The variation of the capacitance must be computed at each iteration of the analog simulation, for accurate prediction of the switching delay. In our implementation of MOS level 3, we use the following model, based on the formulations given in [Fjedly]. The parameter V_{dsat} was given by Equation 3.10.

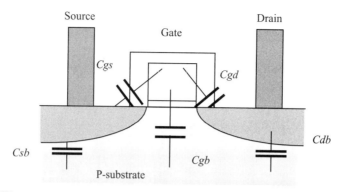

Fig. 3.15   The MOS capacitance considered in MOS model 3

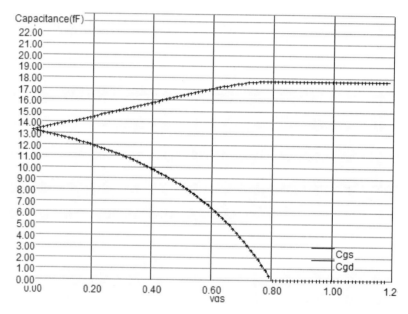

Fig. 3.16   The evolution of MOS capacitance with the drain voltage (W = 10 μm, L = 0.12 μm)

$$C_{GS} = \frac{2}{3} C_i \left[1 - \left(\frac{V_{GS} - V_t - V_{dsat}}{2(V_{GS} - V_t) - V_{dsat}} \right)^2 \right] \quad \text{(Equation 3.15)}$$

$$C_{GD} = \frac{2}{3} C_i \left[1 - \left(\frac{V_{GS} - V_t}{2(V_{GS} - V_t) - V_{dsat}} \right)^2 \right] \quad \text{(Equation 3.16)}$$

$$C_{GB} = 0 \quad \text{(Equation 3.17)}$$

with

$$Ci = W.L. \frac{\varepsilon_0 \varepsilon_r}{TOX} \quad \text{(Equation 3.18)}$$

W = width of the MOS device (m)

L = length of the MOS device (m)

TOX = oxide thickness (m)

The two remaining capacitances C_{DB} and C_{SB} are junction capacitances. Their model is given by Equations 3.19 and 3.20.

$$C_{DB} = W.L_{\text{drain}} \frac{\text{CJ}}{\left(1 - \dfrac{V_{BD}}{\text{PB}}\right)^{\text{MJ}}} \qquad \text{(Equation 3.19)}$$

$$C_{SB} = W.L_{\text{source}} \frac{\text{CJ}}{\left(1 - \dfrac{V_{BS}}{\text{PB}}\right)^{\text{MJ}}} \qquad \text{(Equation 3.20)}$$

where

W is the channel width (m)

L_{drain} is the drain length, according to Figure 3.17 (m)

CJ is around $3 \times 10e\text{-}4$ F/m2

PB is the built-in potential of the junction (around 0.8 V)

MJ is the grading coefficient of the junction (around 0.5)

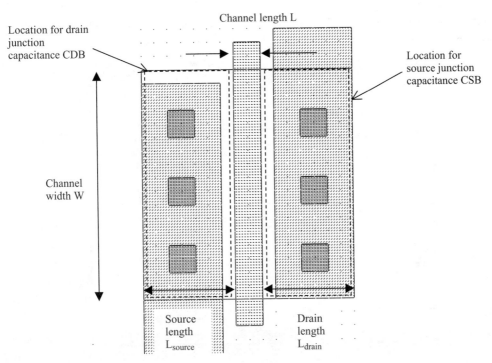

Fig. 3.17   The junction capacitance for drain and source contributes significantly to the MOS capacitance

3.4   The BSIM4 MOS Model

A family of models has been developed at the University of Berkeley for the accurate simulation of sub-micron and deep submicron technologies. The Berkeley Short-channel IGFET Model (BSIM) exists in several versions (BSIM1, BSIM2, BSIM3). The BSIM3v3 [Cheng] version, promoted by the Electronic Industries Alliance (EIA), is an industry standard for deep-submicron device simulation [Eia].

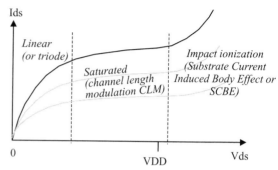

A new MOS model, called BSIM4 [Bsim4], was introduced in 2000. A simplified version of this model is supported by MICROWIND, and recommended for ultra-deep submicron technology simulation. The complete details of BSIM4 are provided in the excellent book [Liu]. BSIM4 still considers the operating regions described in MOS level 3 (linear for low *Vds*, saturated for high *Vds*, sub-threshold for *Vgs<Vt*), but provides a perfect continuity between these regions. BSIM4 introduces a new region wherein the impact ionization effect

Fig. 3.18   The three regions considered in our simplified version of BSIM4

is dominant (Figure 3.18). In that region, *Vds* is very high, over the nominal supply voltage VDD. One of the key features of BSIM4 is the use of one single equation to build the current, valid for all operating modes. Smoothing functions ensure a nice continuity between operating domains.

The number of parameters specified in the official release of BSIM4 is as high as 300. A significant portion of these parameters is unused in our implementation. We concentrate on the most significant parameters for educational purposes. The set of parameters is reduced to around 30.

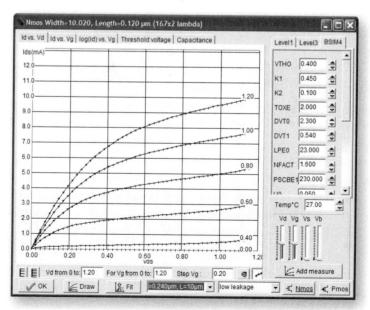

Fig. 3.19   Implementation of BSIM4 within MICROWIND2, based on [Liu]

3.4.1 Effective Channel Length and Width

Once fabricated, the physical length *Leff* and width *Weff* of the MOS device do not correspond exactly to the initial length L and width W drawn using MICROWIND2 (Figure 3.20). The parameters LINT and WINT have been introduced for that purpose, with Equations 3.21 and 3.22.

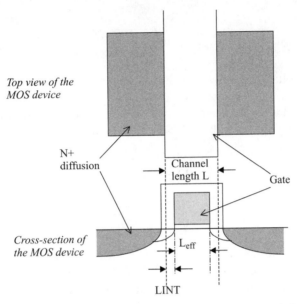

Fig. 3.20   Illustration of the effective channel length Leff

$$Leff = L - 2.LINT \qquad \text{(Equation 3.21)}$$

$$Weff = W - 2.WINT \qquad \text{(Equation 3.22)}$$

3.4.2 Surface Potential and Junction Depth

The surface potential Φ_s and junction depth are basic parameters taken into account in the evaluation of the threshold voltage and the global current. The surface potential Φ_s is defined by Equation 3.23.

$$\Phi_s = 0.4 + vt.In\left(\frac{NDEP}{ni}\right) \qquad \text{(Equation 3.23)}$$

where *vt* is the thermal voltage given by Equation 3.24, NDEP is the channel doping concentration for zero body bias (around 10^{17}cm^{-3} in practice), and *ni* is the intrinsic carrier concentration of silicon (ni = $1.02 \times 10^{10} \text{ cm}^{-3}$ at 300° K). Consequently, the surface potential Φ_s in deep-submicron CMOS process is around 0.85 V.

The thermal voltage is

$$vt = \frac{k_B T}{q} \qquad \text{(Equation 3.24)}$$

k_B = Boldzmann constant = 1.38×10^{-23} J/K

T = temperature (300°K by default)

q = Electronic charge = 1.60×10^{-19} C

The built-in voltage of the source/drain junctions is given by Equation 3.25.

$$V_{bi} = vt. \ \text{In}\left(\frac{NDEP.NSD}{ni^2}\right)$$ (Equation 3.25)

where *vt* is the thermal voltage given by Equation 3.24, *NDEP* is the channel doping concentration for zero body bias (around 10^{17}cm^{-3} in practice), *NSD* is the source/drain doping concentration (around 10^{20}cm^{-3} in practice), and *ni* is the intrinsic carrier concentration of silicon (ni = 1.02×10^{10}cm^{-3} at 300°K). Consequently, the built-in voltage *Vbi* in deep-submicron CMOS process is around 1.0 V.

The depletion depth *Xdep* is computed by Equation 3.26. It corresponds to the thickness of the region near the N+/P– junction interfaces, as illustrated in Figure 3.21.

$$X_{dep} = \sqrt{\frac{2\varepsilon_{rsi}.\varepsilon_0(\Phi_s - Vbs)}{q.NDEP}}$$ (Equation 3.26)

where ε_{rsi} is the dielectric constant of silicon (11.7) ε_0 is the permittivity in vacuum (8.854×10^{-12}F/m), Φs is the surface potential given by Equation 3.23, *NDEP* is the channel doping concentration for zero body bias, *q* is the electronic charge (1.60×10^{-19} C), and *vbs* is the bulk-source potential. The typical value of *Xdep* is 0.5 μm (Figure 3.21).

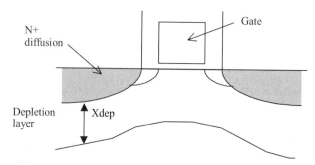

Fig. 3.21   Illustration of the depletion depth Xdep

3.4.3 Threshold Voltage

The main impact of the threshold voltage *Vt* is the *Ioff* parasitic current which exhibits an exponential dependence with *1/Vt*. A high threshold voltage *Vt* leads to a small *Ioff* current, at the price of a low *Ion* current. Low threshold MOS devices consume a very high standby current, which impacts the power consumption of the whole circuit. An accurate prediction of the threshold voltage is a key issue for low power integrated circuit design. The general formulation of the threshold voltage is presented in Equation 3.27.

$$vth = VTHO + K1.\sqrt{\left(\left(\Phi_s - Vbs - \sqrt{\Phi_s}\right)\right)} - K2.Vbs + \Delta Vt_{SCE} + \Delta Vt_{NULD} + \Delta Vt_{DIBL} \quad \text{(Equation 3.27)}$$

where VTHO is the long channel threshold voltage at $Vbs = 0$ (around 0.5 V), K1 is the first order body bias coefficient (0.5 $V^{1/2}$), Φ_s is the surface potential given by Equation 3.23, Vbs is the bulk-source voltage, K2 is the second order body bias coefficient, ΔVt_{SCE} is the short channel effect (SCE) on Vt (detailed in Equation 3.28), ΔVt_{NULD} is the non-uniform lateral doping (NULD) effect explained in Equation 3.29, and ΔVt_{DIBL} is the drain-induced barrier lowering (DIBL) effect of short channel on Vt (detailed in Equation 3.30).

3.4.4 Short Channel Effect

The threshold voltage is not the same for all MOS devices. There is a complex dependence between the threshold voltage and the effective length of the channel. For small channels, the threshold value tends to decrease. Equation 3.28 is proposed, based on an hyperbolic cosine function.

$$\Delta Vt_{SCE} = -\frac{0.5.DVT0}{\cosh\left(DVT1.\dfrac{Leff}{lt} - 1\right)}(Vbi - \Phi_s) \quad \text{(Equation 3.28)}$$

where DVT0 is the first coefficient of the short-channel effect on the threshold voltage (2.2 by default), DVT1 is the second coefficient of the short-channel effect on the threshold voltage (0.53 by default), *Leff* is the effective channel length given in Equation 3.21, and *lt* is the characteristic length, approximated in our implementation to 1/4 of the minimum channel length (0.03 μm for a 0.12 μm), *Vbi* is defined in Equation 3.25, and Φ_s is the surface potential given by Equation 3.23.

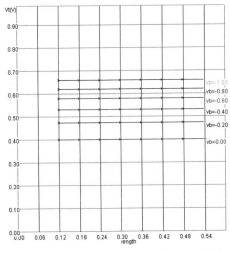

(a) No short channel effect (models 1 and 3)

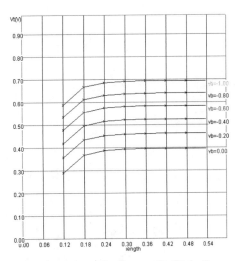

(b) Short channel effect on Vt (Bsim4)

Fig. 3.22 Short channel effect (SCE) on the threshold voltage

The illustration of the effect of ΔVt_{SCE} is proposed in Figure 3.22. Without taking into account the short-channel effect, the threshold voltage is only dependent on *Vbs*. It can be seen that *Vt* increases when *Vbs* decreases. There is no dependence on the length. When we add the contribution of the short-channel effect expressed by Equation 3.28, the threshold voltage is decreased significantly for small length values.

3.4.4   Non-uniform Lateral Doping

The lateral drain diffusion (LDD) is a technique introduced in recent technologies to reduce the peak channel fields in the MOS channel. The location for high-field parasitic effects is illustrated in the process section of Figure 3.23 (a).

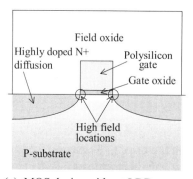

(a) MOS device without LDD structure   (b) MOS device with LDD structure

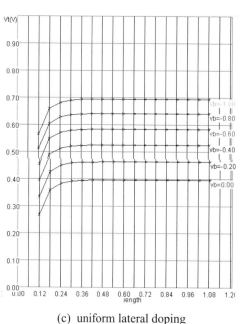
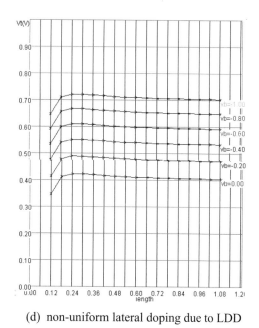

(c) uniform lateral doping   (d) non-uniform lateral doping due to LDD

Fig. 3.23   Effect of non-uniform lateral doping on the threshold dependence with the channel length

In the cross-section of Figure 3.23 (b), the doping concentration at the corner of the gate is reduced thanks to a lightly doped N-type implantation. The Si_3N_4 spacer is grown over the gate before the N+ highly doped implantation is performed. Consequently, the high-field effects are moderated. Unfortunately, the threshold voltage exhibits a complex dependence on the channel length, as illustrated in Figure 3.23 (d), as compared to (c). For a decreasing length, the threshold voltage tends to increase first (due to ΔVt_{NULD}), before decreasing rapidly due to the short channel effect ΔVt_{SCE} described in Equation 3.28.

A simple formulation of the non-uniform lateral doping is ΔVt_{NULD} given below:

$$\Delta Vt_{NULD} = K1\left(\sqrt{1 + \frac{LPE0}{L_{eff}}} - 1\right) \cdot \sqrt{\Phi_s}$$ (Equation 3.29)

3.4.5 Drain-Induced Barrier Lowering

When we apply a positive voltage on the drain of a long-channel nMOS device, we observe no significant change in the value of *Vt*. When we do the same for a short-channel nMOS device, we observe a decrease in the threshold voltage. The physical origin of DIBL is the increase of the depletion layer due to a high value of *Vds* that reduces the equivalent channel length, and consequently decreases the threshold voltage [Liu].

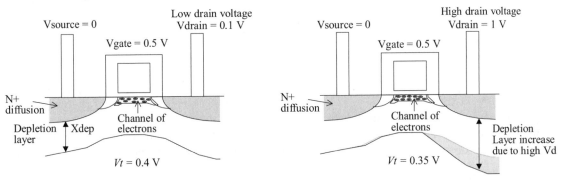

Fig. 3.24 Illustration of the depletion depth Xdep

A simplified model of the DIBL effect on the threshold voltage is proposed in Equation 3.30. The parameter ETA0 is the DIBL coefficient in the sub-threshold region (default value 0.08), and *Vds* is the drain-source voltage.

$$\Delta Vt_{DIBL} = -0.5.ETA0.Vds$$ (Equation 3.30)

3.4.6 Mobility

In this paragraph, we introduce the formulations for the mobility of channel carriers. The generic parameter is U0, the mobility of electrons and holes. The effective mobility μ_{eff} is reduced due to several effects: the bulk polarization and the gate voltage. The equation implemented in MICROWIND2 is one of the mobility models proposed in BSIM4 (Equation 3.31).

$$\mu_{eff} = \frac{U0}{1 + (UA + UC.V_{BS})\left(\dfrac{V_{gsteff} + 2(VTH0 - V_{fb} - \phi_s)}{TOXE}\right)^{EU}}$$ (Equation 3.31)

where

U0 is the low field mobility, in m^2/V–s. Its default value is around 0.06 for n-channel MOS and 0.025 for p-channel MOS.

UA is the first order mobility degradation coefficient, in m/V. Its default value is around 10^{-15}.

UC is the body-effect coefficient of mobility degradation, in m/V^2. Its default value is -0.045×10^{-15}.

VFB is the flat band voltage, in V. It is computed by using Equation 3.32, where Φ_S is derived from Equation 3.23. Its value is around 0.8 V.

$$V_{FB} = \text{VTO} - \Phi_S - \text{K1}\sqrt{\Phi_S} \qquad \text{(Equation 3.32)}$$

TOXE is the oxide thickness, in m. A typical value for TOXE in 0.12 μm is 2 nm (2.10^{-9}m).

V_{BS} is the voltage difference between the bulk and the source (V).

EU is a coefficient equal to 1.67 for n-channel MOS, and 1.0 for p-channel MOS.

The parameter $Vgst_{eff}$ is a smoothing function, to ensure continuity between the sub-threshold region and the linear region.

$$Vgst_{eff} = max\left(\text{VOFF}, \frac{n.vt.ln\left(1 + exp\left(\dfrac{(Vgs - Vth)}{n.vt}\right)\right)}{1 + n.exp\left(\dfrac{-(Vgs - Vth)}{n.vt}\right)} \right) \qquad \text{(Equation 3.33)}$$

$$n = 1 + \text{NFACTOR} \qquad \text{(Equation 3.34)}$$

A specific parameter VOFF is introduced to account for a saturation effect appearing in a short-channel device when *Vgs* is negative. Conventional models predict that the current decrease with an exponential law down to zero with decreasing *Vgs*. For *Vgs* < 0, *Ids* is supposed to be 0.

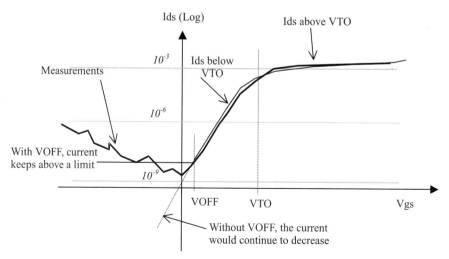

Fig. 3.25 Illustration of the gate-induced drain leakage (GIDL) for negative Vgs

In reality, *Ids* stops decreasing near zero *Vgs*, and even tends to increase with negative *Vgs* (Figure 3.25). This effect is called gate-induced drain leakage (GIDL). Consequently, the leakage current *Ioff* can be significant when *Vgs* is negative (quite frequent in logic cells). The VOFF parameter stops the *Ids* at a certain value, a simplified version of the BSIM4 modelling of the so-called gate-induced leakage current (more information may be found in [Liu]).

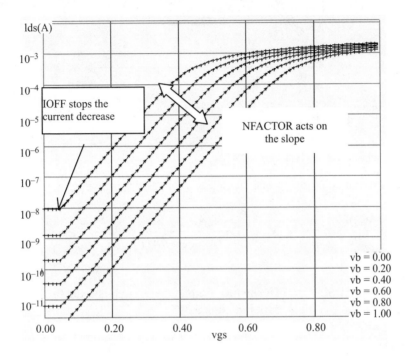

Fig. 3.26   Illustration of the effects of IOFF and NFACTOR in sub-threshold mode

The parameter NFACTOR is usually close to 1, meaning that *n* is close from 2 (Equation 3.34). The effect on NFACTOR is illustrated in the display mode **Id. vs. Vg**, in logarithmic scale, as illustrated in Figure 3.26.

$$Esat = 2\frac{Vsat}{\mu eff} \qquad \text{(Equation 3.35)}$$

$$Vdsat = Esat.L\frac{(Vgsteff + 2.vt)}{(Esat.L + VgstEff + 2.vt)} \qquad \text{(Equation 3.36)}$$

Again, *VdsEff* is defined so as to smoother the evolution from *Vds* to the saturation voltage *Vdsat* (Equation 3.37). The parameter *DELTA* is fixed to 0.01. The effect of *DELTA* is shown in Figure 3.27. With a small value of *DELTA* (0.001 for example), the transition between the linear and saturated regions leads to a discontinuity. Experimental measurements show a gradual transition that is well approximated when *DELTA* = 0.01. A higher value of *DELTA* would lead to an *Ids* curve that is significantly lower than measurements.

$$Vdseff = Vdssat - 0.5\left(Vdsat - Vds - \delta + \sqrt{(Vdsat - Vds - \delta)^2 + 4\delta.Vdsat}\right) \quad \text{(Equation 3.37)}$$

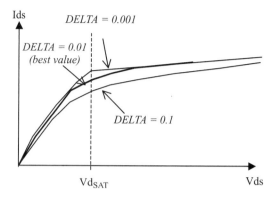

Fig. 3.27 The smoothing function between linear and saturated regions can be modulated by DELTA. In MICROWIND2, DELTA is fixed to 0.01

3.4.7 Current Ids

The current *Ids* is computed by using one single equation, as described below.

$$IdsO = \frac{Weff}{Leff} \mu eff \frac{\varepsilon_0 \, \varepsilon_r}{\text{TOXE}} V_{gsteff} \left(1 - \frac{A_{bulk} V_{dseff}}{(2V_{gsteff} + 4.vt)}\right) \frac{V_{dseff}}{\left(1 + \frac{V_{dseff}}{\varepsilon_{sat} L_{eff}}\right)} \qquad \text{(Equation 3.38)}$$

In our implementation of BSIM4 in **Microwind**, the parameter A_{bulk} is fixed to 1. The final current *Ids* used in analog simulation is computed by Equation 3.39:

$$Ids = IdsO \left(1 + \frac{(V_{ds} - V_{dseff})}{V_{ascbe}}\right)\left(1 + \frac{1}{C_{clm}} ln\left(\frac{V_{ASAT} + V_{ACLM}}{V_{ASAT}}\right)\right) \qquad \text{(Equation 3.39)}$$

Two new terms appear after *IdsO*. The second term of the current equation accounts for impact ionization. It corresponds to a parasitic current at very high *Vds*, created by hot electrons and generating supplementary pairs or electrons and holes, when hitting the drain region after acquiring a high energy level inside the device channel. The parameter V_{ascbe} is a voltage below which the impact ionization becomes significant. If V_{ascbe} is large, *Ids* is almost equal to *IdsO*, meaning that there is no impact ionization effect. If V_{ascbe} is small, the shape of *Ids* is changed for high *Vds*, as illustrated in Figure 3.28.

Two parameters affect the shape of the ionization current: PSCBE1 and PSCBE2. The first parameter can be changed interactively on the screen. The voltage V_{asbe} is determined by using the following equations:

$$V_{ascbe} = \frac{L_{eff}}{\text{PSCBE2}} exp\left(\text{PSCBE1} \frac{litl}{(V_{ds} - V_{dseff})}\right) \qquad \text{(Equation 3.40)}$$

with

$$Litl = \sqrt{\text{XJ} \frac{\varepsilon_{rsi}}{\varepsilon_{rsiO2}} \text{TOXE}} \qquad \text{(Equation 3.41)}$$

XJ is the source/drain junction depth, around 0.1 μm $(10^{-7}$m$)$

TOXE is the oxide thickness, in m (around 3 nm in 0.12 μm)

ε_{rsi} = relative permittivity of silicon (11.7)

$\varepsilon_{rsi\ O2}$ = relative permittivity of silicon oxide (3.9)

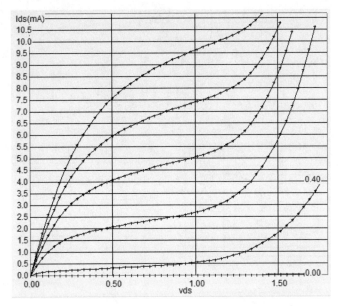

Fig. 3.28   Effect of impact ionization at large *Vds* ($W = 10$ μm, $L = 0.12$ μm)

The third term of Equation 3.39 accounts for the channel length modulation. An illustration of this phenomenon is provided in Figure 3.29. It represents the *Ids* increase with large *Vds*. Graphically, V_{ACLM} is equivalent to an early voltage, i.e. the value for which the *Ids* slope would cross the horizontal axis for negative *Vds*. For long channel devices ($L = 1$ μm for example), the effect of the channel length modulation effect is small, so V_{ACLM} has a very high value (10 V). For very short channels (0.12 μm in the case of Figure 3.30), a significant channel length modulation effect is observed, and V_{ACLM} is small.

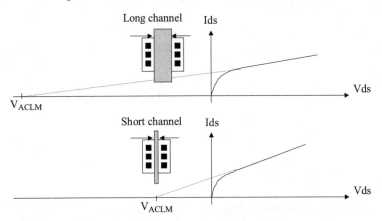

Fig. 3.29   The channel length modulation is significant for short channel devices, and corresponds to the *Ids* increase at high *Vds*

Only one new parameter, PCLM, is introduced in the equations. The parameters V_{ASAT} and V_{ACLM} are detailed below. The original equations from BSIM4 have been significantly simplified, and some fitting parameters have been ignored. See [Liu] for a description and relevant comments about the original equations.

$$C_{clm} = \frac{1}{PCLM \cdot litl} \left(l_{eff} + \frac{V_{dsat}}{\varepsilon_{sat}} \right) \qquad \text{(Equation 3.42)}$$

$$V_{ACLM} = C_{clm} (V_{ds} - V_{dseff}) \qquad \text{(Equation 3.43)}$$

$$V_{ASAT} = (\varepsilon_{sat} \cdot l_{eff} + V_{DSSat}) \left(1 - \frac{A_{bulk} V_{dsat}}{2(V_{gsteff} + 2Vt)} \right) \qquad \text{(Equation 3.44)}$$

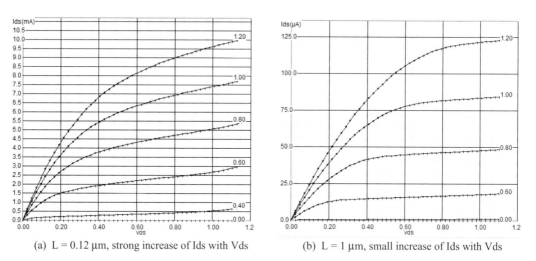

(a) L = 0.12 μm, strong increase of Ids with Vds   (b) L = 1 μm, small increase of Ids with Vds

Fig. 3.30   The channel length modulation effect is significant for short channel devices in 0.12 μm technology

3.4.8   Temperature Effects

Three main parameters are concerned with the sensitivity to temperature: the threshold voltage VTO, the mobility U0 and the slope in sub-threshold mode. Both VTO and *U0* decrease when the temperature increases. The modelling of the temperature effect in BSIM4 is as follows. In MICROWIND2, TNOM is fixed to 300°K, equivalent to 27°C. UTE is negative, and set to −1.8 in 0.12 μm CMOS technology, while KT1 is set to − 0.06 by default.

$$U0 = U0_{(T = 27)} \left(\frac{T + 273}{TNOM} \right)^{UTE} \qquad \text{(Equation 3.45)}$$

$$VT = VTO_{(T = 207)} KT1 \left(\frac{T + 273}{TNOM} - 1 \right) \qquad \text{(Equation 3.46)}$$

A higher temperature leads to a reduced mobility, as UTE is negative. Consequently, at a higher temperature, the current *Ids* is lowered. This trend is clearly illustrated in Figure 3.31. The reduction of the maximum current is 40 per cent between −30°C and 100°C.

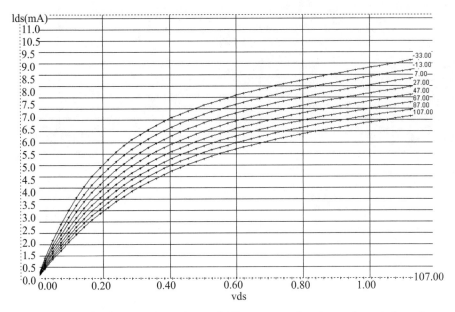

Fig. 3.31 The effect of temperature on the peak Ids current, showing a degradation of current with rising temperature

For a short channel n-channel MOS device ($L = 0.12\ \mu m$), the result of the parametric analysis illustrates the same trend (Figure 3.33). The parametric analysis is conducted as follows: the layout **MosTemperature.MSK** is loaded first. The MOS is polarized with a gate always on, the drain at VSS, and the source at VDD. The parametric analysis is launched. In the new window, select the temperature (upper menu) and the maximum current (lower menu). In Figure 3.33, we observe a significant decrease of the *Ion* current with the temperature.

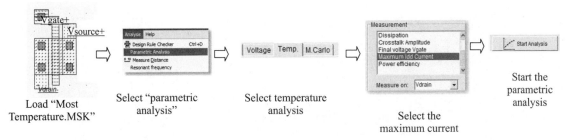

Fig. 3.32 Configuring Microwind to display the variations of Ids vs. temperature

Meanwhile, in an opposite trend, the threshold voltage is decreased, as KT1 is negative (Figure 3.34). Therefore, there exists a remarkable operating point where the *Ids* current is almost constant and independent of temperature variation. In $0.12\ \mu m$ CMOS, the *Vds* voltage with zero temperature coefficient (ZTC) is around 0.9 V, as shown in Figure 3.34.

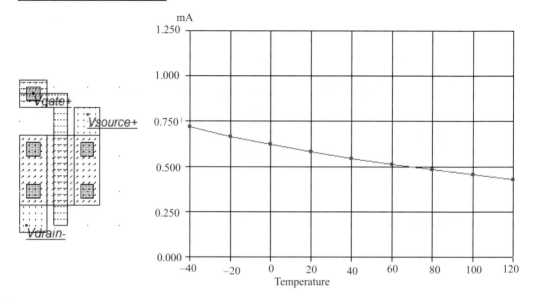

Fig. 3.33 The parametric analysis reveals an important decrease of the maximum current Ids with temperature (MosTemperature.MSK)

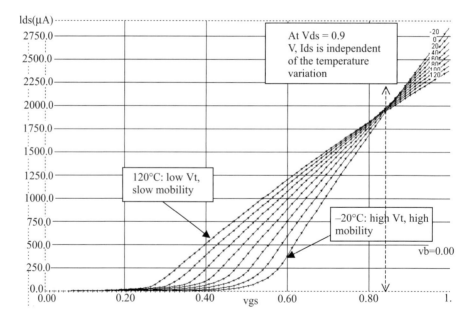

Fig. 3.34 The effect of temperature on the Ids current, showing a zero temperature coefficient (ZTC) operating point

In the sub-threshold region, the impact of temperature is extremely important, as demonstrated in Figure 3.35. At low temperature the current *Ids* decreased rapidly down to 10 nA, corresponding to a small off-leakage current. In contrast, at high temperature, not only is the threshold voltage reduced but the sub-threshold slope is flattened, which signifies an exponential increase of the *Ioff* leakage current (Figure 3.35).

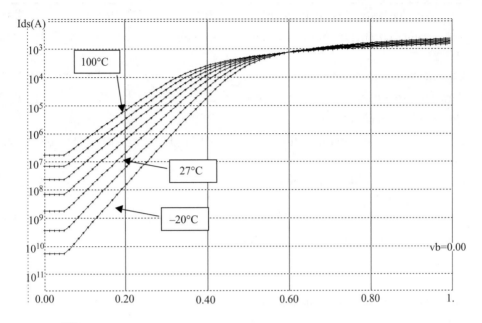

Fig. 3.35 The effect of temperature on the MOS characteristics

Table 3.3 List of user-accessible parameters in the BSIM4 implementation in Microwind

Parameter	Description	NMOS value in 0.12 μm	NMOS value in 0.12 μm	Name in RUL file
DVT0	First coefficient of short-channel effect on threshold voltage	2.2	2.2	B4D0VT
DVT1	Second coefficient of short-channel effect on Vth	0.53	0.53	B4D1VT
ETA0	Drain induced barrier lowering coefficient	0.08	0.08	B4ETA0
LINT	Channel-length offset parameter	0.01^e-6 μm	0.01^e-6 μm	B4LINT
LPE0	Lateral non-uniform doping parameter at Vbs = 0	2.3^e-10	2.3^e-10	B4LPE
NFACTOR	Sub-threshold turn-on swing factor. Controls the exponential increase of current with Vgs	1	1	B4NFACTOR
PSCBE1	First substrate current induced body-effect mobility reduction	4.24e8 V/m	4.24e8 V/m	B4PSCBE1
PSCBE2	Second substrate current induced body-effect mobility reduction	4.24e8 V/m	4.24e8 V/m	B4PSCBE2
K1	First-order body bias coefficient	0.45 V1/2	0.45 V1/2	B4K1

K2	Second-order body bias coefficient	0.1	0.1	B4K2
KT1	Temperature coefficient of the threshold voltage	– 0.06 V	– 0.06 V	B4KT1
NDEP	Channel doping concentration	1.7^e17 cm-3	1.7^e17 cm-3	B4NDEP
PCLM	Parameter for channel length modulation	1.2	1.2	B4PCLM
TOX	Gate oxide thickness	100 nm	100 nm	B4TOX
UA	Coefficient of first-order mobility degradation due to vertical field	11.0e–15 m/V	11.0e–15 m/V	B4UA
UC	Coefficient of mobility degradation due to body-bias effect	–0.04650e-15 V-1	–0.04650e-15 V-1	B4UC
U0	Low-field mobility	0.060 m2/Vs	0.025 m2/Vs	B4U0
UTE	Temperature coefficient for the zero-field mobility U0	–1.8	–1.8	B4UTE
VFB	Flat-band voltage	–0.9	–0.9	B4VFB
VOFF	Offset voltage in sub-threshold region	–0.08 V	–0.08 V	B4VOFF
VSAT	Saturation velocity	8.0e4 m/s	8.0e4 m/s	B4VSAT
VTHO	Long channel threshold voltage at Vbs = 0 V	0.3 V	0.3 V	B4VTHO
WINT	Channel-width offset parameter	$0.01^e–6$ μm	0.01e–6 μm	B4WINT
XJ	Source/drain junction depth	$1.5^e–7$ m	$1.5^e–7$ m	B4XJ

3.5   Specific MOS Devices

New kinds of MOS devices have been introduced in deep-submicron technologies, starting the 0.18 μm CMOS process generation. These MOS devices have specific characteristics which are described in this section.

3.5.1   Low Leakage MOS

The main objective of the low leakage MOS is to reduce the *Ioff* current significantly, that is the small current that flows between the drain and source with a zero gate voltage. The price to be paid is a reduced Ion current. The designer has the possibility to use high speed MOS devices, which have high *Ioff* leakages but large *Ion* drive currents. The symbols of the low leakage MOS and the high speed MOS are given in Figure 3.36.

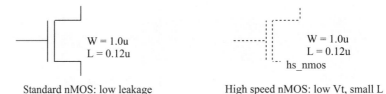

Standard nMOS: low leakage   High speed nMOS: low Vt, small L

Fig. 3.36   The low leakage MOS symbol (left) and the high speed MOS symbol (right) (MosOptions.SCH)

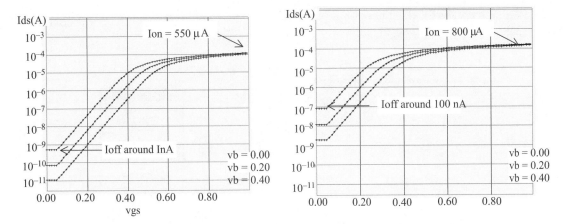

Fig. 3.37 The low leakage MOS offers a low Ioff current (1 nA) but a reduced Ion current (550 μA) as compared to the high speed MOS

In Figure 3.37, the low leakage MOS device (left side) has an *Ioff* current reduced nearly by a factor 100, thanks to a higher threshold voltage (0.4 V rather than 0.3 V) and larger effective channel length (120 nm) as compared to the high speed MOS (100 nm, see Figure 3.30). By default, the MOS device is in low leakage option, to encourage low power design. The *Ion* difference is around 30 per cent. This means that a high speed MOS device is 30 per cent faster than a low leakage MOS. Its use is justified in circuits where speed is critical.

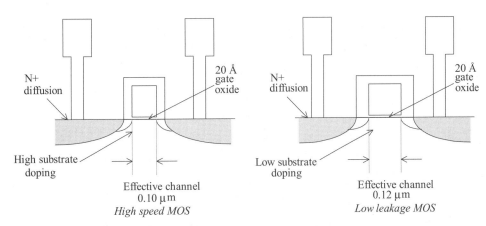

Fig. 3.38 Process section of the high speed (left) and low leakage (right) MOS devices

High speed MOS devices may be found in clock trees, data bus interfaces and central processing units, while low leakage MOS are used whenever possible, for all nodes wherein a maximum switching speed is not mandatory.

3.5.2 MOS Options in Microwind

A specific layer, called option layer, is used to configure the MOS device option. The layer is situated in the upper part of the palette of layers. The bird's eye view of the standard MOS is identical to the high speed MOS, except for the added option layer which surrounds the MOS device. The p-channel MOS device includes an option layer together with the n-well layer, as seen in Figure 3.39.

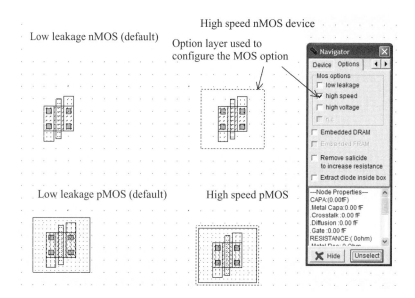

Fig. 3.39 High speed and low leakage MOS layout. The only difference is the option layer configured for the low leakage option

An "ultra-high speed" MOS has been introduced, together with the 90 nm technology. This MOS device has a very narrow channel, nearly half of the technology, which significantly increases the *Ion* current at the price of a very high *Ioff* parasitic leakage current (Figure 3.40).

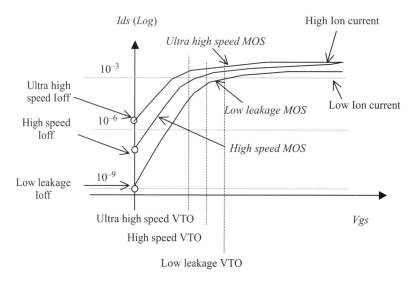

Fig. 3.40 Three types of MOS with different VTO threshold voltage are available in 90 nm technology

3.5.3 High Voltage MOS

Integrated circuits with low voltage internal supply and high voltage I/O interface are becoming common in deep-submicron technology. The internal logic of the integrated circuit operates at very low voltage (typically 1.0 V in 0.12 μm), while the I/O devices operate at standard voltages (2.5, 3.3 or 5 V).

Figure 3.41 shows the evolution of the supply voltage with the technology generation. The internal supply voltage is continuously decreasing. For compatibility reasons, the chip interface is kept at standard voltages, depending on the target application. Consequently, the input/output structures work at high voltages thanks to specific MOS devices with thick oxide called "high voltage MOS", while the internal devices work at low voltage for optimum performances.

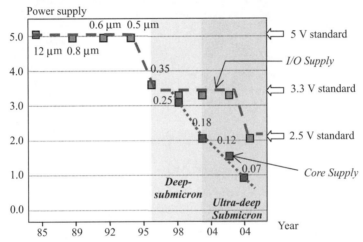

Fig. 3.41   The technology scale down leads to decreased core supply while keeping I/O interfacing compatible with 5 V, 3.3 V and 2.5 V standards

For I/Os operating at high voltage, high voltage MOS devices are commonly used. High-speed or low leakage devices are dangerous to use because of their ultra-thin oxide: a 3 V voltage applied to the gate of a core MOS device would damage the poly/substrate oxide. The high voltage MOS is built by using a thick oxide, two to three times thicker than the low voltage MOS, to handle high voltages as required by the I/O interfaces (Figure 3.42). Furthermore, the length of the channel is 0.25 μm minimum, that is twice the minimum length of the core MOS. The cross-section of the three types of MOS (low leakage, high speed and high voltage) is given in Figure 3.43.

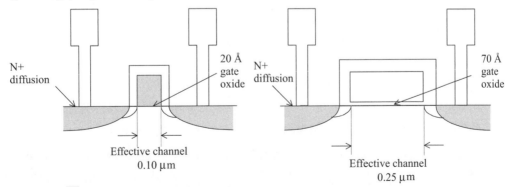

Fig. 3.42   Process section of the high speed and high voltage MOS devices

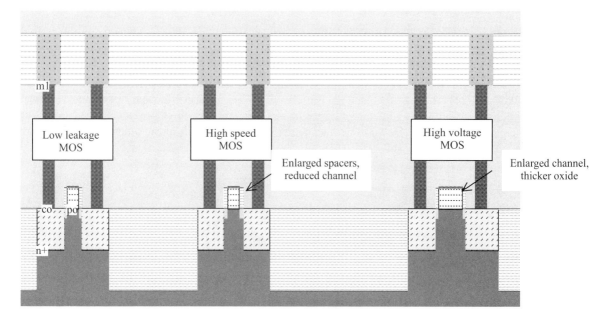

Fig. 3.43 The cross-section of the 3 n-channel MOS options: standard, high speed, and high voltage (lddExplain.MSK)

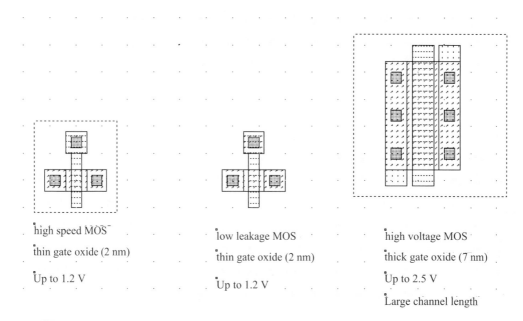

high speed MOS
thin gate oxide (2 nm)
Up to 1.2 V

low leakage MOS
thin gate oxide (2 nm)
Up to 1.2 V

high voltage MOS
thick gate oxide (7 nm)
Up to 2.5 V
Large channel length

Fig. 3.44 High speed, low leakage and high voltage MOS (MosHighVoltage.MSK)

There is no difference between the high speed MOS and the low leakage MOS from a layout point of view (Figure 3.44), expect the option layer for the high speed option. The high voltage MOS has a significantly different layout, due to the enlarged channel length and width.

The I/V characteristics of the high voltage MOS are plotted in Figure 3.45, for *Vgs* and *Vds* up to 2.5 V. The channel length is 0.25 μm and the channel width is 1.2 μm. Due to a large channel length, the current drive is less efficient.

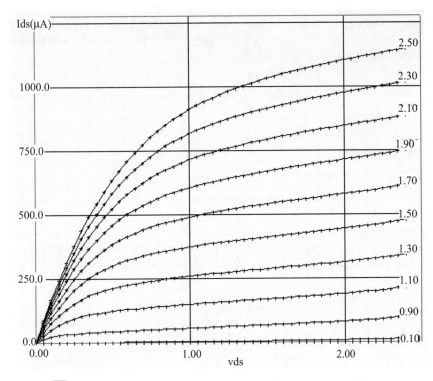

Fig. 3.45 Ids/Vds characteristics of the high voltage MOS

There are two main reasons for keeping a low-voltage supply for the core of the integrated circuit. The first one is low power consumption, which is of key importance for integrated circuits used in cellular phones or any portable devices. Low supply strongly reduces power consumption by reducing the amplitude of signals, thus reducing the charge and discharge of each elementary node of the circuit. Equation 3.47 gives an approximation of power consumption. We deduce that even a small reduction of Vdd has a very positive impact on the reduction of power consumption.

$$P = k.C.f.V_{DD}^{2}$$ (Equation 3.47)

where

P = power consumption (Watts)

K = technology factor, close to 0.5

C = total active capacitance of electrical nodes (F) (not taking into account decoupling capacitance)

f = operational frequency of the integrated circuit (Hz)

Vdd = supply voltage (V)

3.5.4 Oxide Breakdown

The second reason for internal low voltage operation is the oxide breakdown. Increased switching performances have been achieved by a continuous reduction of the gate oxide thickness. In 0.12 μm technology, the MOS device has an ultra thin gate oxide, around 0.002 μm, that is 2 nm or 20 Å. Knowing that the molecular distance of SiO_2 oxide is around 2Å, 20 Å means 10 atoms. The oxide may be destroyed by a voltage higher than a maximum limit *Vcrit*, called oxide breakdown voltage. A first order estimation is 0.1V/Å [Wang], which is expressed in Equation 3.47.

$$Vcrit = \frac{K}{tox}$$
(Equation 3.47)

With

K = breakdown coefficient (Close to 1 V.nm)

Tox = oxide thickness in nm

$Vcrit$ = critical breakdown voltage (V)

Consequently, in 0.12 μm, the breakdown voltage is around 2.0 V, that is less than twice the nominal VDD (1.2 V). An illustration of the breakdown voltage is proposed in Figure 3.46. If we display the Id/Vd characteristics with *Vg* higher than VDD, (for example 2.5 V instead of 1.2 V), the oxide damage is represented by dotted lines (here for *Vg* > 2.0V). The MOS polarization should always be fixed in such a way that the gate voltage is lower than the breakdown voltage limit.

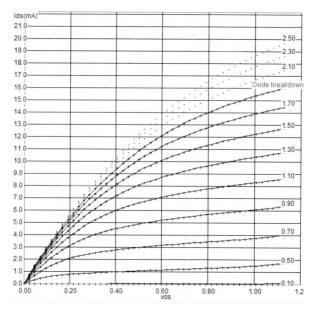

Fig. 3.46 Illustration of the breakdown voltage for a low leakage nMOS device, with a very high voltage applied on the gate

The oxide may be damaged by a 2 V gate voltage. If the gate voltage is further increased, the physical destruction of the oxide may be observed, which usually results in a permanent conductive path between the gate and the source (Figure 3.47).

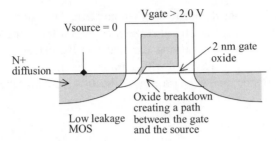

Fig. 3.47 Oxide breakdown for gate voltage appears to be significantly higher than the nominal supply voltage

3.5.5 Microwind Configuration

A set of specific parameters is used for each MOS option to configure the BSIM4 and Level 3 models. The industrial approach usually consists in describing each MOS device in a completely separated set of model parameters. Consequently, MOS model cards may include several thousands of parameters. We are trying to be as practical and didactic as possible, at the cost of a poor matching between the measured and simulated MOS characteristics. In Table 3.4, the list of the main varying parameters includes the gate oxide, the effective channel length parameter, and the threshold voltage.

Table 3.4 BSIM4 parameters variation depending on the MOS option

Parameter	Description	nMOS value in 0.12 μm	nMOS value in 0.12 μm	Name in RUL file
TOX	Gate oxide thickness (low leakage) (high speed) (high voltage)	20 Å 20 70	20 Å 20 70	B4TOX B4T2OX B4T3OX
LINT	Channel-length offset parameter (low leakage) (high speed) (high voltage)	0.0 nm 10 0.0	0.0 nm 10 0.0	B4LINT B4L2INT B4L3INT
VTHO	Long channel threshold voltage (low leakage) (high speed) (high voltage)	0.40 V 0.30 0.50	0.40 V 0.30 0.50	B4VTHO B4V2THO B4V3THO

3.6 Process Variations

The simulated results should not be considered as absolute values. Due to unavoidable process variations during the hundreds of chemical steps for the fabrication of the integrated circuit, the MOS characteristics are never exactly identical from one device to another, and from one die to an other. It is very common to measure 5 per cent to 20 per cent electrical difference within the same die, and up to 30 per cent difference between separate dies. One varying parameter is the effective channel length. In Figure 3.48, though both devices have been designed with a drawn 2 lambda, the result is a 0.11 μm length for the MOS situated on the left side, and 0.13 μm for the MOS situated on the right side.

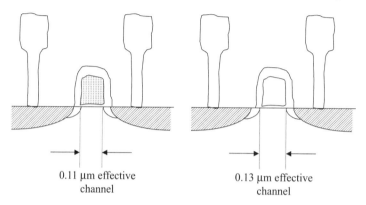

Fig. 3.48   The same MOS device may be fabricated with an important effective channel variation

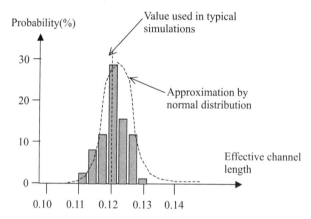

Fig. 3.49   The effective channel length may vary significantly with the process

If we cumulate several measurements on a wide number of devices, we can plot the probability of occurrence versus the measured effective length. The curve is usually a normal distribution with a centre close to the default parameter given in the electrical rules (Figure 3.49).

3.6.1   Simulation with Microwind

The menu **Simulate → Simulation parameters** gives a simple access to minimum/typical/maximum parameter sets (Figure 3.50). The industrial approach usually consists in providing a separate set of model parameters for each case, which represents a huge amount of model parameters. In Microwind, the approach has consisted in altering two main parameters: the threshold voltage (20 per cent variation) and the mobility (20 per cent variation). All other parameters are supposed to be constant.

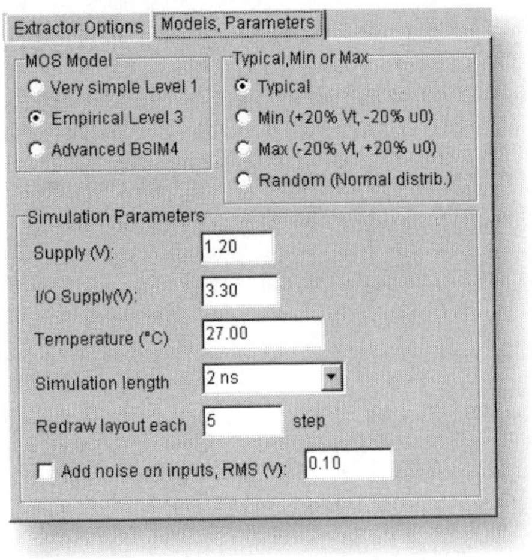

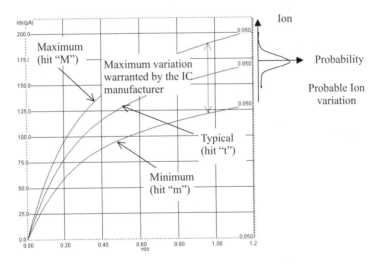

Fig. 3.50   Access to minimum, typical, maximum model parameters or random simulation

A comparative simulation of the *Id/Vd* curve in typical, maximum and minimum scenarios shows a very large variation of performances (Figure 3.51). The user may automatically switch from one parameter set to another with the press of a key ("M" for maximum, "m" for minimum,"t" for typical).

Fig. 3.51   The MOS Id/vd curve in Min, Typ, Max modes

In order to superimpose the three curses, click the small brain icon (Enable Memory), and increase the **Step Vg** to 1.2 to draw only one curve for each mode. Notice the important variation between the minimum Ion and maximum Ion (from 125 µA to 200 µA). In reality, the MOS characteristics vary in a normal distribution around the typical case. Consequently, the Ion current of this MOS device is very likely to reach the typical value. The min/max simulation is very interesting to validate the design in extreme situations. The min/max simulation should also consider the temperature: the worst current is obtained at high temperature and with a minimum set of parameters, while the highest current is obtained at low temperature and with a maximum set of parameters.

3.7 Concluding Remarks

The number of parameters required for various MOS models is shown in Figure 3.52. It can be seen that the trend is to increase the number of parameters, in order to take into account various effects linked to the device scale down.

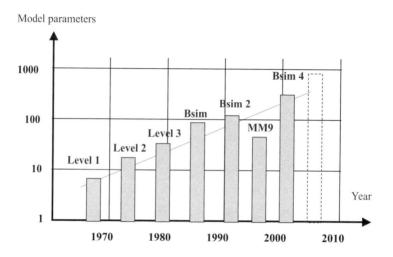

Fig. 3.52 Increased number of parameters in the MOS models

Even with advanced models, the resulting models may not fit well in all operating regions, for all device sizes. This is why the industrial approach for building model parameters is based on error minimization algorithms. In deep-submicron technology, the model parameters have a strong variation with the device size. For example, the threshold voltage and mobility vary significantly with the device length, and the equations cannot always handle these dependencies properly. One solution is called binning. It consists in breaking the width-length space into several regions, as illustrated in Figure 3.53. In each region, a specific set of model parameters is set up and optimized.

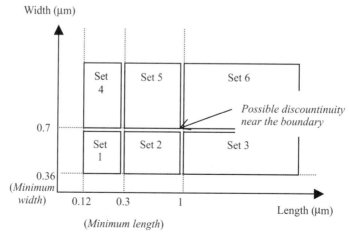

Fig. 3.53 Using six sets of parameters to accurately cover the whole range of width and length

Binning severely complicates the procedure for parameters extraction. In the case of Figure 3.53, six sets of parameters are required. Notice that set 1 covers small length and small width. Set n°2 covers a wider length interval, while set n°3 is valid for any length greater than 1 µm. This is because long length devices are easier to model than short length devices, where many second order effects appear. Binning is used in industry to increase the analog simulation accuracy, at the cost of several drawbacks: the simulation time cost due to model complexity, and discontinuities in the current prediction that may be observed at the boundary of two sets. These limitations are the fuel for constructing more complex models that fit well for the whole range of width and length. In the future, nano-scale technologies may require MOS models with up to 1000 parameters, requiring a degree of expertise never attained up to now.

References

[Schockley] W. Shockley "A Unipolar field effect transistor", Proceedings of IRE, Vol. 40, Nov. 1952, pp. 1365-1376.

[EIA] *http://www.eia.org/eig/CMC*

[Shichman] H. Shichman, D. Hodges, "Modelling and simulation of insultaed-gate field effect transistor switching circuits", IEEE J. Solid State Circuits, Vol. 3, pp 285–289, 1968.

[Tsividis] Y. P. Tsividis "Operating and Modelling of the MOS transistor", McGraw-Hill, 1987, ISBN 0-07-065381-X

[Sze] S. M Sze "Physics of semiconductor devices", John-Wiley, 1981, ISBN 0-471-05661-8

[Cheng] Y. Cheng, C. Hu "MOSFET Modelling and BSIM3 user's guide", Kluwer Academic Publishers, 1999.

[Bsim4] BSIM4 website *www-device.eecs.berkeley.edu*

[Weste] N. Weste, K. Eshraghian "Principles of CMOS VLSI design", Addison Wesley, ISBN 0-201-53376-6, 1993.

[Lee] K. Lee, M. Shur, T.A Fjeldly, T. Ytterdal "Semiconductor device modelling for VLIS", Prentice-Hall, 1993, ISBN 0-13-805656-0.

[Liu] W. Liu "Mosfet Models for SPICE simulation including Bsim3v3 and BSIM4", Wiley and Sons, 2001, ISBN 0-471-39697-4.

[Wang] Albert Z.H. Wang "On-chip ESD protection for Integrated Circuits", An IC Design Perspective, Kluwer Academic Publishers, 2002, ISBN 0-7923-7647-1.

[TSMC] *http://www.tsmc.com*

EXERCISES

1. Configure Microwind in 0.35 µm technology (File **cmos035.RUL**) using the command **File →**
   Select Foundry. Compare simulation and measurement in 0.35 µm for an nMOS device, W = 10 µm,
   L = 0.4 µm (File **Na10x0ES**). Evaluate the mismatch between level 1, level 3 and BSIM4. In which
   domain is model 3 poorly fitted?

2. Configure Microwind in 0.18 µm technology (File **cmos018.RUL**). Compare simulation and
   measurement in 0.18 µm for an nMOS device, W = 4 µm, L = 0.2 µm, low leakage option (File
   Nc4x0.2.mes). Evaluate *Ron*, *Ion*, *Ioff*, and *Vt*. Perform the same evaluation for the high speed

MOS, same size (File **NcHS4x0.2.mes**). Perform the same evaluation for the high voltage MOS, same size (File **NcHV4x0.2.mes**).

3. In low power applications, the n-well can be connected to a voltage *Vwell* different from VDD. This unusual polarization aims at modifying the threshold of the p-MOS transistor. Under which condition for *Vwell* the threshold of the PMOS transistor (Low Leakage option) is decreased to 0.2 V in 0.12 μm?

4. Design an MOS device with *Ion* =10 mA, minimum gate length in 0.12 μm. What is *Ioff*? How can one obtain a MOS with *Ion* = 10 mA but twice less *Ioff*?

5. Compare the value of V_{GS_T0} (the value of *Vgs* for which *Ids* is independent of temperature) between MOS options in 0.12 μm.

6. Show that four MOS devices (Wn, Ln) connected in parallel have approximately the same Ion than a single MOS device with a width equal to 4 Wn. What is the origin of the mismatch?

7. What is the channel size of the following MOS device? Does Microwind correctly extract the channel size of that device?

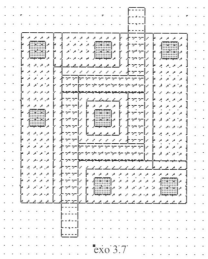

exo 3.7

8. In the following picture of a 50 nm MOS device, locate the gate, drain and source, field oxide and gate oxide.

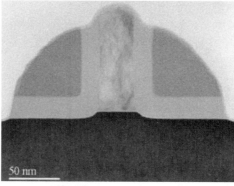

TSMC www.tsmc.com

4

The Inverter

The inverter is probably the most important basic logic cell in circuit design. This chapter introduces the logical concepts of the inverter, its layout implementation, the link between the transistor size, and the static and analog characteristics. The manual design of the inverter is detailed. The performances of the inverter are analyzed in terms of static transfer function, switching speed, MOS options influence, and power consumption.

4.1 Logic Symbol

Two logic symbols are often used to represent the inverter: the "old style" inverter (left of Figure 4.1), and the IEEE symbol (right of Figure 4.1). In DSCH, we preferably use traditional symbol layout. As the logic truth table of Figure 4.1 shows, the cell inverts the logic value of the input *In* into an output *Out*.

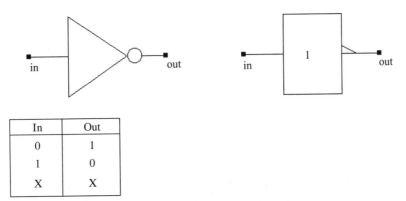

In	Out
0	1
1	0
X	X

Fig. 4.1 Symbols used to represent the logic inverter

In the truth table, the symbol 0 represents 0.0 V while 1 represents the logic supply, which is 1.2 V in 0.12 μm. The symbol X means "undefined". This state is equivalent to an undefined voltage, just like in the case of a floating input node without any input connection. The undefined state appears in gray in the simulations and chronograms.

4.2   CMOS Inverter

The CMOS inverter design is detailed in Figure 4.2. Here one p-channel MOS and one n-channel MOS transistors are used as switches. Notice that the size of each device is plotted (W accounts for the width, L for the length). The channel width for pMOS devices is set to twice the channel width for nMOS devices. The reason is described in detail in the next chapters.

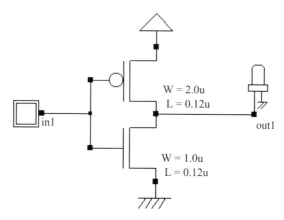

Fig. 4.2   The CMOS inverter is based on one n-channel and one p-channel MOS device

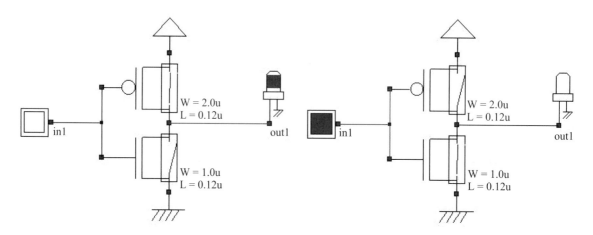

Fig. 4.3   Logic simulation of the CMOS inverter (CmosInv.sch)

When the input signal is logic 0 (Figure 4.3 left), the nMOS is switched off while the pMOS passes VDD through the output, which turns to 1. When the input signal is logic 1 (Figure 4.3 right), the pMOS is switched off while the nMOS passes VSS to the output which goes back to 0. In that simulation, the MOS is considered as a simple switch. The n-channel MOS symbol is a device that allows the current to flow between the source and the drain when the gate voltage is "1".

In order to simulate the inverter at logic level, start the software DSCH 2, load the file "CmosInv.SCH", and launch the simulation by the command **Simulate → Start Simulate**. Click inside the button *in1*.

The result is displayed on the output *out1*. The red value indicates logic 1, while the black value means a logic 0.

Click the button **Stop simulation** of the simulation menu to return to the schematic editor. Click the **chronogram** icon to get access to the chronograms of the previous simulation (Figure 4.4). As seen in the waveform, the value of the output is the logic opposite of that of the input.

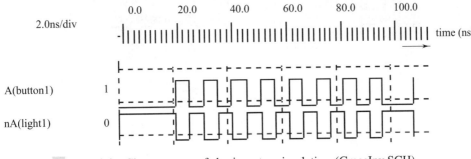

Fig. 4.4   Chronograms of the inverter simulation (CmosInv.SCH)

4.3   Inverter Layout

In this paragraph, details of the layout of a CMOS inverter are provided. The simplest way to create a CMOS inverter is to generate both n-channel MOS and p-channel MOS devices using the cell generator provided by **Microwind**. The advantage of this approach is to avoid any design rule error. The corresponding menu is reported below. You can generate an n-channel or p-channel device. A double gate device may also be created for EEPROM memory devices (see the memory chapter in book II). By default, the proposed length is the minimum length available in the technology (2 lambda), and the width is 10 lambda. In 0.12 μm technology, where lambda is 0.06 μm, the corresponding size is 0.12 μm for the length and 0.6 μm for the width.

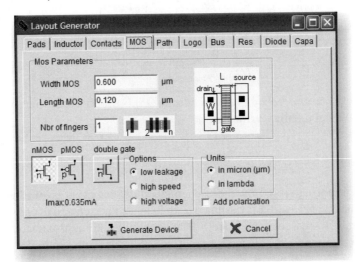

Fig. 4.5   Using the MOS generator to add n-channel and p-channel MOS devices on the layout

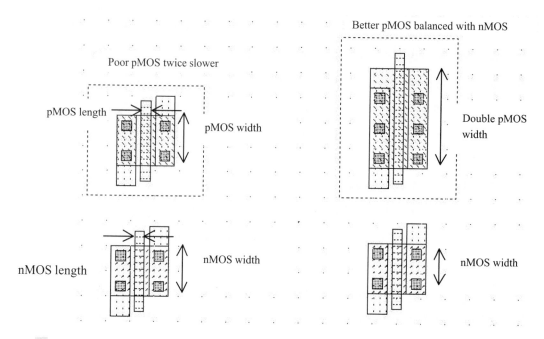

Fig. 4.6   The layout of one nMOS and one pMOS to build the CMOS inverter (invSizing.MSK)

The design starts with the implementation of one nMOS and one pMOS, as shown in Figure 4.6. Using the same default channel width (0.6 μm in CMOS 0.12 μm) for nMOS and pMOS is not the best idea, as the p-channel MOS switches half the current of the n-channel MOS. The origin of this mismatch can be seen in the general expression of the current delivered by n-channel MOS devices (Equation 4.1) and p-channel MOS devices (Equation 4.2).

$$Ids(Nmos) \approx \varepsilon_0 \varepsilon_r \frac{\mu_n}{TOX} \frac{W_{Nmos}}{L_{Nmos}} f(Vd, \, Vg, \, Vs, \, Vb) \qquad \text{(Equation 4.1)}$$

$$Ids(Nmos) \approx \varepsilon_0 \varepsilon_r \frac{\mu_p}{TOX} \frac{W_{Pmos}}{L_{Pmos}} f(Vd, \, Vg, \, Vs, \, Vb) \qquad \text{(Equation 4.2)}$$

If WnMOS = WpMOS and LnMOS = LpMOS, *Ids*(*nMOS*) is proportional to μn while *Ids* (pMOS) is proportional to μp. Typical mobility values are:

$$\mu_n = 0.068 \text{ m}^2/\text{v.s} \quad \text{for electrons}$$

$$\mu_p = 0.025 \text{ m}^2/\text{v.s} \quad \text{for holes}$$

Consequently, the current delivered by the n-channel MOS device is more than twice that of the p-channel MOS. Usually, the inverter is designed with balanced currents to avoid significant switching discrepancies. In other words, switching from 0 to 1 should take approximately the same time as switching from 1 to 0. Therefore, balanced current performances are required.

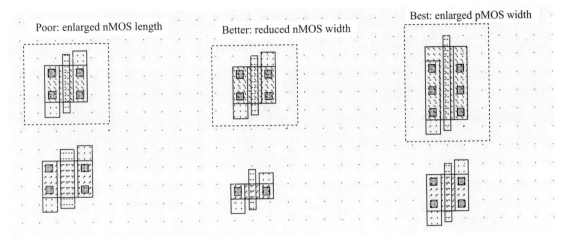

Poor: enlarged nMOS length   Better: reduced nMOS width   Best: enlarged pMOS width

Fig. 4.7   Three techniques to compensate the lower hole mobility (invSizing.MSK)

There are several techniques to counter-balance the intrinsic mobility difference: increase the nMOS channel length (left of Figure 4.7), decrease the nMOS channel width (middle), or increase the pMOS channel width. The main drawback of the design of Figure 4.7(left) is the spared silicon area. The design in the middle is equivalent, but consumes less silicon space. However, reducing the nMOS width slows down the switching. The best approach (right) consists in enlarging the pMOS width. Its *Ion* current is doubled, and becomes comparable with the nMOS current. The behaviour will be balanced in terms of switching speed.

4.3.1   Connection between Devices

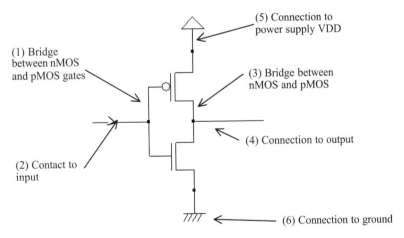

Fig. 4.8   Connections required to build the inverter (CmosInv.SCH)

Within CMOS cells, metal and polysilicon are used as interconnects for signals. Metal is a much better conductor than polysilicon. Consequently, polysilicon is only used to interconnect gates, such as the bridge (1) between pMOS and nMOS gates, as described in the schematic diagram of Figure 4.8. Polysilicon is rarely used for long interconnects, except if a huge resistance value is expected.

In the layout shown in Figure 4.9, the polysilicon bridge links the gate of the n-channel MOS with the gate of the p-channel MOS device. The polysilicon serves as the gate control and the bridge between the MOS gates.

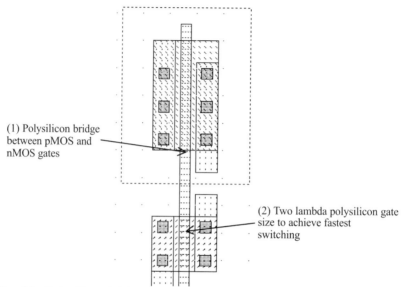

(1) Polysilicon bridge between pMOS and nMOS gates

(2) Two lambda polysilicon gate size to achieve fastest switching

Fig. 4.9 Polysilicon bridge between nMOS and pMOS devices (InvSteps.MSK)

4.3.2 Useful Editing Tools

The following commands may help you in the layout design and verification processes:

Table 4.1 A set of useful editing tools

Command	Icon/Short cut	Menu	Description
UNDO	CTRL+U	Edit menu	Cancels the last editing operation.
DELETE	CTRL+X	Edit menu	Erases some layout included in the given area or pointed by the mouse.
STRETCH		Edit menu	Changes the size of one box, or moves the layout included in the given area
COPY	CTRL+C	Edit menu	Copies the layout included in the given area
VIEW ELECT-RICAL NODE	CTRL+N	View menu	Verifies the electrical net connections.
2D CROSS-SECTION		Simulate menu	Shows the aspect of the circuit in vertical cross-section.

4.3.3 Metal-to-polysilicon

As polysilicon is a poor conductor, metal is preferred to interconnect signals and supplies. Consequently, the input connection of the inverter is made with metal. Metal and polysilicon are separated by an oxide which prevents electrical connections. Therefore, a box of metal drawn across a box of polysilicon does not allow an electrical connection (Figure 4.10). In order to build an electrical connection, a physical contact is needed. The corresponding layer is called "contact". You may insert a metal-to-polysilicon contact in the layout using a direct macro situated in the palette.

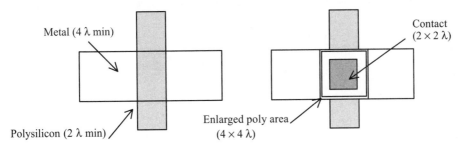

Metal (4 λ min)
Contact (2 × 2 λ)
Polysilicon (2 λ min)
Enlarged poly area (4 × 4 λ)

Fig. 4.10   Physical contact between metal and polysilicon

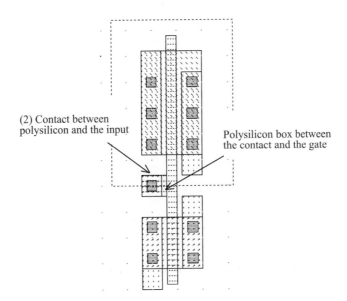

(2) Contact between polysilicon and the input

Polysilicon box between the contact and the gate

Fig. 4.11   Physical contact between metal and polysilicon (InvSteps.MSK)

The *Process Simulator* shows the vertical aspect of the layout, as when fabrication has been completed. This feature is a significant aid to understand the circuit structure and the way layers are stacked on top of each other. A click of the mouse on the left side of the n-channel device layout and the release of the mouse at the right side give the cross-section shown in Figure 4.13.

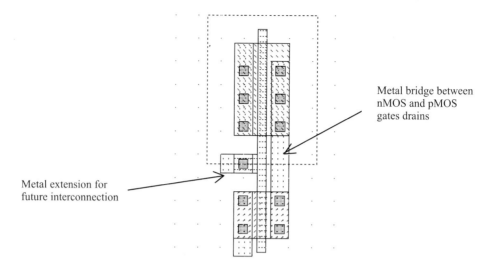

Metal bridge between nMOS and pMOS gates drains

Metal extension for future interconnection

Fig. 4.12 Adding a poly contact, poly and metal bridges to construct the CMOS inverter (InvSteps.MSK)

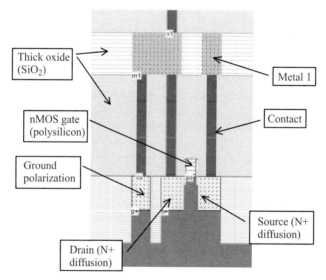

Thick oxide (SiO$_2$)

Metal 1

nMOS gate (polysilicon)

Contact

Ground polarization

Source (N+ diffusion)

Drain (N+ diffusion)

Fig. 4.13 The 2D process section of the inverter circuit near the nMOS device (InvSteps.MSK)

4.3.4 Supply Connections

The next design step consists in adding supply connections, that is the positive supply VDD and the ground supply VSS. In Figure 4.14, we use the metal 2 layer (second level of metallization) to create horizontal supply connections. Notice that the metal connections have a large width. This is because a strong current may flow within these supply interconnects. Enlarging the supply metal lines reduces the resistance and avoids electrical overstress called electromigration (more details are given in Chapter 5 dedicated to interconnects).

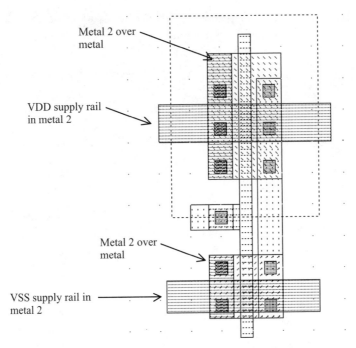

Fig. 4.14 Adding metal 2 supply lines and the appropriate vias (InvSteps.MSK)

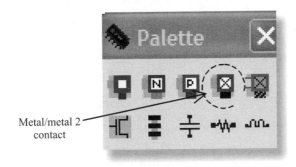

Fig. 4.15 The metal/metal 2 contact in the palette

The metal layers are electrically isolated by a SiO_2 dielectric. Consequently, the metal 2 supply line floats over the inverter cell and no physical connection exists down to the MOS source region. The simplest way to build the physical connection is to add a metal/metal 2 contact that may be found in the palette (Figure 4.15).

As seen in Figure 4.16, the connection is created by a plug called "via" between the metal 2 and metal layers.

The final layout design step consists in adding polarization contacts. These contacts convey the VSS and VDD voltage supply close to the bulk regions of the device. We have seen that the MOS behaviour is influenced by the bulk polarization. For example, *Ion* is directly dependent on V_{BS}, which represents the voltage difference between the bulk and the source (see for example Equation 3.31). See also the characteristics *Ids* versus *Vgs* for varying bulk voltage like in Figure 3.26. If we ensure a clean supply polarization near each device (VSS for nMOS, VDD for pMOS), we avoid such variations. Remember that the n-well region should always be polarized to a high voltage to avoid a short-circuit between VDD and VSS.

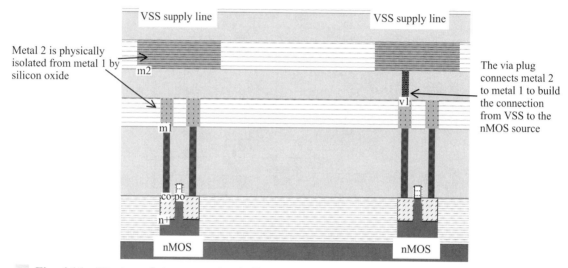

Fig. 4.16 2D view of the connection built near the nMOS region to connect the source to the VSS supply line

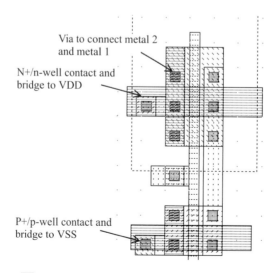

Fig. 4.17 Adding polarization contacts

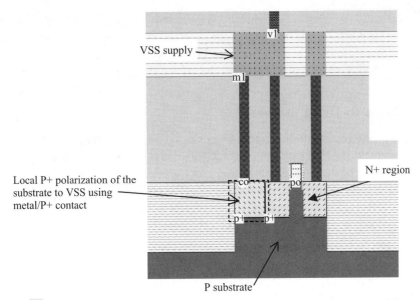

Fig. 4.18 A 2D view of the VSS polarization built near the nMOS source

More details about the vertical aspect of the VSS polarization are given in Figure 4.18. When adding the metal/P+ contact, we create a VSS supply path to the P substrate. Consequently, the surrounding of the n-MOS device is firmly tied to the VSS supply voltage. We also illustrate the VDD polarization near the pMOS channel in Figure 4.19 using an N+ diffusion. The n-well region cannot be left without polarization.

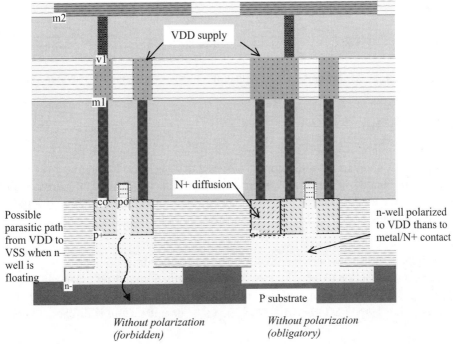

Fig. 4.19 A 2D view of the VDD polarization built near the pMOS source

Adding the VDD polarization in the n-well region is a very strict rule. The local polarization built with a metal/N+ diffusion contact, as shown in Figure 4.19, is efficient to avoid a floating n-well region, which may result in a parasitic current path from the PMOS source down to the P substrate usually tied to VSS. The current path may be strong enough to damage the chip. This effect is called latchup.

4.3.5   Process Steps to Build the Inverter

At that point, it might be interesting to illustrate the steps of fabrication as they would be sequenced in a foundry. Microwind includes a 3D process viewer for that purpose. Click **Simulate → Process steps in 3D**. The simulation of the CMOS fabrication process is performed, step-by-step by a click on **Next Step**. In Figure 4.20, the picture on the left represents the nMOS device, pMOS device, common polysilicon gate and contacts. The picture on the right represents the same portion of the layout with the metal layers stacked on top of the active devices.

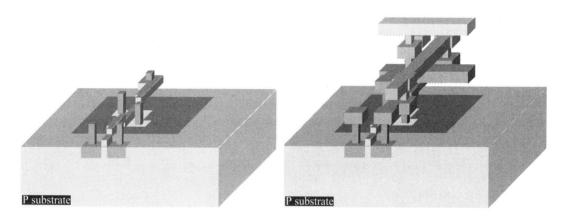

Fig. 4.20   The step-by-step fabrication of the inverter circuit (InvSteps.MSK)

4.4   Inverter Simulation

The inverter simulation is conducted as follows. Firstly, a VDD supply source (1.2 V) is fixed to the upper metal 2 supply line, and a VSS supply source (0.0 V) is fixed to the lower metal 2 supply line. The properties are located in the palette menu. Simply click the desired property, and click on the desired location in the layout. Add a clock on the inverter input node (the default node name *clock1* has been changed into *Vin*) and a visible property on the output node (the default name *out1* has been changed into *Vout*).

The expected behaviour is shown in Figure 4.22. The basic phenomenon is the charge and discharge of the output parasitic capacitor *Cout*, which is the sum of junction and wire capacitance. When *In1* is equal to 0, the pMOS device is on, and the capacitor *Cout* is charged until its voltage rises to VDD. When *In1* is equal to 1, the nMOS device is on, and the capacitor *Cout* is discharged until its voltage reaches VSS.

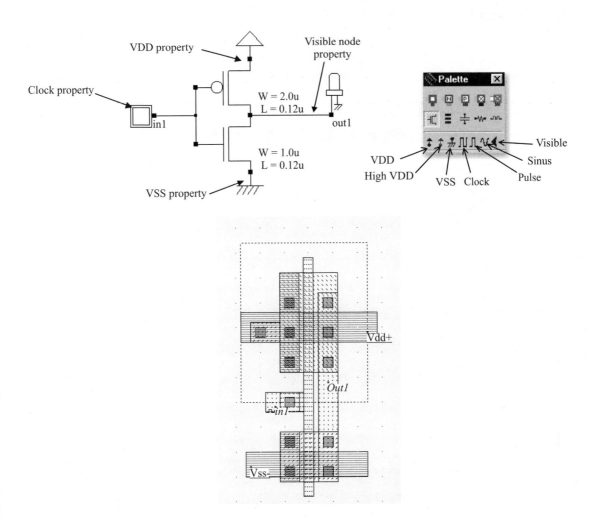

Fig. 4.21 Adding simulation properties (InvSteps.MSK)

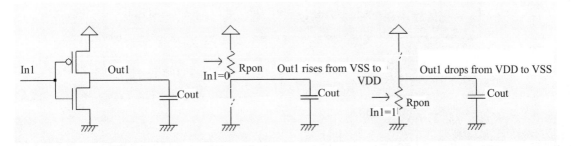

Fig. 4.22 Expected behaviour of the CMOS inverter (InverterLoad.SCH)

4.4.1 Starting Simulation

The command **Simulate** → **Run Simulation** gives access to four simulation modes: **Voltage vs. time**, **Voltage and current vs. Time, Static Voltage vs. voltage and Frequency vs. time**. All these simulation modes are applicable to the inverter simulation.

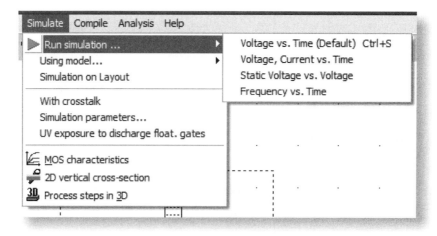

Fig. 4.23 The four simulation modes in Microwind

Due to the fact that the layout **InvSteps.MSK** not only includes the inverter correctly polarized, but also several other MOS devices without any simulation properties, a warning window appears prior to the analog simulation, as shown in Figure 4.24. In this case, you may click **Simulate as it**. In normal cases, all n-well regions should be stuck at VDD.

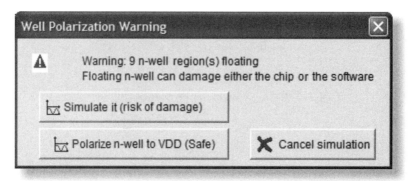

Fig. 4.24 Missing polarization in n-well regions provokes a warning prior to simulation (InvSteps.MSK)

4.4.2 Voltage vs. Time

Select the simulation mode **Voltage vs. Time**. The analog simulation of the circuit is performed. The time domain waveform, proposed by default, details the evolution of the voltages *in1* and *out1* versus time. This mode is also called transient simulation, as shown in Figure 4.25.

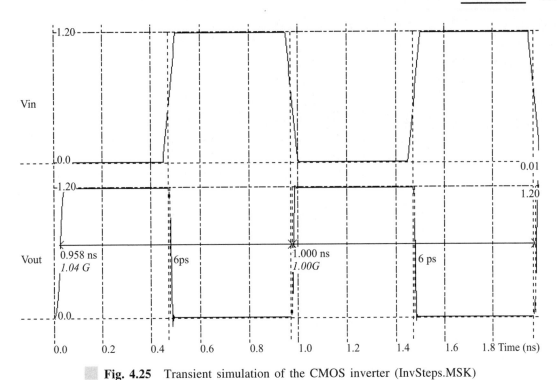

Fig. 4.25 Transient simulation of the CMOS inverter (InvSteps.MSK)

The truth table is verified as follows. A logic zero corresponds to a zero voltage and a logic 1 to a 1.20 V. When the input rises to 1, the output falls to 0, with a 6 Pico-second delay (6.10^{-12} second).

In	Out
0	1
1	0

Logic table

In (V)	Out (V)
0.0	1.2
1.2	0.0

Analog voltage table

4.4.3 Current vs. Time

The inverter consumes power during transitions, due to two separate effects. The first is short-circuit power arising from momentary short-circuit current that flows from *VDD* to *VSS* when the transistor functions in the incomplete-on/off state (Figure 4.26). The second is the charging/discharging power, which depends on the output wire capacitance. With small loading, the short-circuit power loss is dominant. With a huge loading, that is a large output node capacitance, the loading power is dominant.

The power consumption occurs briefly during transitions of the output, either from 0 to 1 or from 1 to 0 (Figure 4.27). The simulation contains the supply currents in the upper window, and all voltage waveforms in the lower window. The current consumption is important only during a very short period corresponding to the charge or discharge of the output node. Without any switching activity, the current is almost equal to zero.

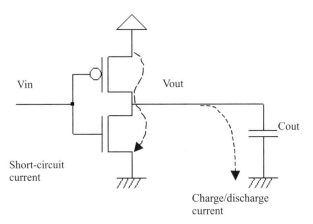

Fig. 4.26 Short circuit current in CMOS inverters (InverterLoad.SCH)

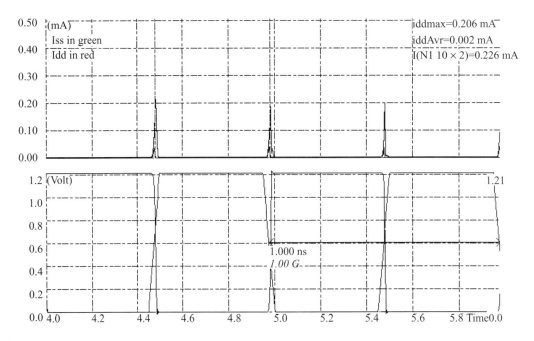

Fig. 4.27 Simulation of the current peaks appearing between VDD and VSS in the CMOS inverter at each output transition (InvSteps.MSK)

4.4.4 Inverter Delay

As the number of gates connected to the inverter output node increase, the load capacitance increases. The fanout corresponds to the number of gates connected to the cell output. Physically, a large fanout means a large number of connections (Figure 4.28), that is, a large load capacitance.

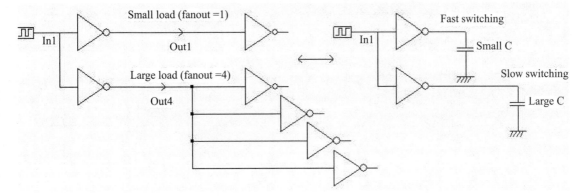

Fig. 4.28 One inverter connected either to a single inverter or to four inverters in parallel (InverterLoad.SCH)

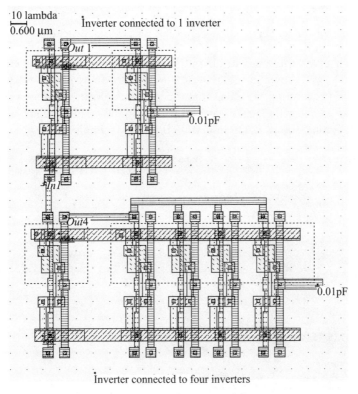

Fig. 4.29 One inverter connected either to a single inverter or to four inverters in parallel (InvFanout.MSK)

An inverter circuit is simulated by using different clock, fanout and supply conditions. The initial configuration is based on one inverter controlled by a 2 GHz clock, with its output connected either to a single inverter or to four inverters (Figure 4.30). The supply voltage is 1.2 V, with a 0.12 µm CMOS technology.

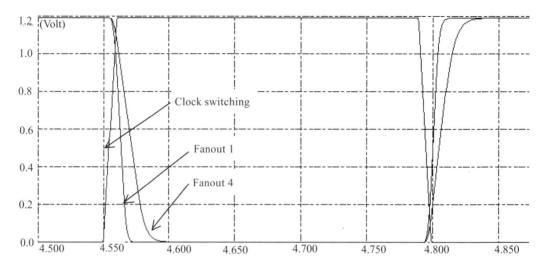

Fig. 4.30   Influence of the output capacitance on the current and switching response (InvFanout.MSK)

Now, we connect four inverter circuits to the output node, thus increasing the charge capacitance. In the simulation chronograms shown in Figure 4.30, the inverter delay is significantly increased. When we investigate the delay variation with the output capacitance load, we observe the curve shown in Figure 4.31. It can be seen that the gate delay variation with the loading capacitance is quite linear. A 100 fF load leads to around 300 ps delay in CMOS 0.12 μm technology.

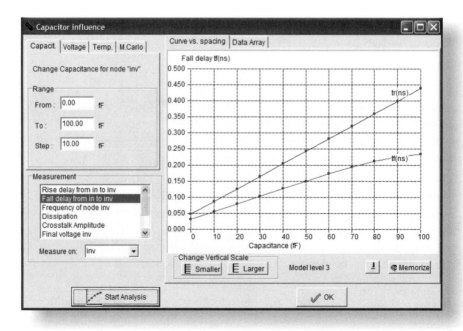

Fig. 4.31   Inverter delay increases with the output capacitance (InvCapa.MSK)

In Microwind, we may directly obtain this type of screen thanks to the command **Parametric Analysis**. Load the file InvCapa.MSK, invoke the command **Parametric Analysis**, click in the output node, and click **Start Analysis**. By default, the capacitance of the output node is increased step-by-step from its default value C_{def} to $C_{def}+100$ fF. For each value of the output capacitance, the analog simulation is performed, and the last computed rise time is plotted, appearing as one single red dot in the graphs. The complete graph is built once all analog simulations have been completed. The memory button enables us to store one curve (evaluation of the rise time for example) prior to a new parametric simulation, for comparison purposes. Three main parameters may vary in the parametric analysis: the capacitance as in Figure 4.31, voltage, or temperature. Several analog parameters may be monitored: rise and fall delay, oscillating frequency, power consumption, final voltage of a node, crosstalk, etc.

4.5   Power Consumption

The power consumption P is computed by MICROWIND as the average product of the supply voltage VDD and the supply current IDD, computed at each iteration step. In other words:

$$P = \frac{\sum I_{DD} \cdot V_{DD}}{steps}$$

(Equation 4.3)

Three main factors contribute to the power consumption P: the load capacitance C, the supply voltage *VDD* and the clock frequency f. For a CMOS inverter, this relation is usually represented by the first-order approximation below. Equation 4.4 shows a linear dependence of the power consumption P with the total capacitance C and the operating frequency f. The power consumption is also proportional to the square of the supply voltage *VDD*.

$$P = \frac{1}{2}\eta \cdot C\, V_{DD}^2\, f$$

(Equation 4.4)

Where:

k: technological factor (close to 1)

C: Output load capacitance (Farad)

VDD: supply voltage (V)

f: Clock frequency (Hz)

η: switching activity factor (Between 0 and 1)

4.5.1   Frequency Dependence

We can verify the linear dependence of the power consumption with the operating frequency by simulating a CMOS inverter circuit. At each time-domain analog simulation, we get a value of the power consumption, which is computed by MICROWIND as the average product of the supply voltage VDD and the supply current IDD (Equation 4.3).

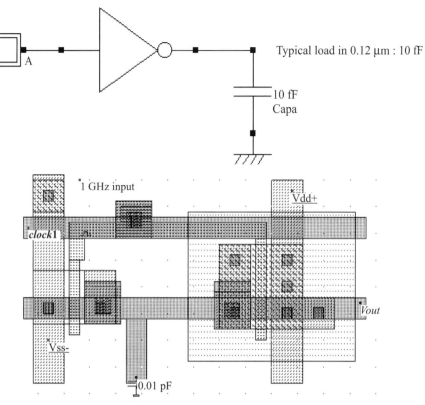

Typical load in 0.12 µm : 10 fF

10 fF
Capa

1 GHz input

Vdd+

clock1

Vss-

0.01 pF

Fig. 4.32   CMOS inverter set-up used to simulate the effect of the clock frequency on the power consumption. A 10 fF load is added to the output to represent a typical loading condition in 0.12 µm (CmosLoad.MSK)

In the case of Figure 4.32, a 1 GHz switching of the inverter induces a circuit power dissipation of 15.7 µW. When we change the frequency, we observe a linear increase of P with the clock frequency, as forecast in Equation 4.4 (Figure 4.33).

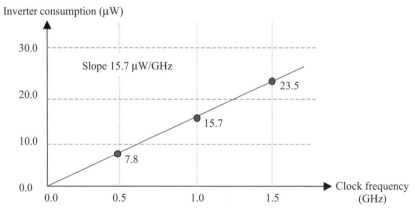

Inverter consumption (µW)

Slope 15.7 µW/GHz

30.0

20.0

23.5

10.0

15.7

7.8

0.0

0.0   0.5   1.0   1.5   Clock frequency
                                                    (GHz)

Fig. 4.33   Power consumption increase with the clock frequency, for an inverter with a 10 fF load, in 0.12 µm CMOS technology (CmosLoad.MSK)

As the power consumption is linearly proportional to the clock frequency, a usual metric found in most cell libraries is the µW/GHz. In the case of the simple inverter and its 10 fF load, we get 15.7 µW/GHz.

4.5.2 Supply Voltage Dependence

It can be considered, as a first-order approximation that the average power consumption is proportional to VDD^2 (Equation 4.4). We use the parametric analysis tool in MICROWIND to control the incremental change of the supply voltage, from 0.5 to 2.0 V. The supply voltage step is 0.1 V. In the measurement window, the item "Dissipation" is selected. The result plotted in Figure 4.44 shows a non-linear dependence of the power dissipation with VDD. The square law fits with the experimental data from 0.8 to 1.5 V. We notice a very important rise of the power consumption over 1.5 V, due to the avalanche effects in n-channel MOS devices. This simulation demonstrates the interest for a minimum supply operation to achieve optimum low-power operation.

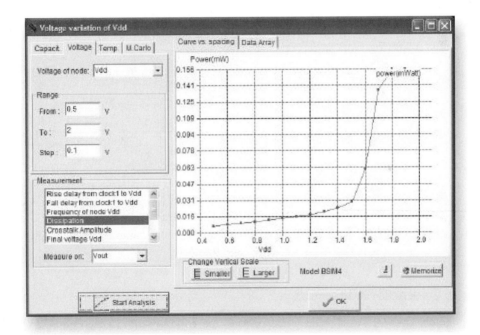

Fig. 4.34 Analysis of the power consumption increase with the supply voltage VDD (CmosLoad.MSK)

4.5.3 Minimum Supply Voltage

The question is: what is the supply voltage below which the inverter does not work any more? The answer can be given by the parametric analysis, focusing this time on the inverter delay dependence versus the supply voltage. Load the file **CmosLoad.MSK** for this study. Invoke the command **Parametric Analysis** of the Analysis menu. Click the layout region corresponding to the node VDD. Verify that the

Voltage menu is selected in the parametric analysis window. Verify that the node "VDD" is selected. Modify the VDD voltage range from 0.5 to 1.5 V, step 0.1 V. Finally, in the measurement menu, select the item Rise **delay** and click **Start Analysis**.

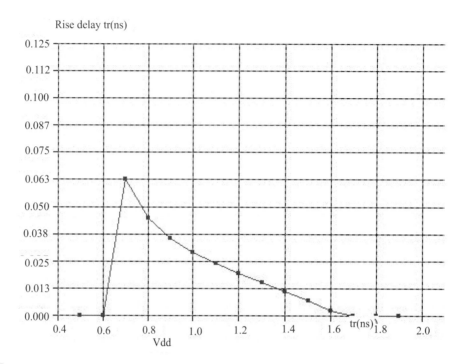

Fig. 4.35 Switching delay dependence with the supply voltage VDD (CmosLoad.MSK)

We observe that the delay is significantly increased as we decrease VDD from its nominal value 1.2 V down to 0.6 V. Below 0.7 V, the inverter delay is higher than the default transient simulation time (10 ns) so that the delay evaluator does not work anymore.

4.6 Static Characteristics

The static characteristics of the inverter correspond to the variation plot of the output voltage versus the input voltage. The simulation involves a step-by-step increase of *Vin*, and the monitoring of *Vout*. In the simulation window, the static characteristics are obtained by a click on the item **Voltage vs Voltage** situated in the selection menu, at the bottom of the chronograms. The curve shown in Figure 4.36 appears.

When *Vin* is low, *Vout* is high, which corresponds to one logic state of the inverter. When *Vin* increases, *Vout* starts to decrease slowly, and suddenly crosses the VDD/2 boundary. At that point, the value of Vin is the commutation point of the inverter, called *Vc*. Then, when *Vin* rises to *VDD*, *Vout* reaches 0, which corresponds to the other logic state of the inverter.

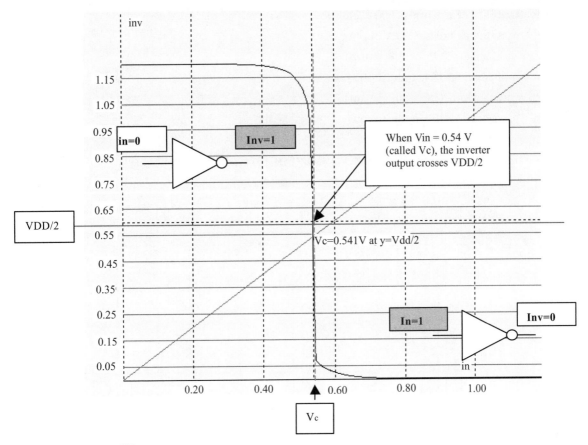

Fig. 4.36 The static characteristics of the inverter (Inv.MSK)

4.6.1 Modify the Commutation Point

Several theoretical formulations of the commutation voltage versus layout parameters exist. A simple formula derived from MOS model 1 is reported in Equation 4.5 [Baker]. Although based on an obsolete model, this formulation may be applied for first-order hand calculations. The verification must be performed by simulation.

$$Vc = \frac{K.V_{TN} + V_{DD} - V_{TP}}{1 + K} \qquad \text{(Equation 4.5)}$$

with

$$K = \sqrt{\frac{\mu_n \dfrac{Wn}{Ln}}{\mu_p \dfrac{Wp}{Lp}}}$$

μn = mobility of electrons (600 V.cm^{-2})

μp = mobility of holes (270 V.cm^{-2})

Wn = n-channel MOS width (in μm)

Ln = n-channel MOS length (in μm)

Wp = p-channel MOS width (in μm)

Lp = p-channel MOS length (in μm)

V_{DD} = supply voltage (1.2 V)

V_{TN} = threshold voltage of n-channel device (0.30 V)

V_{TP} = threshold voltage of p-channel device (0.30 V)

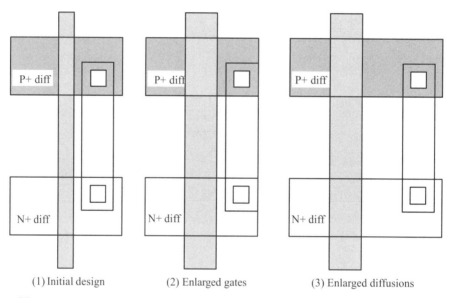

(1) Initial design   (2) Enlarged gates   (3) Enlarged diffusions

Fig. 4.37 Layout modifications that do not change the commutation point

As predicted by the formulation, the sizing of the n-channel and p-channel MOS devices has a strong influence on the commutation point *Vc*. Enlarging both the nMOS and pMOS channels does not change the commutation voltage, nor the supplementary diffusion area (Figure 4.37). As the ratio between the nMOS and pMOS sizes has an effect on *Vc*, only one device should be modified.

In Figure 4.38, we have designed three inverters with almost identical characteristics. The only change is the n-channel or p-channel sizing. The inverter on the left uses the default MOS size, that is *Wp* = 16, *Lp* = 2 lambda, *Wn* = 6, *Ln* = 2 lambda. The large width for the pMOS device compensates the low mobility of holes as compared to electrons, in order to achieve a balanced inverter in terms of switching performances. As a result, the static characteristics are almost symmetrical. In other words, when *Vin* is VDD/2, *Vout* is nearly VDD/2 (curve *inv_1* in Figure 4.39).

For the inverter situated in the middle of the layout, the width and length of the n-channel MOS are identical to those of the p-channel MOS. The result is a lower commutation point, as shown in curve *inv_2* of Figure 4.39. Thanks to this modification, the nMOS device can drive stronger currents and moves the whole curve towards lower voltages. Now, if we enlarge the n-channel MOS channel to reduce its current, the opposite result is achieved, with a commutation point shifted to higher voltages (curve *inv_3* in Figure 4.39).

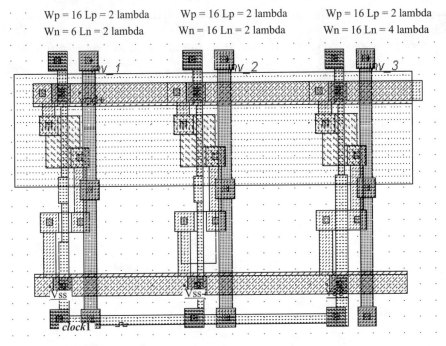

Fig. 4.38   Three different inverter sizings used to investigate its influence on the commutation point Vc (InvSizing.MSK)

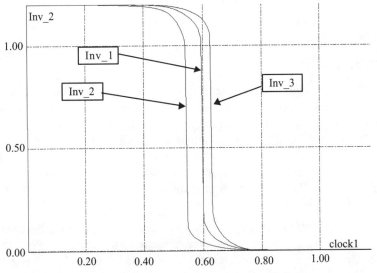

Fig. 4.39   Influence of the inverter sizing on the commutation point (InvSizing.MSK)

4.6.2   Influence of the Model

Using the analog simulation with various models, we may obtain significantly different estimations of the switching characteristics. In Figure 4.40, we superimpose the static characteristics of the same inverter using model 3 and BSIM4. While the simulation with model 3 gives $Vc = 0.6$ V, the simulation with BSIM4 gives Vc = 0.63 V with smooth transitions from 1.2 V down to 0.0 V. This difference is not significant as far as logic behaviour is concerned, but may lead to wrong performance estimation in the case of analog design.

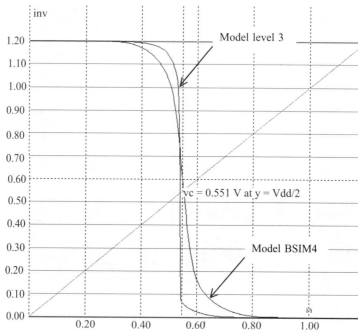

Fig. 4.40   Influence of the model on the simulation (Inv.MSK)

4.7   Random Simulation

As explained in Chapter 3, unavoidable process variations may occur during the integrated circuit fabrication, which may impact the static and dynamic characteristics of the inverter. In the menu **Simulate → Simulation parameters** the default set of parameters corresponds to the "typical" case. We may simulate the inverter in "minimum" or "maximum" case, as we did for the MOS device. An interesting alternative consists in using the "random" mode, also called "Monte-Carlo" analysis, wherein the threshold voltage and the mobility are chosen in a random way, following a Guassian distribution as illustrated in Figure 4.41. There is a high probability that VTO is close to the typical value, and almost no chance that VTO is higher than 0.44 V or lower than 0.36 V.

The simulations of the transient response may be cumulated by a press of the **Reset** button. A new button, **Memory**, appears in the simulation window, at the right lower corner. Press this button to draw all simulations together without refreshing the grid. Each time the Reset button is activated, a new set of threshold and mobility parameters is used to conduct the simulation. The accumulation of ten successive transient simulations is represented in Figure 4.43.

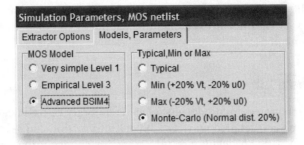

Fig. 4.41   Access to random simulation using an arbitrary set of MOS model parameters (Inv.MSK)

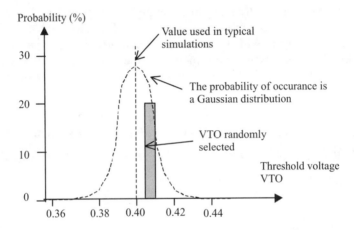

Fig. 4.42   Random selection of *Vt*, with a normal probability

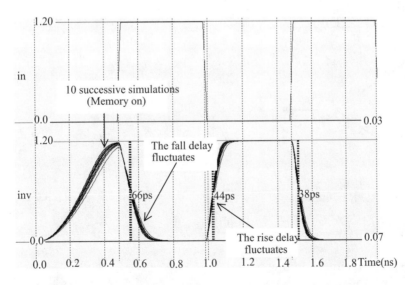

Fig. 4.43   The Monte-Carlo simulation of the inverter transient characteristics, using random VTO and UO parameters (Inv.MSK)

This is a strong probability of the inverter behaving close to the typical value. In some rare cases, the switching performances vary significantly. The min/max simulation is also very interesting to simulate the inverter in extreme situations.

4.8   The Inverter as a Library Cell

Generally speaking, the integrated circuit design relies on a library of basic cells. In this library, each basic cell is described in a very detailed way. The layout information and several static and dynamic aspects are usually included. Such details are important for choosing the appropriate cell, and for evaluating the circuit size, standby parasitic current and switching performances.

The data-sheet of the inverter usually looks like that given in Figure 4.44. Firstly, the header gives the cell name. The mask level file and symbol files are also listed. The truth table recalls the logic behaviour of the cell. The symbol is also provided. In the case of complex cells such as latches, where numerous versions and options co-exist beyond the same name, the truth table is of key importance. The node capacitance is useful for propagation of delay prediction, as the switching performance is linked with the capacitance load.

The operating point recalls the value of the supply with which the characterization has been conducted. Some information is also given for low supply voltage (see also the switching characteristics at 0.9 V).

Name: INVERTER   Technology: 0.12 μm CMOS   Layout: INV.MSK   Symbol: NOT.SYM

Operating point: VDD=1.2 V, Temperature=25°C

Truth table:

In	Out
0	1
1	0

Symbol:

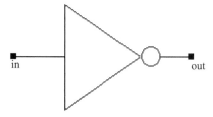

Capacitance:

- Input: 0.5 fF
- Output 0.5 fF

Drive: 1x

Cell Area: 1.26 μm × 4.3 μm (5.14 μm²)

Power consumption: 1.02 μW/MHz typical

Standby current: 100 pA

Inverter	Rise time (ps)				Fall time (ps)			
Input slope	0.01 ns (fast)		0.1 ns (slow)		0.01 ns (fast)		0.1ns(slow)	
Load (fF)	10 fF	100 fF	10 fF	100 fF	10 fF	100 fF	10 fF	100 fF
Delay In → Out	42	340	61	416	35	288	49	338
Delay In → Out (VDD = 0.9 V)			86				65	
Delay In → Out (max, – 40°)			70				50	
Delay In→Out (min, 120°C)			122				98	

Inverter	Peak current (μA)			
Input slope	0.01 ns (fast)		0.1 ns (slow)	
Load (fF)	10 fF	100 fF	10 fF	100 fF
Peak current (typ)			138	
Peak current (max, – 40°)			189	
Peak current (min, 120°C)			105	
Peak current (typ, VDD = 0.9 V)			79	

Fig. 4.44 The library information for the basic inverter (Inv.MSK)

The power consumption is usually described in μW/MHz. To characterize this value, a 1 MHz clock is connected to the input and the total power consumption is computed from the integral of the current. In MICROWIND, we preferably use a 1 GHz clock, and consequently divide the power estimation by 1000. The cell consumption increases linearly with frequency. The standby current is a key information in low-power circuits, wherein the standby parasitic current should be as small as possible. Depending on the MOS option (normal, high-speed, low leakage), the standby current may vary in a very significant way.

The keyword "1x" refers to the inverter strength. A cell with 1x strength is designed with small output MOS devices, usually close to the minimum length and width. A cell with 2x has medium size MOS devices, a cell with drive 4x is used for high speed signals. Cells with 8x drive or even 16x drive may exist, to propagate very fast signals such as clocks and bus. However, using cells with high drive means a high power consumption and high risks of signal integrity problems.

The delay between signals *in* and *out* is strongly dependent on the slope of the input signal, the capacitance connected to the output signal, the temperature, the power supply and the process variations. The goal of the switching delay table is to summarize the delay in typical conditions as well as in extreme conditions. Several design tools such as the timing analyzer and the power consumption extractor will use these data to guess all possible cases of loading conditions, temperature variation, etc.

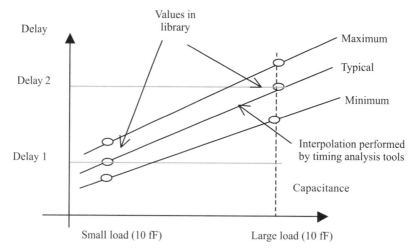

Fig. 4.45   Delay parameters are used by timing analysis tools to predict the cell switching performances for any capacitance

The current peak details are used to evaluate the power consumption of the circuit. The value of the current changes significantly depending on the loading conditions, temperature and supply voltage, as expected.

4.9   3-State Inverter

Until now all the symbols produced the value logic '0' and logic '1'. However, if several inverters share the same node, such as bus structures (Figure 4.46), conflicts will rise. In order to avoid multiple access at the same time, specific circuits called 3-state inverters are used, featuring the possibility of remaining in a 'high impedance' state when access is not required.

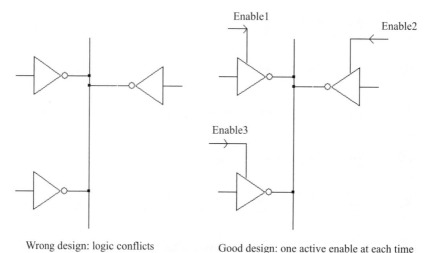

Fig. 4.46 If multiple access is required on a single node, 3-state inverters are used for interfacing (Inv3state.SCH)

The 3-state inverter symbol consists of the logic inverter and an enable control circuit. The output remains in 'high impedance' (logic symbol 'X') as long as the enable *En* is set to level '0'. The truth table is reported below.

NOTIF1

In	En	Out
0	0	X
0	1	1
1	0	X
1	1	0
x	0 or 1	X
0 or 1	X	X

The internal structure of the 3-state inverter is shown in Figure 4.47. The basic CMOS inverter is no longer connected to the supply lines VDD and VSS directly. In contrast, pass nMOS and pMOS devices are inserted to disconnect the inverter when the cell is disabled.

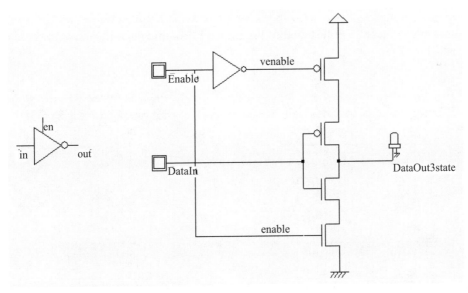

Fig. 4.47 Schematic diagram and logic symbol for the 3-state inverter (CmosInv3State.SCH)

Unfortunately, a supplementary inverter is needed to generate the /*venable* signal required to control the pMOS device. The logic simulation shown in Figure 4.48 illustrates two basic situations: one where *enable* is inactive, and the output is in high-impedance state as no path exists to VDD or VSS, and the other where the circuit is equivalent to an inverter, as the upper and lower pass transistors are enabled.

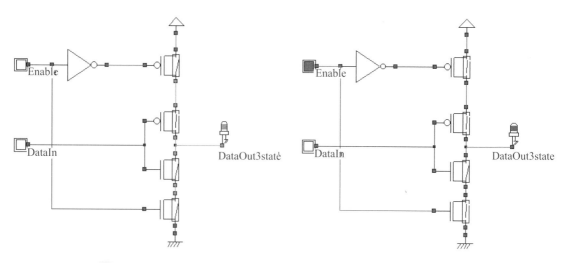

Fig. 4.48 Simulation of the 3-state inverter (CmosInv3State.SCH)

Two versions of the layout are proposed in Figure 4.49, and they correspond to the same design. The cell situated on the left is the direct implementation of the schematic diagram of the 3-state inverter. The layout implementation is not optimal as we loose some silicon area due to severe diffusion design rules which require a 4 lambda spacing. The new arrangement, shown on the right of Figure 4.49, is significantly more compact, thanks to the horizontal flip of the *Enable* inverter, and the sharing of the ground and supply contacts, as illustrated in Figure 4.50. Continuous diffusions always lead to more compact and faster designs.

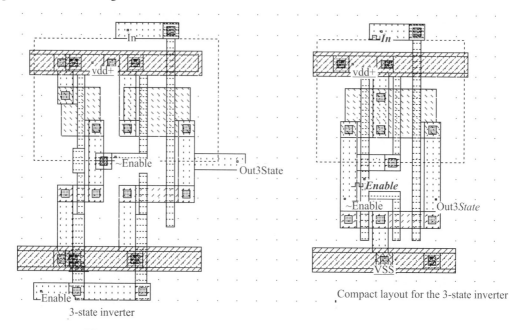

Fig. 4.49 The layout of the 3-state inverter (Inv3State.MSK)

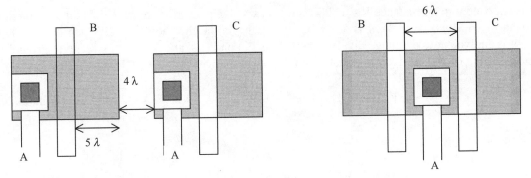

Fig. 4.50 A permutation technique to achieve more compact layout

The analog simulation reported in Figure 4.51 gives an interesting view of the high impedance state. From the chronograms, we see that when *Enable* = 1, the cell acts as a regular CMOS inverter, while when *Enable* = 0, the output "floats" in an unpredictable voltage value, which tends to fluctuate at the switching of the input, mainly due to parasitic leakage and couplings.

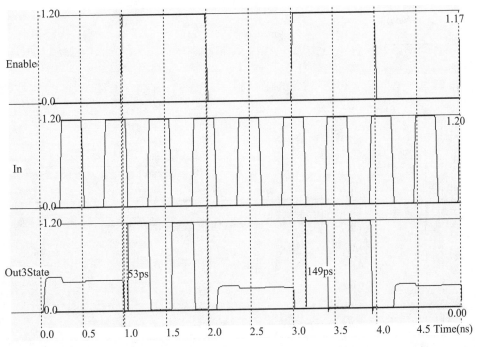

Fig. 4.51 Analog simulation of the 3-state inverter (INV3STATE.MSK)

4.10 All nMOS Inverters

Several other circuits exist to build the logic inverter function [Baker]. One popular inverter is shown in Figure 4.52. It consists of a normal n-channel MOS device *N1* and of another n-channel MOS device *N2* connected as a simple load. Due to the permanent connection of the gate to VDD, the nMOS device

N2 is equivalent to a resistance R_{on_N2}. When *Clock* = 0, a path exists to raise the output through the resistance. The final value V_{high} is $VDD\text{-}V_{tn}$, where V_{tn} is the threshold of *N2*. When *clock* = 1, the circuit is equivalent to a voltage divider where *N2* still conducts, and *N1* creates a path to ground. An approximation of V_{low} is given in Equation 4.6.

$$V_{low} = VDD \frac{R_{on\_N2}}{R_{on\_N1} + R_{on\_N2}} \qquad\text{(Equation 4.6)}$$

Fig. 4.52   An inverter only made with nMOS devices (InvNmos.SCH)

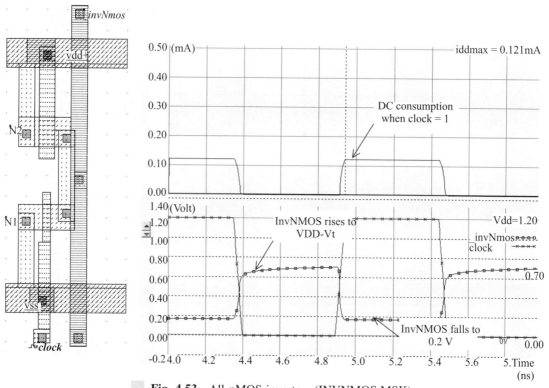

Fig. 4.53   All nMOS inverters (INVNMOS.MSK)

The simulation waveforms given in Figure 4.53 are quite unusual as the low state (*Clock* = 1) of the output leads to a standby current which had not appeared until now in CMOS circuits. This dc power waste is a major drawback for this kind of design. Furthermore, the logic level 0 corresponds to 0.3 V while the logic level 1 corresponds to 0.8 V. The switching is slow, specifically from 0 to 1, due to a weak nMOS device. However, no pMOS device is required, which simplifies both the design and the process. The device *N2* is sized with a large length and a small width to increase the resistance *Ron_N2*, to lower the output voltage when *clock* = 1. All nMOS inverters were used before CMOS technology was made available.

4.11 Ring Oscillator

The ring oscillator made from five inverters has the property of oscillating naturally. We observe in the circuit of Figure 4.54 the oscillating outputs and measure their corresponding frequency.

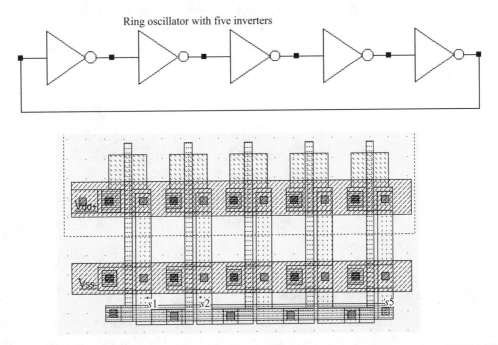

Fig. 4.54 Schematic diagram and layout of the ring oscillator used for simulation (INV5.MSK)

The ring oscillator circuit can be simulated easily at the layout level with MICROWIND using various technologies. The time-domain waveform of the output is shown in Figure 4.55 for 0.8, 0.12 μm and 70 nm technologies. Although the supply voltage (VDD) has been reduced (VDD is 5 V in 0.8 μm, 1.2 V in 0.12 μm, and 0.7 V in 70 nm), the gain in frequency improvement is significant.

By default, the software is configured with 0.12 μm technology. Use the command **File → Select Foundry** to change the configuring technology. For example, select **cmos08.RUL** which corresponds to the CMOS 0.8 μm technology, or the file **cmos90n.RUL** which configures MICROWIND to the CMOS 90 nm technology. When you run the simulation again, you may observe a change of VDD and a significant change in the oscillating frequency.

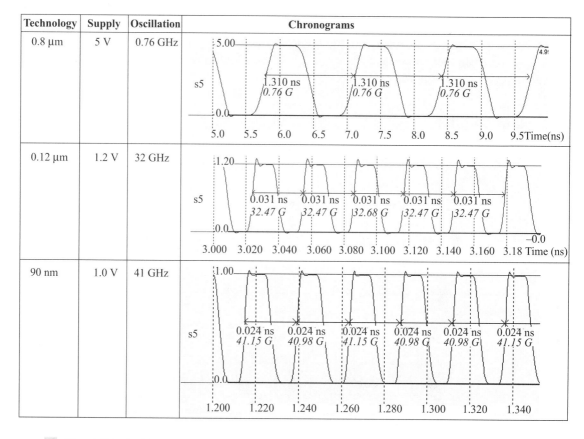

Technology	Supply	Oscillation	Chronograms
0.8 μm	5 V	0.76 GHz	
0.12 μm	1.2 V	32 GHz	
90 nm	1.0 V	41 GHz	

Fig. 4.55   Oscillation frequency improvement with the technology scale down (Inv5.MSK)

4.11.1   High Speed vs Low Leakage

Let us consider the ring oscillator with an enable circuit, where one inverter has been replaced by a NAND gate to enable or disable oscillation (Inv5Enable.MSK). The schematic diagram is shown in Figure 4.56, as well as its layout implementation. We analyze the switching performances in high speed and low leakage mode, by changing the properties of the option layer which surrounds all devices.

Ring oscillator with 4 inv and one NAND gate

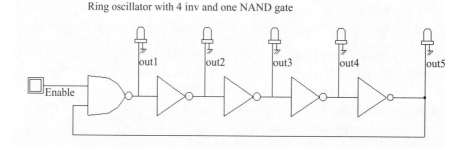

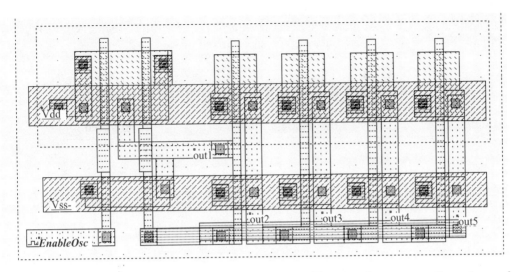

Fig. 4.56 Schematic diagram and layout of the ring oscillator used to compare the analog performances in high speed and low leakage mode (INV5Enable.MSK)

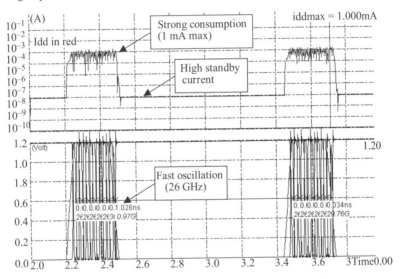

Fig. 4.57 Simulation of the ring oscillator in high speed mode, using the BSIM4 model. The oscillating frequency is fast but the standby current is high (Inv5Enable.MSK)

The option layer which surrounds the oscillator is set to high speed mode by a double click inside that box. In high speed mode, the circuit works fast (26 GHz) but consumes a lot of power (1 mA) when on, and a significant standby current when off (10 nA), as shown in the simulation of the voltage and current given Figure 4.56. Notice the tick in front of "Scale I in log" to display the current in logarithmic scale.

In contrast, the low leakage MOS features slower oscillation (20 GHz in Figure 4.57, that is, approximately a 25 per cent speed reduction), but with 40 per cent less current when on, and more than one decade less standby current when off (1 nA). In summary, low leakage MOS devices should be used whenever possible. High speed MOS devices should be used only when speed is critical, such as communication bus, critical path, etc. The analog performances are summarized in Table 4.2.

Table 4.2 Comparative performances of the ring oscillator (Inv5Enable.MSK)

Parameter	Low leakage mode	High speed mode
Imax	0.6 mA	1.0 mA
I standby	<1 nA	>10 nA
Oscillating frequency	20 GHz	26 GHz

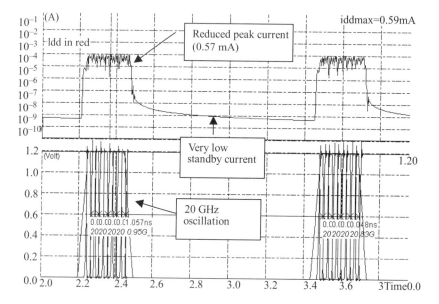

Fig. 4.58 Simulation of the ring oscillator in low voltage mode, using the BSIM4 model. The oscillating frequency is slower but the standby current is very low (Inv5Enable.MSK)

4.11.2 Temperature Effects

The main consequence of a temperature increase is a decreasing mobility of the electrons and holes of the MOS channel, leading to slower transient performances. Thus, the propagation delay due to the logic gate is increased, as illustrated in Figure 4.58, which concerns the switching characteristics of the 5-inverter ring oscillator. In MICROWIND, you can get access to temperature using the command **Simulate → Simulation Parameters**. The temperature is given in °C.

In the simulation in Figure 4.59, we used the BSIM4 model for a temperature set to 25°C and 120°C. We can observe a 10 per cent decrease in the switching speed, which finds its origin in the mobility degradation, which is computed by the following formulation:

$$U0 = U0_{(T=27)} \left(\frac{T + 273}{300} \right)^{-1.8}$$

(Equation 4.7)

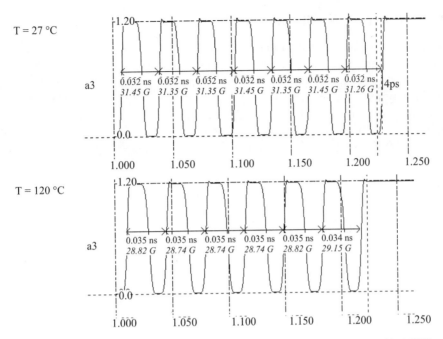

Fig. 4.59 Propagation delay increases with temperature (Inv5Enable.MSK)

We may conduct the parametric analysis of the temperature influence on the oscillating frequency, in order to obtain the results given in Figure 4.60. It may be seen that the frequency variation from −100°C to +100°C is kept below 15 per cent. The reason for this reduced dependence is that the mobility reduction is compensated by the threshold voltage decrease, also strongly dependent on the temperature, which tends to limit the overall effects of temperature variations.

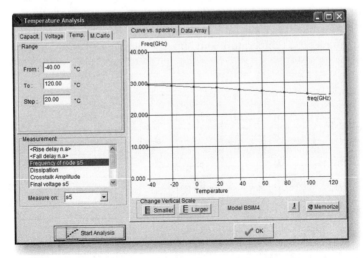

Fig. 4.60 Performances of the ring oscillator versus temperature increase (Inv5Enable.MSK)

4.11.3   A 2.5 GHz Ring Oscillator

The previous ring oscillator operated around 30 GHz, which is of no practical use. In contrast, the 2.5 GHz frequency is widely used for a variety of wireless network applications. In this paragraph, we investigate several possibilities of slowing down the oscillator frequency to 2.5 GHz. One immediate idea consists in designing a ring oscillator with more inverter stages (around 70). This is a power-consuming and silicon area-consuming approach. A more attractive solution consists in the reduction of the MOS current capabilities. Then a new question rises: how should we proceed, as we should increase the channel length, decrease the channel width, or add parasitic capacitance in each switching node (Figure 4.60)?

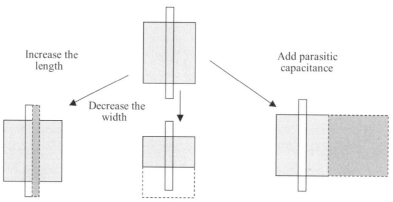

Fig. 4.61   The current of the MOS device may be reduced by increasing the length or decreasing the width

One solution is proposed in Figure 4.62. It combines the channel length increase, the width decrease to its minimum value, and the enlarging of drain areas, whenever possible, to increase the parasitic junction surface, and consequently its parasitic capacitance. All these layout modifications have a sufficient impact in reducing the oscillating frequency to around 2.5 GHz. Notice that this frequency is very sensitive to process parameters, temperature and supply voltage variation.

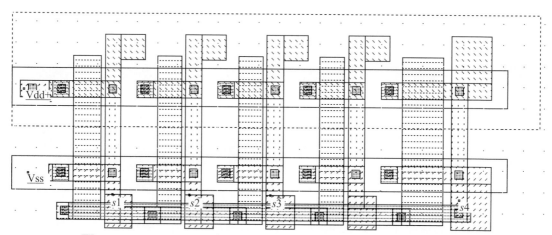

Fig. 4.62   The 5-inverter oscillator tuned to 2.5 GHz (Inv2,5GHz.MSK)

4.12 Latch-up Effect

The latch-up effect is a parasitic shortcut between VDD and VSS that can lead to the destruction of the integrated circuits. The origin of latch-up is the activation of a parasitic N/P/N/P device (also called thyristor) appearing in the vertical cross-section of the nMOS and pMOS structures as shown in Figure 4.63.

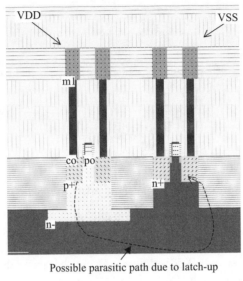

Possible parasitic path due to latch-up

Fig. 4.63 Origin of latch-up

4.12.1 Limiting the Latch-up Effect

The latch-up effect is almost eliminated if the substrate is locally polarized to ground, and the n-well is locally polarized to VDD (Figure 4.64). In the upper layout (Figure 4.64a), the situation is extremely dangerous as the n-well region is floating. If the n-well potential drops around VDD/2 and the local substrate voltage rises to VDD/2, the latch-up phenomenon is initiated. Most layout tools alert the designer in the case of floating n-well regions. The good approach consists in inserting a polarization diode N+/n-well and sticking it to the highest possible potential, typically VDD.

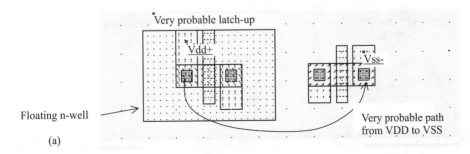

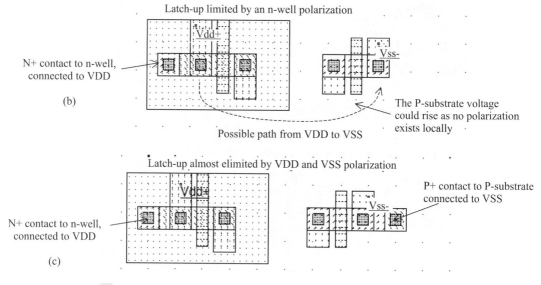

Fig. 4.64   Limiting the latch-up effect by polarization diodes

Many designers consider that there exists an "automatic" polarization of the substrate to ground and forget to add a local P+/P-substrate contact to ground, near the nMOS device (Figure 4.64b). This might be a dangerous assumption which can cause latch-up: in 0.12 μm technologies, several manufacturers use a highly resistive P-doped substrate. In that case, the electrical link between the physical ground (back of the IC) and the local nMOS area is equivalent to a resistor of several K Ohm. Consequently, what is supposed to be a good 0 V reference is a very weak 0 V, that can easily fluctuate and turn on the N/P/N/P device, which may lead to latch-up and possible destruction. This is why it is highly recommended to also add a P+/P–substrate polarization to ground, which protects the logic cell from latch-up (Figure 4.64c).

4.13   Conclusion

This chapter has described the CMOS inverter, from a logic and analog point of view. The mobility difference between electrons and holes has been counter-balanced at the layout level to obtain symmetrical static and dynamic characteristics. The effect of MOS model and temperature on the simulation results have also been investigated. The 3-state inverter, and all n-MOS inverter and ring-oscillator circuits have been designed and simulated. Finally, we have presented the basic polarization techniques to avoid the parasitic latch-up effect.

References

[Weste] N. Weste, K. Eshraghian "Principles of CMOS VLSI design", Addison Wesley, ISBN 0-201-53376-6, 1993.

[Baker] R.J. Baker, H. W. Li, D.E. Boyce "CMOS circuit design, layout and simulation", IEEE Press, ISBN 0-7803-3416-7, 1998.

EXERCISES

1. Create the layout and compare the static characteristics of the following three inverters:

   - Wn < Wp
   - Wn = Wp
   - Wn > Wp

   Which one seems to be well balanced? Justify your answer.

2. We consider the two inverters of Figure 4.65. We define t_{PLH1} as the delay from a low to high value of the output inv1 (Figure 4.65). We define t_{PLH2} as the delay from a low to high value of the output inv2. Find the relation between the propagation delays t_{PLH1} and t_{PLH2}.

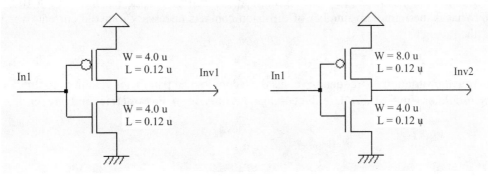

Fig. 4.65 Compared performances of two inverters with different sizing (ch42.sch)

Answer: $t_{PLH2} = 2/3 * t_{PLH1}$

3. Using Microwind in 0.12 μm, draw an inverter (W_{PMOS} = 2 μm, W_{NMOS} = 1 μm) connected to a 100 fF capacitor. Find *RON* and *ROFF* of each transistor and calculate t_{PHL} and t_{PLH}. Compare their values with the simulated results. Using BSIM4 model.

   Answer: RON_P = 1.2 kΩ, $ROFF_P$ =, RON_n = , $ROFF_n$ = <Sonia>

4. Configure Microwind in 90 nm and design a ring oscillator using an odd number of inverters (for example, 11). Based on the analog simulation using BSIM4, find the inverter propagation delay (t_p) and the oscillator frequency (f_{osc}). Extract t_{PHL} and t_{PLH} and deduce the theoretical oscillator frequency f_{osc_theory}. Compare it with the simulation (f_{osc}). Playing with the available MOS options in 90 nm technology, analyze the variation of f_{osc}.

5. Using Microwind, design two inverters: INV1 (L_{min}, WN_{min}, WP_{min}), INV2 (L_{min}, 4 x WN_{min}, 4 x WP_{min}). What is the input capacitance of INV1? Simulate the following four configurations using the same input clock (clock switching: $tr = tf = 10$ ps) and extract the switching delays (out put from 0 to 1). Would the switching delay be different with a larger tr and tf?

   - INV1 alone
   - INV1 connected to INV1
   - INV1 connected to INV2

- INV1 connected to a 10fF load

Answer: 1.73 fF, 43 ps, 8 ps, 15 ps, 29 ps

6. Using Microwind in 0.12 μm, draw an inverter (L = 0.12 μm, W_{PMOS} = 0.6 μm, W_{NMOS} = 0,3 μm).

- Simulate the PMOS characteristics to find its I_{ON} current.

- Use the time domain simulation in mode **voltage and current** to compare I_{CCmax} to I_{ON}. Extract the power dissipation.

- Add a capacitor on the inverter output. For the two following values (1 fF, 100 fF) compare I_{Ccmax} and the power dissipation.

- Design a new inverter (L = 0.12 μm, W_{PMOS} = 6 μm, W_{NMOS} = 3 μm) with a 100 fF load. What is the minimum number of diffusion contacts necessary to avoid current overstress in the inverter?

Answer: 275 μA, 111 μA

7. Analyze the variation of frequency versus the technological parameter variation, for the 5-inverter ring oscillator (Inv5.MSK). Use the command **Analysis → Parametric analysis** for this study.

Interconnects

5.1 Introduction

The role of interconnects in integrated circuit performances has considerably increased with the technology scale down. Figure 5.1 shows the evolution of the aspect of the integrated circuit. In 0.12 μm, 6 to 8 metal layers are available.

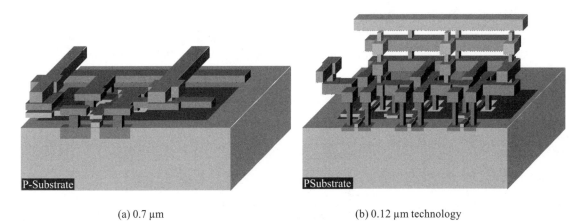

(a) 0.7 μm   (b) 0.12 μm technology

Fig. 5.1 Evolution of interconnect between 0.7 μm technology and 0.12 μm technology (Inv3.MSK)

5.2 Metal Layers

In the previous chapter, we designed the CMOS inverter using two layers of metal. However, up to six metal layers are available for signal connection and supply purpose. A significant gap exists between the 0.7 μm 2-metal layer technology and the 0.12 μm technology in terms of interconnect efficiency.

Firstly, the contact size is 6 lambda in 0.7 μm technology, and only 4 lambda in 0.12 μm. This features a significant reduction of device connection to metal and metal 2, as shown in Figure 5.2. Notice that an MOS device generated by using 0.7 μm design rules is still compatible with 0.12 μm technology. But an MOS device generated by using 0.12 μm design rules would violate several rules if checked using 0.7 μm design rules.

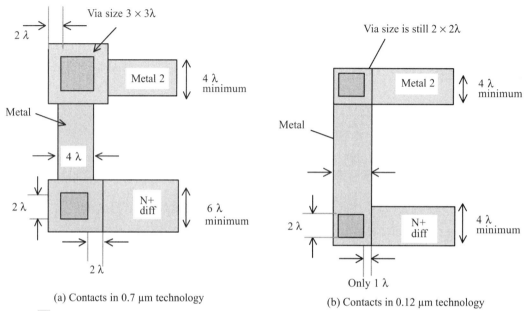

(a) Contacts in 0.7 μm technology

(b) Contacts in 0.12 μm technology

Fig. 5.2   Contacts in 0.7 μm technology require more area than in 0.12 μm technology

Secondly, the stacking of contacts is not allowed in micron-range technologies. This means that a contact from poly to metal 2 requires a significant silicon area (Figure 5.3a) as contacts must be drawn in a separate location. In deep-submicron technology (starting from 0.35 μm and below), stacked contacts are allowed (Figure 5.3b).

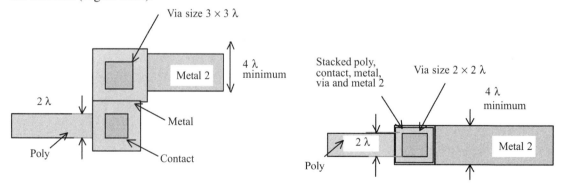

(a) Poly to metal 2 contact in 0.7 μm technology

(b) Poly to metal 2 contact in 0.12 μm technology

Fig. 5.3   Stacked vias are allowed in 0.12 μm technology, which saves a significant amount of silicon area as compared to 0.7 μm design style

Metal layers are labelled according to the order in which they are fabricated, from the lower level 1 (metal 1) to the upper level (metal 6 in 0.12 μm). Each layer is embedded into a silicon oxide (SiO_2) which isolates layers from each other. A cross-section of a 0.12 μm CMOS technology is shown in Figure 5.4.

Fig. 5.4   Cross-section of a 0.12 μm technology (*Courtesy:* Fujitsu)

5.3   Contact and Vias

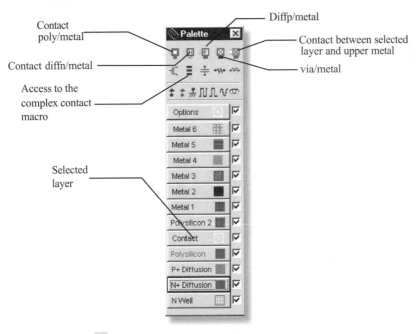

Fig. 5.5   Access to basic contact macros

The connection material between diffusion and metal is called "contact". The same layer is also used to connect poly to metal, or poly 2 to metal. The connection material between metal and metal 2 is called "via". By extension, the material that connects metal 2 to metal 3 is "via 2", metal 3 to metal 4 "via 3", etc.

In MICROWIND, specific macros are accessible to ease the addition of contacts in the layout. These macros may be found in the palette, as shown in Figure 5.6. As an example, you may instantiate a design-error free poly/metal contact by a click on the upper left corner icon in the palette. You may obtain the same result by drawing one box of poly (4 × 4 lambda), one box of metal (4 × 4 lambda) and one box of contact (2 × 2 lambda), according to design rules.

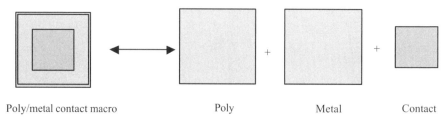

| Poly/metal contact macro | Poly | Metal | Contact |

Fig. 5.6   Access to basic contact macros

Additionally, an access to complex stacked contacts is proposed thanks to the icon "complex contacts" situated in the palette, second row, second column. The screen shown in Figure 5.7 appears. By default, you create a contact from poly to metal 1, and from metal 1 to metal 2. Add new ticks to build more complex stacked contacts.

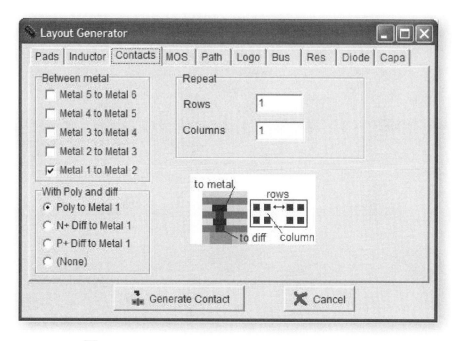

Fig. 5.7   Access to complex stacked contact generator

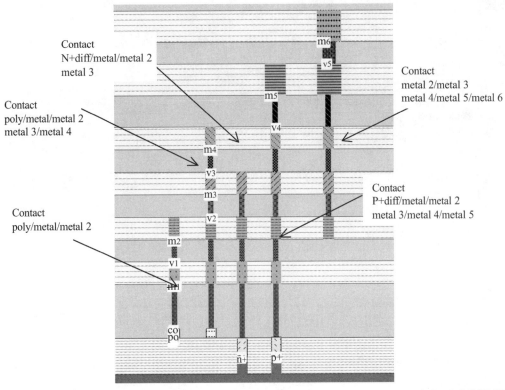

Fig. 5.8 Examples of layer connection using the complex contact command from MICROWIND (Contacts.MSK)

5.3.1 Direct Layer Connection

A convenient command exists in MICROWIND to add the appropriate contact between two layers. Let us imagine that we need to connect two signals, one routed in polysilicon and another in metal 3. Rather than invoking the complex macro command, we may just select the icon "connect layers". As a result, a stack of contacts is inserted at the desired location to connect the lower layer to the upper layer. An illustration of this command is shown in Figure 5.9.

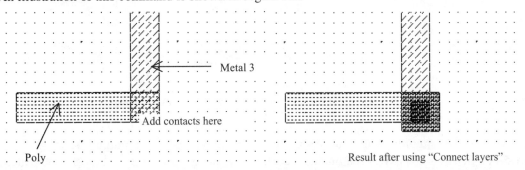

Fig. 5.9 The command "connect layers" inserts the appropriate stacked contacts to build the connection between the desired layers (ConnectLayers.MSK)

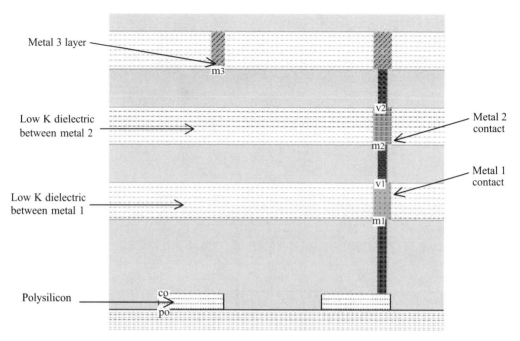

Fig. 5.10 A 2-D cross-section of the layout before and after connecting poly and metal 3 layers (ConnectLayers.MSK)

5.4 Design Rules

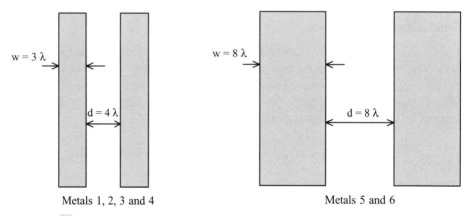

Fig. 5.11 Minimum width w and distance d between metal layers

In 0.12 μm technology, the metal layers 1, 2, 3 and 4 have almost identical characteristics. As regards the design rules, the minimum size w of the interconnect is 3 λ. The minimum spacing is 4 λ (Figure 5.11). In MICROWIND, each interconnect layer is drawn with a different colour and pattern. Examples of minimum width and distance interconnects are given in Figure 5.12.

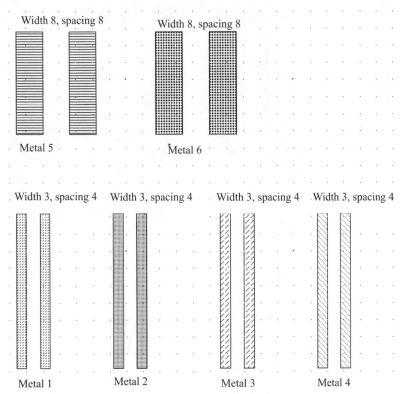

Fig. 5.12 Metal layers, associated rules and patterns as appearing in the Microwind editor (DesignRulesMetal.MSK)

These minimum width and spacing dimensions are critical. They define the limit below which the probability of manufacturing error rises to an unacceptable level. If we draw metal 1 lines with 2 λ width and 2 λ spacing, interconnect interruptions or short-cuts may appear, as illustrated in Figure 5.13. Prior to fabrication, the design rules must be checked to ensure that the whole circuit complies with the width and spacing rules, to avoid unwanted interruptions or bridges in the final integrated circuit. The Microwind command to get access to the design rule checker is **Analysis →
Design Rule Checker**. Still, there exists a significant probability of manufacturing error even if the circuit complies with all design rules. A wafer of 500 integrated circuits has a typical yield of 70 per cent in mature technologies, which means that 30 per cent of the circuits have a fabrication error and must be rejected. The yield may drop to as low as 20 per cent, as in the case of state-of-the art technologies with all process constraints pushed to their limits, and very large silicon dies.

The practical design width for metal interconnects is usually a little higher than the minimum value. In MICROWIND, the routing interconnects are drawn in 4 λ width. The pitch is the usual distance that separates two different interconnects. In 0.7 μm, due to severe constraints in the contact size, the pitch has been fixed to 10 λ. In deep-submicron technology, improvements in contact sizing may reduce that pitch to 8 λ. In 0.12 μm technology, this routing pitch is equivalent to 0.48 μm.

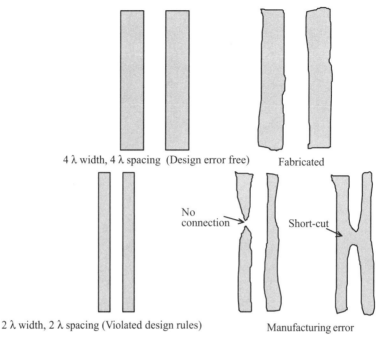

4 λ width, 4 λ spacing (Design error free)   Fabricated

No connection →   Short-cut →

2 λ width, 2 λ spacing (Violated design rules)   Manufacturing error

Fig. 5.13 The manufacturing of interconnects which violate the minimum width and distance may result in interruptions or short-cuts, which may have catastrophic consequences on the behaviour of the integrated circuit

Table 5.1   Conductor parameters vs. technology

Technology	Filename	Upper metal			Lower metal		
		Width	Thickness	Distance	Width	Thickness	Distance
0.8 μm	CMOS08. RUL	1.2	0.70	1.20	3.20	0.70	6.00
0.6 μm	CMOS06. RUL	0.9	0.70	0.90	2.40	0.70	4.50
0.35	CMOS035.RUL	0.6	0.70	0.60	1.60	0.70	3.00
0.25	CMOS025.RUL	0.38	0.60	0.50	1.00	0.70	1.88
0.18	CMOS018.RUL	0.30	0.50	0.40	0.80	1.00	0.80
0.12	CMOS012.RUL	0.18	0.40	0.24	0.48	0.80	0.48
90 nm	CMOS90n.RUL	0.15	0.35	0.20	0.40	1.60	0.40
65 nm	CMOS65n.RUL	0.10	0.30	0.14	0.28	1.00	0.52

Integrated circuit manufacturers usually specify this routing pitch in their non-confidential technological descriptions, as one of the commercial arguments for designing compact (and behind this, low cost) integrated circuits. The common industrial pitch in the 0.12 μm CMOS process is around 0.4 μm. The design styles in 0.7 μm and 0.12 μm are illustrated in Figure 5.14.

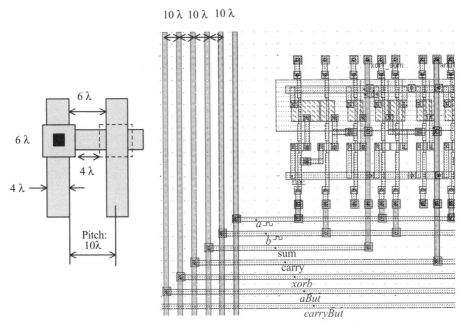

Fig. 5.14 Illustration of the routing pitch in 0.7 μm, set to 10 λ due to the large size of the contacts

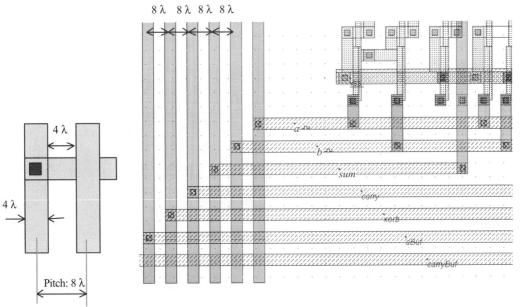

Fig. 5.15 Illustration of the routing pitch in 0.12 μm, set to 8 λ thanks to the reduced size of the contact

5.5 Capacitance Associated with Interconnects

Interconnect lines exhibit the property of capacitance, as they are able to store charges in the metal interface with oxide. The capacitance effect is not simple to describe and to modellize. This is due to the fact that interconnects are routed very close to each other, as shown in the example in Figure 5.16. The capacitance effects are represented by a set of capacitors which link interconnects electrically.

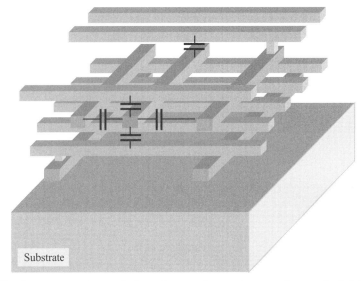

Fig. 5.16   One interconnect is coupled to other conductors in several ways, both lateral and vertical

5.5.1 Large Plates

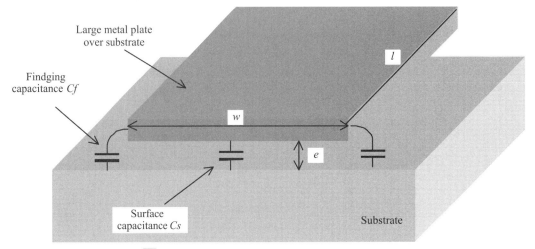

Fig. 5.17   Large plate of metal above the substrate

In the case of a large metal area (width w, length l) separated from the substrate or other metal areas by an oxide with a thickness e, the Equation 5.1 is quite accurate. We neglect the fringing capacitance Cf in that case. Large plates of metal are used in pads (see the chapter on input/output in Book II) and supply lines.

$$Cs = \varepsilon_0 \varepsilon_r \frac{w \cdot l}{e}$$ (Equation 5.1)

$\varepsilon_0 = 8.85\ e^{-12}$ (Farad/m)

$\varepsilon_r = 3.9$ for SiO_2

w = conductor width (m)

l = conductor length (m)

e = dielectric thickness (m)

5.5.2 Conductor above a Plane

Several formulations have been proposed [Sakurai] [Delorme] to compute the capacitance of a conductor when the width is comparable to the oxide thickness, which is the case with a large majority of conductors used to transport signals. We give the formulation of the total capacitance which consists in the sum of Cs and twice the fringing capacitance Cf [Delorme].

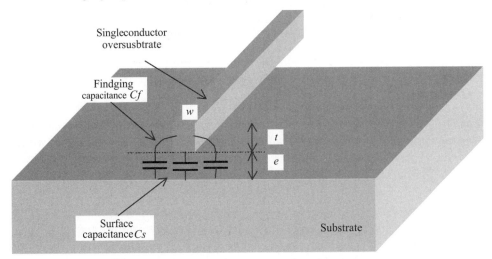

Fig. 5.18 One conductor above a ground plane

$$C = C_s + 2 \cdot C_f = \varepsilon_0 \varepsilon_r + \left(1,13 \cdot \frac{w}{e} + 1,44 \cdot \left(\frac{w}{e} \right)^{0,11} + 1,46 \cdot \left(\frac{t}{e} \right)^{0,42} \right)$$ (Equation 5.2)

C = total capacitance per metre (Farad/m)

Cs = surface capacitance (Farad/m)

Cf = fringing capacitance (Farad/m)

ε_0 = 8.85 e^{-12} (Farad/m)

ε_r = relative permittivity 3.9 for SiO$_2$

w = conductor width (m)

t = conductor thickness (m)

e = dielectric thickness (m)

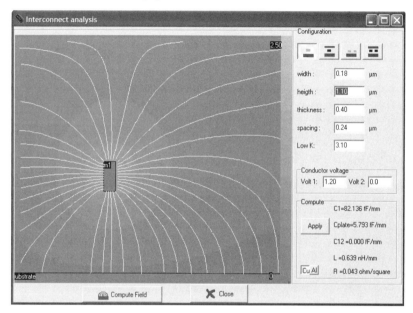

Fig. 5.19   A 2D simulation of the electrical field lines between the conductor and the ground

The command **Analysis → Interconnect Analysis** with FEM gives some interesting information about the electric coupling between the conductor and the ground (Figure 5.19). The palette of colours indicates the potential in the oxide region, in Volt. The field lines are the white wires which link the conductor to the ground. Some field lines are directly coupled to the substrate ground, while some other field lines go to the free space. In 0.12 μm technology, the metal conductor is 1.1 μm above the ground plane, which corresponds to the simulation shown in Figure 5.19. Its minimum width is 0.18 μm (equivalent to 3 λ) while its thickness is 0.4 μm. These physical dimensions are listed in the parameter menu situated on the right side of the window. The formulations given in (Equation 5.1) produce an underestimated value for *Cplate*, equal to 6 fF/mm. The formulations proposed in (Equation 5.2) give a more reliable result for *C1*, equal to 82 fF/mm.

5.5.3   Two Conductors above a Ground Plane

When a conductor is routed close to another conductor, a crosstalk capacitance, defined as *C12* is created between the two conductors (Figure 5.20). In 0.12 μm technology, a specific dielectric with a low permittivity (this parameter, called Low K, is approximately 3.0 instead of 3.9) is used to fill the gaps between interconnects. This is an efficient technique to reduce the crosstalk capacitance while keeping the upper and lower capacitance almost unchanged. Consequently, the oxide stack alternates between high K and low K materials, as shown in the cross-section. Low K dielectrics were introduced with 0.18 μm technology. Air gaps are the ultimate low K materials, with a lowest possible K = 1. Intensive research is being conducted on the enclosure of air gaps between coupled connectors, and could become a standard in future technologies.

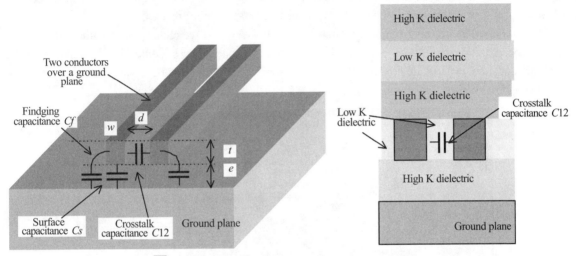

Fig. 5.20   Two conductors above a ground plane

$$C = C_s + C_f = \varepsilon_0 \varepsilon_r \left(1.10 \frac{w}{e} + 0.79 \left(\frac{w}{e} \right)^{0.1} + 0.46 \cdot \left(\frac{t}{e} \right)^{0.17} \left(1 - 0.87 e^{\left(\frac{-d}{e} \right)} \right) \right) \qquad \text{(Equation 5.3)}$$

$$C_{12} = \varepsilon_0 \varepsilon_{\text{rlowK}} \left(\frac{t}{d} + 1.2 \left(\frac{d}{e} \right)^{0.1} \cdot \left(\frac{d}{e} + 1.15 \right)^{-2.22} + 0.253 ln \left(1 + 7.17 \frac{w}{d} \right) \cdot \left(\frac{d}{e} + 0.54 \right)^{-0.64} \right) \quad \text{(Equation 5.4)}$$

C = conductor capacitance to ground per meter (Farad/m)

Cs = surface capacitance (Farad/m)

Cf = fringing capacitance (Farad/m)

$C12$ = crosstalk capacitance (Farad/m)

ε_0 = 8.85 e^{-12} (Farad/m)

ε_r = relative permittivity 3.9 for SiO_2

$\varepsilon_{\text{rlowK}}$ = relative permittivity of low dielectric material (around 3.0 in 0.12 μm)

w = conductor width (m)

  t = conductor thickness (m)

  e = dielectric thickness (m)

  d = conductor distance (m)

In the default 0.12 µm technology, two coupled interconnects with the minimum width and distance routed with the first level of metallization have a cross-section as shown in Figure 5.21. The benefits of the low permittivity dielectric appear clearly in the value of the coupling capacitance *C12* (75 fF/mm), which is comparable to the ground capacitance *C1* (51 fF/mm). If we change the parameter Low K to the silicon dioxide permittivity equal to 4, the coupling capacitance becomes much higher than the ground capacitance. More details about the crosstalk effect, including simulations and measurements, are given at the end of this chapter.

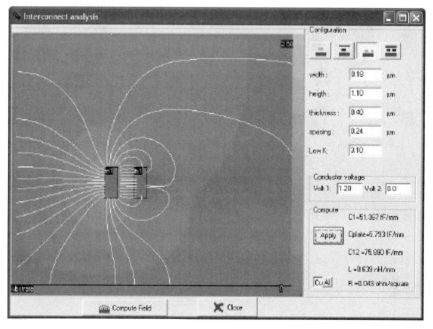

Fig. 5.21 A 2D simulation of the field lines between two conductors and the ground

5.5.4 Real Case Conductors

In practice, the accurate extraction of the capacitance of each node is based on a 2D partitioning of the layout, and the 3D computing of capacitance for each elementary configuration. Even a simple interconnect configuration, such as the one shown in Figure 5.21, is equivalent to a large set of 3D configurations, with each of them requiring the use of a 3D static field solver.

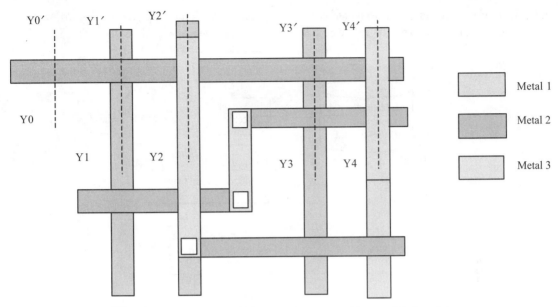

Fig. 5.22 Example of interconnects routed in metal 1, 2 and 3

Examples of standard configurations that may be found in the layout of Figure 5.22 are shown in Figure 5.23, with numerical values corresponding to 0.12 µm technology, metal wire width of 4 lambda (0.24 µm), and wire spacing of 4 lambda (0.24 µm).

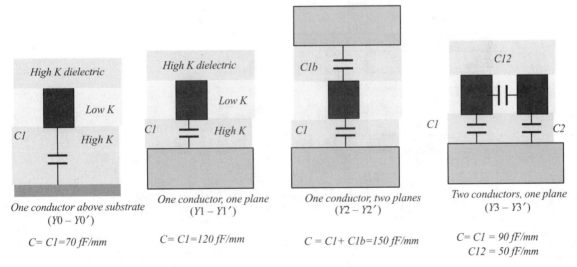

Fig. 5.23 Basic configurations illustrating the cross-sections of the interconnect network (0.12 µm)

The extraction of the layout is conducted according to the flow described in Figure 5.24. The MOS devices and interconnects are extracted separately. In order to compute the crosstalk capacitance, the interconnect network is parsed into elementary structures, with each one having a fixed stack of layers.

An iterative procedure permits us to locate vertical and horizontal elementary crosstalk contributions and to compute the sum which appears in the SPICE netlist and the electrical node properties (Figure 5.24).

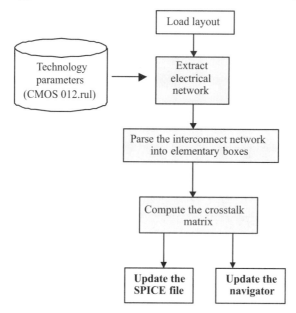

Fig. 5.24 Crosstalk capacitance extraction steps in MICROWIND

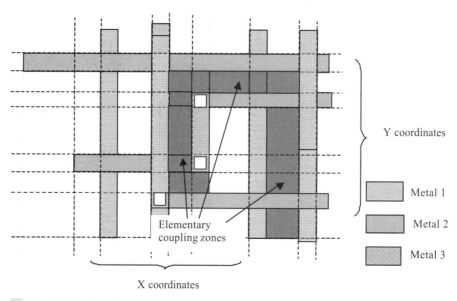

Fig. 5.25 Identification of coupling zones in the test case presented in Figure 5.22

Figure 5.25 shows where the elementary coupling zones have been detected. Only lateral coupling between identical layers, close enough to generate a significant crosstalk contribution, are considered. The other couplings are neglected in this extraction phase. In the test case proposed in Figure 5.22, couplings occur in metal 1, metal 2 and metal 3, as seen in Figure 5.25. Each contribution is very small (less than 1femto-Farad). However, the sum of elementary coupling contributions may result in hundreds of femto farad, which may be comparable to the ground capacitance, and thus create significant crosstalk coupling effects.

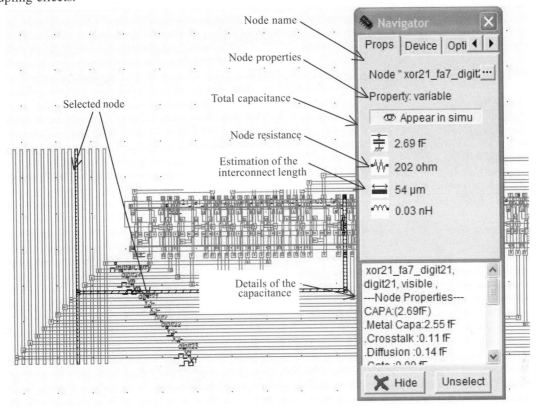

Fig. 5.26 Extraction of a real-case interconnect capacitance using the command "View Electrical Node"

The command **View →View Electrical Node** or the above icon launches the extraction of ground and crosstalk capacitance for each electrical net of the layout. The ground capacitance of the node and the crosstalk capacitance are detailed in the navigator window, as illustrated in Figure 5.26. Notice that the global capacitance is split into metal, crosstalk, gate and diffusion capacitance. The selected net has weak crosstalk coupling as compared to ground coupling.

5.6 Resistance Associated with Interconnects

The resistivity of interconnect materials used in CMOS integrated circuits is listed in Table 5.2. Conductors have very low resistivity, while semiconductor materials such as highly doped silicon have a moderate resistivity. In contrast, the intrinsic silicon resistivity is very high.

Table 5.2   Resistivity of several materials used in CMOS circuits

Symbol	Description	Material Used for	Resistivity at 25°C
ρ_{cu}	Copper resistivity	Signal transport	1.72×10^{-6} Ω.cm
ρ_{al}	Aluminum resistivity	Signal transport	2.77×10^{-6} Ω.cm
ρ_{Ag}	Gold resistivity	Bonding between chip and package	2.20×10^{-6} Ω.cm
$\rho_{tungsten}$	Tungsten resistivity	Contacts	5.30×10^{-6} Ω.cm
ρ_{Ndiff}	Highly doped silicon resistivity	N+ diffusions	0.25 Ω.cm
ρ Nwell	Lightly doped silicon resistivity	n-well	50 Ω.cm
ρ si	Intrinsic silicon resistivity	Substrate	2.5×10^{5} Ω.cm

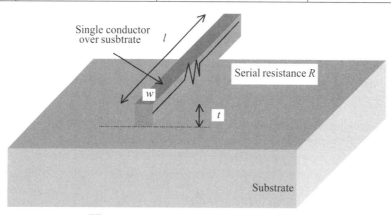

Fig. 5.27   Resistance of a conductor

If a conductor with a resistivity has length *l*, width *w*, and thickness *t*, then its serial resistance *R* (Figure 5.27) can be computed by using the following formula:

$$R = \rho \frac{l}{w \cdot t}$$

(Equation 5.5)

where

R = serial resistance (ohm)

ρ = resistivity (ohm.m)

w = conductor width (m)

t = conductor thickness (m)

l = conductor length (m)

d = conductor distance (m)

5.6.1 Resistance per Square

When designing interconnects, a very useful metric is the "resistance per square". We assume that the width is equal to the length, that is:

$$R_{square} = \rho \frac{w}{w \cdot t} = \frac{\rho}{t} \qquad \text{(Equation 5.6)}$$

The interconnect material used has for long been aluminum as it is an easy-to-process material. Unfortunately, the resistivity of this material is quite high. Recently, copper has replaced aluminum for the manufacturing of interconnects, with a significant gain in terms of resistance, as the intrinsic resistivity of copper is almost two times lower than that of aluminum (see Table 5.2). Tungsten is used in several CMOS technologies to fabricate contact plugs. Its ability to fill narrow and deep holes compensates its high resistance. Gold is only used to connect the final chip to its packaging.

For a copper interconnect ($1.72 \ 10^{-6} \ \Omega.cm$) and a 0.4 µm thickness, the resistance is around 0.043 ohm/square. The measured square resistance is higher than this theoretical value as the conductor is not homogenous. The titanium barriers located on both sides of the conductor have an important resistance, which reduces the effective section of the conductor. In the CMOS 0.12 µm process files, a value of 50 m Ohm is used.

The square resistance is used to rapidly estimate the equivalent resistance of an interconnect by splitting its layout into elementary squares. The sum of square is then multiplied by R_{square} in order to evaluate the global interconnect resistance. We illustrate this concept in the layout shown in Figure 5.28. We assume a resistance per square R_{square} of 50 mΩ. The resistance from A to B can be approximated by 10 squares, that is a resistance of 0.5 Ω.

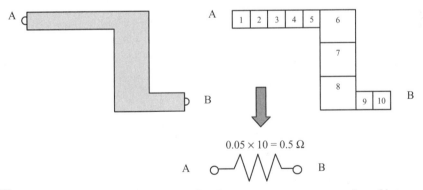

Fig. 5.28   Applying the concept of resistance per square to a portion of interconnect

The conductor resistivity is usually considered as a constant value. However, it depends on temperature in a complex manner [Hastings]. A usual approximation consists in considering the linear temperature coefficient of resistivity (TCR) expressed in parts per million per degree (ppm/°C). The copper and aluminum materials exhibit similar behaviour, with a TCR of around 4000 ppm/°C. The resistance at a temperature T is given by the following equation:

$$R_T = R_{T0} [1 + 10^{-6} \ TCR \ (T - T0)] \qquad \text{(Equation 5.7)}$$

where

R_T = serial resistance at temperature T (ohm)

R_{T0} = serial resistance at reference temperature $T0$ (ohm)

TCR = temperature coefficient of resistivity (ppm/°C)

T = temperature (°C)

$T0$ = reference temperature (Usually 25°C)

Considering a typical 4 λ wide metal interconnect, we observe that its cross-section is dramatically reduced with the scale down, due to a decrease of both the lateral and vertical dimensions of the elementary conductors. The elementary resistance *Rsquare* is increased at each technology generation, as shown in Figure 5.30. The introduction of copper in high performance CMOS process has lowered the square resistance almost by 50 per cent, but significantly increased the process complexity and overall fabrication cost.

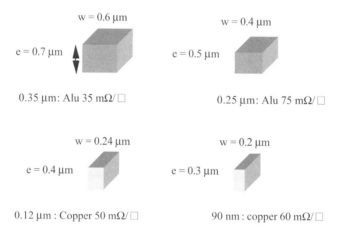

Fig. 5.29   Evolution of interconnect resistance with the technology scale down

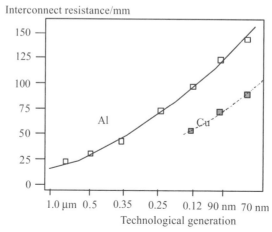

Fig. 5.30   Evolution of interconnect resistance with the technology scale down

5.6.2 Via Resistance

Each contact and via has a significant resistance. Typical values for these resistances are given in Table 5.3 for three technologies: 0.7 µm, 0.12 µm and 90 nm.

Table 5.3 Typical resistance of contacts and vias

Technology	0.7 µm	0.12 µm	90 nm
Contact resistance	0.5 Ω	15 Ω	20 Ω
Via	0.3 Ω	4 Ω	8 Ω
Upper via	—	1 Ω	3 Ω

Globally, the contact resistance and vias increase with the technology scale down, due to the continuous reduction in the contact plug section, resulting in a reduced path for current. The contact resistance from active regions to the first metal layer is very important due to the thick oxide (around 1.0 µm) that separates the active MOS device altitude from the metal 1 altitude. A thick oxide is necessary to insert several optional materials such as double gate MOS devices for EEPROM memories, or large storage capacitances for DRAM memories. The upper via resistance is quite small as the size of that via is very large as compared to the lower via.

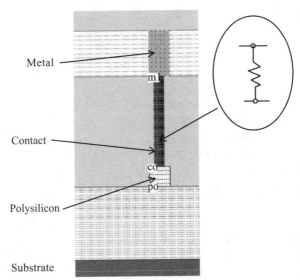

Fig. 5.31 The contact is equivalent to a resistance (Contacts.MSK)

5.7 Signal Transport

The signal transport from one logic cell to another logic cell uses metal interconnects. Depending on the distance between the emitting cell and the receiving cell, the interconnect may be considered as a simple parasitic capacitance or a combination of capacitance and resistance. By default, Microwind only considers the parasitic capacitance. This assumption is valid for short to medium length interconnects. In 0.12 μm, interconnects with a length of up to 1000 μm can be considered as a pure capacitance load *Cl* (Figure 5.32). For interconnects larger than 1000 μm, the serial resistance *Rl* should be included into the model. The usual way consists in splitting the capacitance *C l* into two equivalent capacitances and placing the serial resistance in between.

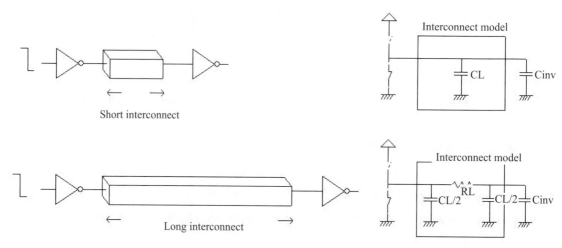

Short interconnect

Long interconnect

Fig. 5.32   C versus RC model for interconnects (RcModels.SCH)

5.7.1   Simulation of the RC Effect

The RC delay within interconnects can be simulated by using MICROWIND2 as follows. Consider the circuit shown in Figure 5.33. It represents a buffer (left lower corner) which drives a long interconnect, and then a loading inverter. From a layout point of view, the interconnect is usually straight, but for simplicity's sake, we use here a serpentine to emulate the RC effect without the need for a very large silicon area. The 4 mm interconnect is obtained by connecting small portions of metal lines. The layout shown in Figure 5.33 is based on 40 bars of metal 4, with a length of 100 μm. The serial resistance   *Rl* is equal to 820 ohm, and the capacitance *C l* is around 130 fF, divided into two parts, shared on each side of the virtual resistance.

Although both the capacitance and resistance of each node are extracted, the simulator considers, by default, only the capacitance and ignores the resistance. We add a "virtual" resistance using the resistance icon situated

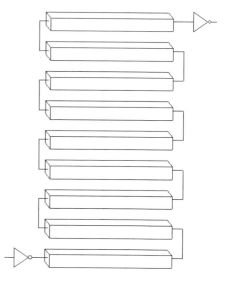

Fig. 5.33   Emulation of the RC effect in a 4 mm interconnect using serpentine (RcEffect.SCH)

in the palette (Figure 5.34). This resistance is placed directly in the interconnect, as detailed in the layout of Figure 5.35, to force MICROWIND to handle the equivalent resistance of the interconnect. We assign to the resistor the value of the metal interconnect resistance, appearing as *R* (metal). This value is updated during the electrical network extraction phase.

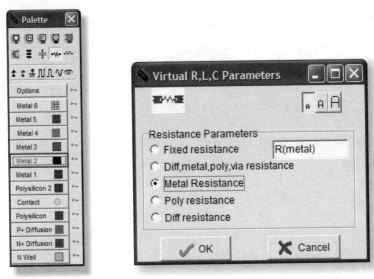

Fig. 5.34 Inserting a virtual resistance to take into account the serial resistance of the interconnect (RcModels.SCH)

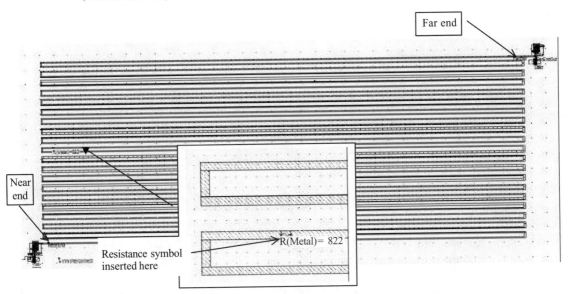

Fig. 5.35 Adding a virtual resistance within the layout to simulate the resistance (RcEffect.MSK)

The RC effect of the interconnect appears very clearly in the analog simulation (Figure 5.36). The initial phase runs from time 0.0 to time 1.0 ns, which is not significant. At time t = 1.0 ns, the input shifts from a high to a low state. The near end of the line switches within approximately 200 ps. The far end of the line reaches VDD/2 after twice this delay. The same delay effects are observed at the fall edge of the clock (*t* = 1.5 ns).

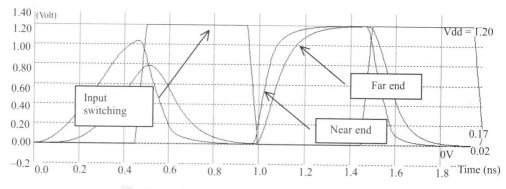

Fig. 5.36   RC delay simulation (RCEffect.MSK)

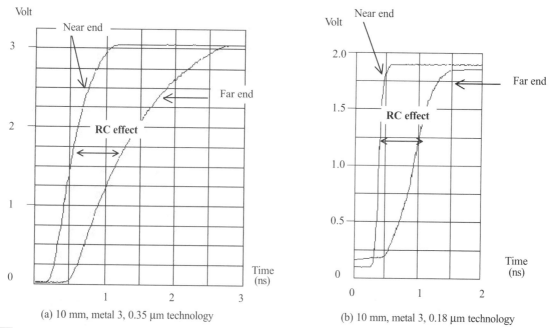

(a) 10 mm, metal 3, 0.35 μm technology   (b) 10 mm, metal 3, 0.18 μm technology

Fig. 5.37   Measurement of RC effect in a very long interconnect (10 mm) in metal 3, experimented in 0.35 μm and 0.18 μm technologies [Bendhia]

The RC effect can be measured thanks to an on-chip oscilloscope approach [Bendhia] at the near end and far end of metal interconnects. In the left part of Figure 5.37, the measurement concerns a 10 mm interconnect routed in metal 3, fabricated in a 0.35 μm 5-metal layer process from ST-Microelectronics [ST]. The observed waveforms confirm the important impact of the serial resistance, estimated in this particular case at around 300 ohm. The waveform is similar in 0.18 µm technology, 6-metal layers, but the near end of the line is prompt to switch within 100 ps, while the far end of the line needs more than 500 ps to pass the VDD/2 limit. Notice the supply voltage difference between these two technologies 3 V and 1.8 V.

5.7.2 Limit between C and RC Models

In 0.12 µm technology, the interconnect length limit above which the resistance should not be neglected is around 1 mm. This limit can be illustrated by implementing one configuration with the capacitance model, and another configuration with the RC model. In order to implement a long interconnect, MICROWIND offers a bus generation command, with an interface shown in Figure 5.38. This menu is accessible by the command **Edit → Generate → Metal bus**. Rather than drawing a straight line with a 300, 800 or 2000 µm length, we split the metal wire into several portions. This makes the layout more compact, and equivalent to the straight line. In the example given in Figure 5.38, we split the 2 mm line into ten portions of interconnects with a length of 200 µm each.

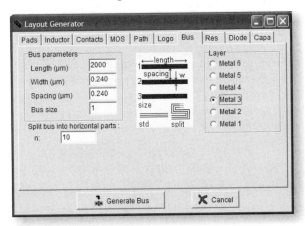

Fig. 5.38 Generating a long metal interconnect using the bus generator command

In this study, three interconnect lengths are investigated: 300 µm, 800 µm and 2 mm. Each configuration is implemented with and without the resistance to compare the simulation with the C model alone or the RC model.

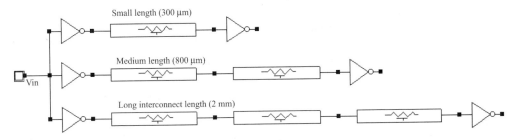

Fig. 5.39 Simulation of three interconnect configurations to investigate the impact of C/RC models (RcModels.SCH)

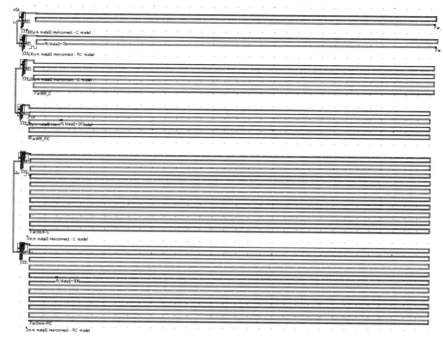

▨ **Fig. 5.40** Layout of the three interconnect configurations (300 μm, 800 μm and 2 mm) to investigate the impact of C/RC models (RCModel.MSK)

Each wire is implemented twice: one version is used to simulate the capacitor model, while the other one includes a small resistance fixed approximately half way between the near and far ends of the interconnect, to force the simulator to take into account the serial resistance of the interconnect.

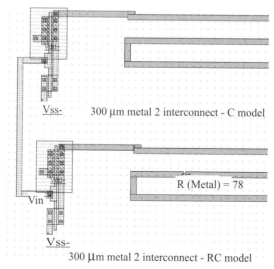

▨ **Fig. 5.41** Portion of the layout showing the two versions of the 300 μm configuration, with and without resistance (RCModel.MSK)

In order to handle the resistance effect of the interconnect, we place a virtual resistance approximately in the middle of the metal path. In the menu, we choose **Metal resistance**, which indicates that the extracted metal resistance should be used as the interconnect resistance. In the layout, the text appearing in the *R* symbol will be *R* (*metal*). The corresponding value appears after extraction or simulation. When the technology is changed, the value of the resistance will be updated according to the new sheet resistance parameters.

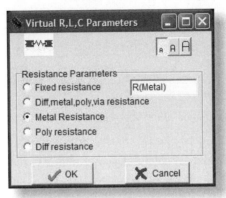

Fig. 5.42 Configuring the virtual resistance to handle the metal serial resistance effect in simulation (RCModel.MSK)

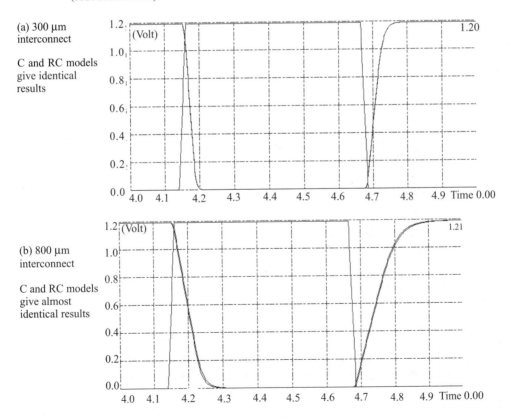

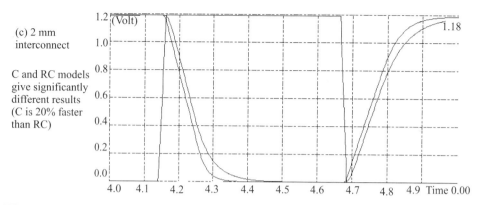

(c) 2 mm interconnect

C and RC models give significantly different results (C is 20% faster than RC)

Fig. 5.43   Comparative simulation of the C and RC models in signal propagation (RCModel.MSK)

The analog simulation of the signal transport with the 300 μm, 800 μm and 2 mm interconnects are given in Figure 5.43. The simulated propagation with C and RC models gives no visible difference below 1 mm. Above 1 mm, the C model gives an optimistic prediction of the delay as compared to the RC model. When we plot the delay versus the interconnect length (Figure 5.44), we will see that the RC model is preferable for interconnects with a length greater that 1 mm, while the C model is sufficient below 1 mm.

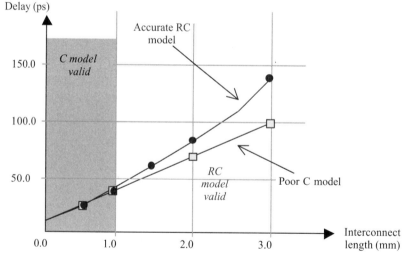

Fig. 5.44   Below 1 mm, the C model is valid. Above 1 mm, the RC model should be considered in 0.12 μm CMOS technology

5.8   Improved Signal Transport

The basic layout techniques used to reduce the signal transport within interconnects are detailed in this paragraph. Two approaches are considered: improving the drive of the switching inverters, and inserting repeaters. We shall also discuss the crosstalk effect and its possible reduction through technology and design improvements.

5.8.1 Increased Current Drive

The simplest approach to reduce the gate delay consists in connecting MOS devices in parallel. The equivalent width of the resulting MOS device is the sum of each elementary gate width. Both nMOS and pMOS devices are designed by using parallel elementary devices. Most cell libraries include so-called *x1, x2, x4, x8* inverters. The *x1* inverter has the minimum size, and is targeted for low speed, low power operations. The *x2* inverter uses two devices *x1* inverters, in parallel. The resulting circuit is an inverter with twice the current capabilities. The output capacitance may be charged and discharged twice as fast as with the basic inverter (Figure 5.45), because the *Ron* resistance of the MOS device is divided by two. The price to be paid is a higher power consumption. The equivalent *Ron* resistance of the *x4* inverter is divided by four.

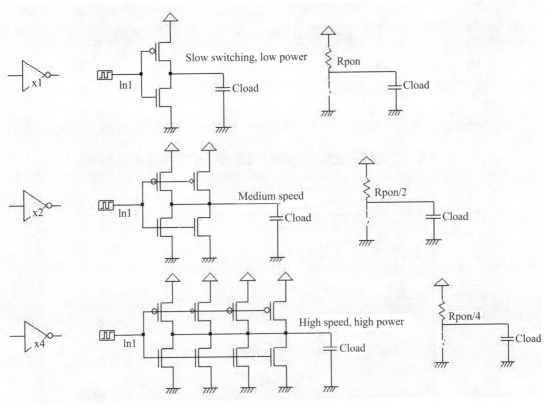

Fig. 5.45   The *x1, x2* and *x4* inverters (Invx124.SCH)

We may use the parametric analyzer included in MICROWIND to investigate the delay in increase with the capacitance load on the output node, for the *x1, x2* and *x4* inverters. In a first approximation, the capacitance increase is similar to the interconnect length increase. Remember that below 1 mm, the interconnect is basically a parasitic capacitance, with a value of 80 fF/mm approximately in 0.12 µm. What the tool does during the parametric analysis is to modify the output node capacitance step-by-step, according to the desired range of study, to perform the simulation, and to plot the desired delay in information.

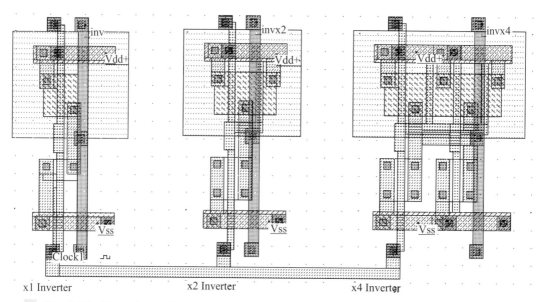

Fig. 5.46   Three sizes of inverters used to investigate the delay vs capacitance load (Invx124.MSK)

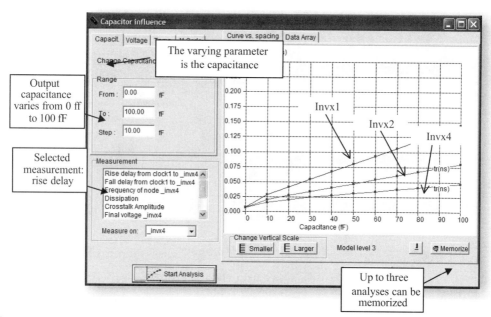

Fig. 5.47   Three sizes of inverters used to investigate the delay vs capacitance load (Invx124.MSK)

From the simulation trends obtained in Figure 5.47, it can be seen that a standard inverter delay (*x1*) increases rapidly with the capacitance. The inverter with a double drive (*x2*) has a switching delay divided by 2 while the inverter with a quadruple drive (*x4*) has a switching delay divided by 4. As interconnects may be assimilated to capacitance, driving long interconnects requires large buffers.

High drive buffers keep the propagation delay short, at the price of a proportionally higher current consumption. The clock signals, bus, ports and long wires with severe time constraints use such high drive circuits.

A fixed ratio is maintained between the p-channel MOS width and the n-Channel MOS width to balance the rise and fall time performance. It is much easier and safer to design the logic cells with similar rise and fall time performances, otherwise the timing analyzer would have to consider the rise and fall time cases separately.

5.9 Repeaters for Improved Signal Transport

Long distance routing means a huge loading due to a series of RC delays, as shown in Figure 5.48. A long line may be considered as a series of RC element where R is the serial resistance and C, the ground capacitance. For example, one RC cell represents one millimeter of interconnect. If a very long interconnect is implemented between an emitter and a receiver inverter, the delay is increased according to n^2, where n is the number of RC cells, as given in Equation 5.6.

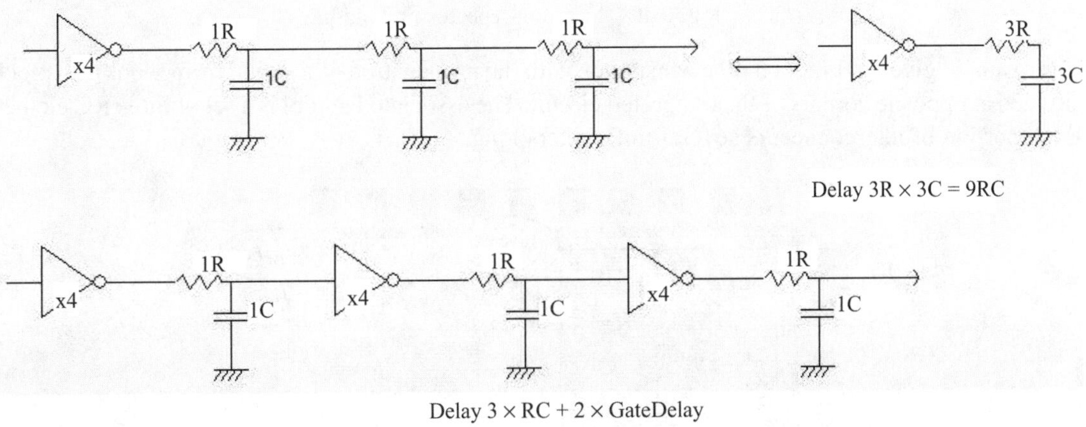

Fig. 5.48 The propagation delay of a long line with one inverter can be longer than with three inverters (RcLines.SCH)

In the case of a long line driven by a single inverter, the propagation delay on a line modellized by RC cells is given by:

$$t_{dly} = t_{gate} + nR \cdot nC = t_{gate} + n^2RC \qquad \text{(Equation 5.6)}$$

The propagation delay on a long line is not linearly dependent on the number of cells n, but proportionally dependent on the square of n. A good alternative is to use repeaters, by splitting the line into several pieces. Why can this solution be a better one in terms of delay? Because the gate delay is quite small compared to the RC delay. If two repeaters are inserted, the delay becomes:

$$t_{dly} = 3t_{gate} + 3RC \qquad \text{(Equation 5.7)}$$

Consequently, if the gate delay is much smaller than the RC delay, repeaters improve the switching speed performances, at the price of a higher power consumption. In the case of very long interconnects (several mm), it is interesting to place repeaters on the path of the interconnects to limit the slowing down effect of the interconnect resistance.

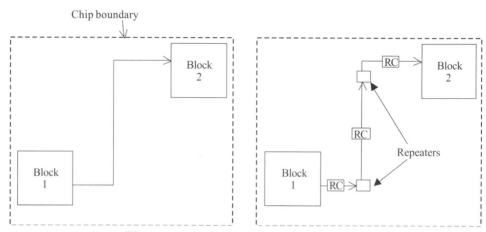

Fig. 5.49   Inserting repeaters in long lines

The example given in Figure 5.49 corresponds with the propagation of a signal from block 1 to block 2, situated at opposite corners of the integrated circuit. The associated model is a set of three RC elements. Each portion of interconnect is several millimeters long.

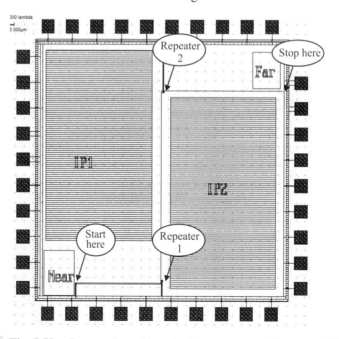

Fig. 5.50   Propagation with and without repeaters (Repeater.MSK)

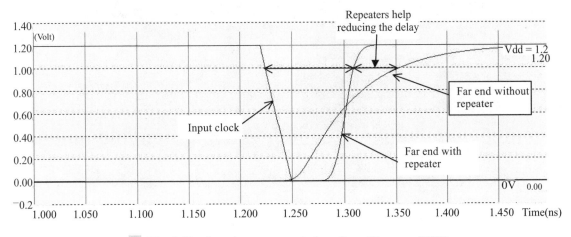

Fig. 5.51   Inserting repeaters in long lines (Repeater.MSK)

5.10   Crosstalk Effects in Interconnects

5.10.1   Coupling Increased with Scale Down

The crosstalk coupling represents the parasitic transient voltage induced by a switching interconnect on a neighbour interconnect. The disturbance may be high enough to create a temporary erroneous state on an interconnect which is supposed to be constant. Over the past recent years, the crosstalk effect has been the focus of active research, in terms of modelling and technology. The main reason for this interest is illustrated in Figure 5.52. The aspect of interconnects has dramatically changed with the technology scale down.

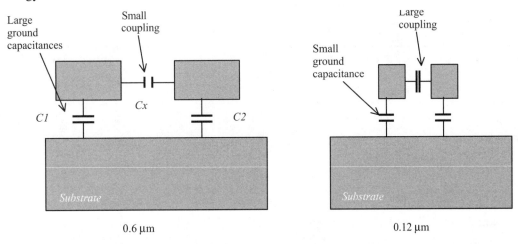

Fig. 5.52   The scale down tends to increase lateral coupling and to decrease vertical coupling

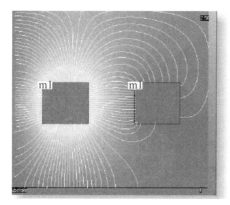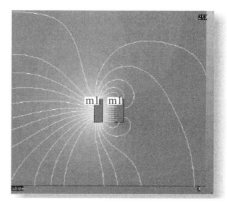

Fig. 5.53   Coupling between adjacent interconnects in 0.6 μm and 0.12 μm technology showing a very strong coupling increase with the scale down

Using the command **Analysis → Interconnect analysis** with FEM, we compare the coupling effects within two conductors, between the 0.6 μm and the 0.12 μm technologies. The field lines are computed by a clock on **Compute Field**. In Figure 5.53(a), the field lines link the left conductor mainly to the ground, with about one-third of the lines to the right conductor, which creates the coupling effect. In Figure 5.53(b), the number of field lines have been reduced, because of reduced conductor surfaces. Only 15 field lines couple to the ground, which means a very low capacitance to the ground. However, the coupling field lines are still numerous and very short, which means a very high coupling.

Some details about the size and electrical properties of the interconnects are given in Table 5.4. The metal width and spacing are scaled according to the lithography improvement, but the interconnect thickness is not reduced with the same trend. Starting at 0.18 μm, copper has been proposed as an alternative to aluminum, for its lower resistivity. In order to decrease the crosstalk coupling capacitance, low permittivity (Low K) dielectrics have been introduced, with 0.18 μm technology.

Table 5.4   Evolution of interconnect parameters with the technology scale down

Techno-logy	Metal layers	Lower metal width (μm)	Metal spacing (μm)	Thickness (μm)	Low K	Interconnect material	MICROWIND2 file
1.2 μm	2	1.8	2.4	0.8		Al	CMOS12.rul
0.7 μm	2	1.2	1.6	0.7		Al	CMOS 07.rul
0.6 μm	3	0.75	1.0	0.7		Al	CMOS 06.rul
0.35 μm	5	0.6	0.8	0.7		Al	CMOS 035.rul
0.25 μm	6	0.5	0.6	0.6		Al	CMOS 025.rul
0.18 μm	6	0.3	0.4	0.5	3.1	Al, Cu	CMOS 018.rul
0.12 μm	6-8	0.18	0.24	0.4	2.8	Al, Cu	CMOS 012.rul
90 nm	8-10	0.15	0.2	0.35	2.5	Cu	CMOS 90n.rul
65 nm	8-12	0.1	0.14	0.3	2.0	Cu	CMOS 70n.rul

5.10.2   Simulation of the Crosstalk Effect

The simulation of the crosstalk effect is based on two inverters, one considered as the affecting signal, and the other as the victim signal. The inverters are connected to long interconnects routed with the minimum distance. The victim is connected to a weak inverter, and surrounded by two aggressor lines connected to a very powerful inverter, to create the maximum crosstalk effect (Figure 5.54).

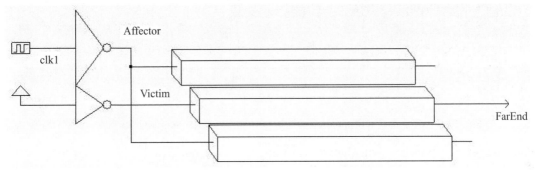

Fig. 5.54   The coupling configuration used to simulate the crosstalk effect (Crosstalk.SCH)

The serpentine shown in the layout of Figure 5.55 corresponds to approximately 1 mm of interconnect. Virtual resistance symbols are added in the middle of the interconnect to handle the RC effect in the simulation. Notice the unbalanced inverter size to create the worst case conditions for parasitic coupling.

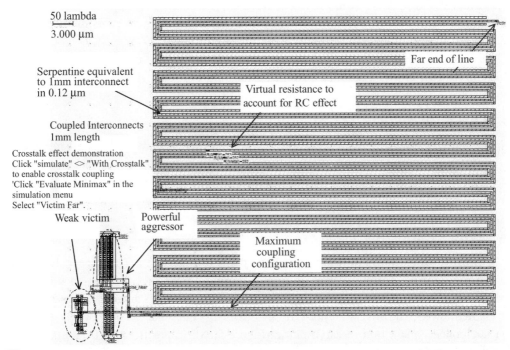

Fig. 5.55   Implementation of strongly coupled lines in worst case configuration (Crosstalk.MSK)

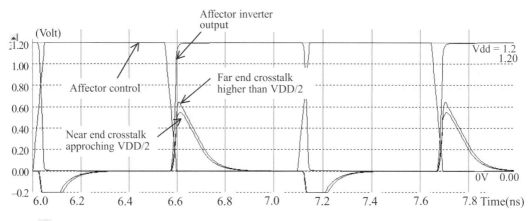

Fig. 5.56 Simulation of the crosstalk coupling in a 1 mm interconnect (Crosstalk.MSK)

When the aggressor lines are switching, the coupling is strong enough to increase the voltage at the far end of the victim line, higher than the switching threshold of logic gates (which is around VDD/2), which may provoke a permanent logic fault (Figure 5.56). The noise is quite impressive. Remember that the line is only 1 mm long, which is very common in circuit design. However, the situation where the 1 mm interconnect is driven by a very low drive inverter is not usual. Nevertheless, crosstalk is very dangerous and almost uncontrollable when dealing with millions of interconnects, as may be found in high complexity designs. One solution to avoid crosstalk is to avoid routing long interconnects. The critical routing length is the limit above which a crosstalk fault may occur. This metric has recently been introduced in design guidelines. In 0.12 μm CMOS technology, the critical routing length is 1 mm. It means that interconnects longer than 1 mm could suffer from crosstalk noise in worst case conditions. In the case of very long routing, repeaters should be used.

5.10.3 Low K Dielectrics

When investigating the maximum crosstalk amplitude versus the technology for a given interconnect length, we observe a severe increase in the coupling effect, as a direct consequence of lithography improvements. In ultra-deep submicron technologies (with lithography lower than 0.18 μm) the permittivity of the lateral oxide that fills the spacing between adjacent interconnects is reduced (low K dielectric with a permittivity of around 3.0), while the oxide that separates vertical layers is kept with a high permittivity (around 4.0 for SiO_2).

The main effect is the decrease in lateral coupling effects (Figure 5.57). The introduction of low K materials may reduce the coupling effect up to a certain limit. Several CMOS compatible materials exist, which must be compatible with the CMOS process. Low K oxides are sometimes called SiOLK (Silicon Oxide Low K). The SiOLK permittivity should ideally be 1, that is, corresponding to an air gap between interconnects, still in a research phase (Figure 5.58). In practice, for 0.12 μm technology, SiOLK ε_r is around 3.0.

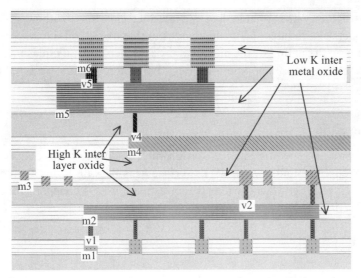

Fig. 5.57   Low dielectric permittivity between lateral metal interconnects reduces the crosstalk effect (Metals.MSK)

Fig. 5.58   Air-gap between interconnects to cut by half the crosstalk coupling (*Courtesy:* ST-Microelectronics)

5.11 Antenna Effect

During the fabrication of interconnects, charges may accumulate and endanger the MOS gate by forcing current through the gate oxide [Hastings]. This effect, called the antenna effect, appears on long metal interconnects connected to small gate oxide areas, without any path to diffusion. The antenna effect is particularly important in deep-submicron technology: without any possible discharge path, the interconnect accumulates sufficient charges to rise its potential to several volts, positive or negative, depending on the nature of the chemical process step. Usually, plasma etching charges the interconnect with electrons, corresponding to a negative charge with respect to the substrate ground voltage.

5.11.1   Antenna Rule

With the 0.35 μm process, specific antenna design rules have been introduced. If we consider an interconnect with a length Li, and a width Wi, its surface Ai is $Wi \times Li$. If this interconnect is connected to an MOS device with a length L and width W, corresponding to a channel surface A, the following rule should be verified:

$$Ai \leq R_{\text{antenna}} \cdot A \qquad\qquad \text{(Equation 5.10)}$$

where

$$A = W \times L = \text{MOS channel surface (m}^2)$$

$$Ai = Wi \times Li = \text{interconnect surface (m}^2)$$

$$R_{\text{Antenna}} = \text{antenna ratio (around 100)}$$

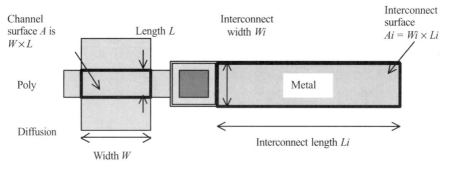

Fig. 5.59   Antenna rules related to the surface of the interconnect with respect to the surface of the gate

5.11.2   Design Example

An example of valid and invalid interconnect design is given in Figure 5.60. The upper layout complies with the antenna rules as the interconnect surface is less than 100 times the gate surface. In contrast, design (b) is dangerous as the interconnect surface is more than 100 times larger than the gate surface, which has been designed very small. The antenna rules are not yet verified by the design rule checker of MICROWIND.

One solution consists in creating a discontinuity by using a bridge with a higher metal, so that when the lower metal is etched, the main part of the interconnect will not be connected to the gate (Figure 5.60(d)). An alternative way to avoid the antenna effect is to build a discharge path to evacuate the parasitic charges accumulated during fabrication. A simple diode is an efficient solution (Figure 5.61). Its parasitic capacitance is quite small, and the diode has no important electrical effect on the nominal signal transport.

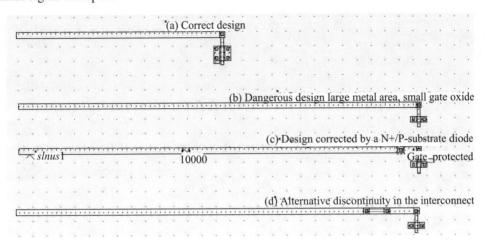

Fig. 5.60   Illustration of the antenna design rule (AntennaRules.MSK)

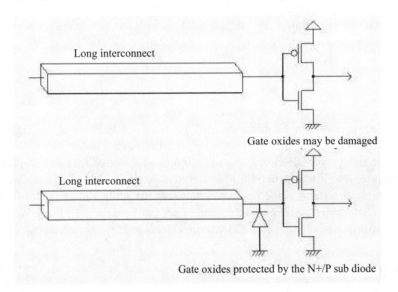

Fig. 5.61   Inserting a diode to discharge the interconnect during plasma etching (AntennaRules.SCH)

5.11.3 Simulation

The N+/P-substrate diode inserted near the gate (Figure 5.60(c)) turns on when the interconnect voltage is negative, or higher than the reverse Zener voltage, as seen in the simulation shown in Figure 5.62. Notice that the BSIM4 model has been used to handle the Zener effect. The Y axis has been changed to −2 to 24 V, thanks to the Y scale cursors situated in the left upper part of the voltage chronograms. By default, MICROWIND does not extract diodes. This is why an option layer surrounds the N+ area near the gate, with a tick in front of the **Extract Diode inside box**.

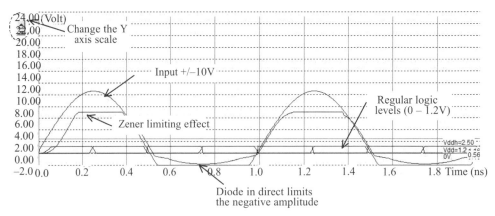

Fig. 5.62 The diode clamps the negative charges, and limits the positive amplitudes (AntennaRules.MSK)

The serial resistance used for simulation is very high (10 K Ohm). Removing that resistance would completely hide the clamping effect of the diode. In reality, the charging of the interconnect is not equivalent to a perfect voltage source, as Microwind does with the sinusoidal property. The high serial resistance accounts for the weak charging process, which can be counter-balanced by a small discharge diode.

5.12   Inductance

The inductance effect is usually not significant in signal transport because of the high serial resistance of interconnects. This is why we have not paid much attention to the inductance value and its possible consequence on the delay estimation or crosstalk amplitude. A lot of research has been dedicated in recent years to the extraction and handling of inductance. The inductor is described in detail in the Part II of the book, as a stand-alone passive component for application in radio-frequency circuits. In that case, the inductance is no longer a parasitic effect but a voluntary effect. We give here a brief evaluation and illustration of the parasitic inductance effect in deep-submicron interconnects.

5.12.1   Parasitic Inductance Formulation

The wire inductance formulation is based on the estimation of a cylinder for which a very simple formulation exists [Lee]. A well-known rule of thumb consists in approximating the serial inductance to 1 nH/mm, which is close to reality in the case of bonding wires. The wire has a cylindrical shape and is situated far from the ground plane.

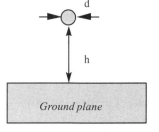

$$L = \frac{\mu 0}{2\pi} \ln\left(4\frac{h}{d}\right)$$

(Equation 5.7)

with

$$\mu_0 = 1.257e^{-6} \text{ H/m for most materials (Al, Cu, Si, SiO}_2 \text{ and Si}_3\text{N}_4)$$

d = wire diameter (m)

h = height of the wire vs. ground (m)

In the case of metal interconnects, Equation 5.7 is adapted with an approximation of the interconnect diameter, based on the conductor width and thickness. The serial parasitic inductance of the conductor appears in the navigator menu, after extraction, together with the capacitance and resistance (Figure 5.63). A metal interconnect exhibits an inductance of around 0.5 nH/mm. Notice that the lineic inductance value is also provided in the **interconnect analysis** window, accessible from the **Analysis** menu.

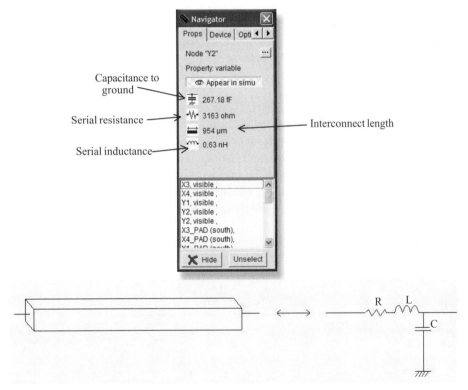

Fig. 5.63   An evaluation of the parasitic serial inductance is given in the navigator menu (Rlcg.SCH)

5.12.2 Simulation with/without Inductance

Let us compare the signal propagation of a logic signal within a 500 μm interconnect with and without the serial inductance. In MICROWIND, the inductance must be added through a virtual inductor symbol that may be found in the palette. The inductor is placed at the beginning of the line. Handling the simulation of the inductance is not simple in Microwind, which has been optimized for RC networks. Several problems rise at simulation: the initial ringing is very high, which is not realistic at all. This numerical instability is due to the initialization of the inductance, which disturbs the circuit in the early nanoseconds. Secondly, the simulation step by default is not small enough to ensure a correct simulation. In most cases, keeping the default time step of 0.3 ps will create the oscillations of all floating nodes with several volts of amplitude. The only solution consists in reducing the simulation time step, at least to 0.03 ps or even less. After some nanoseconds, the simulation becomes stable, as displayed in Figure 5.64. We see no significant difference between the RC and RLC models.

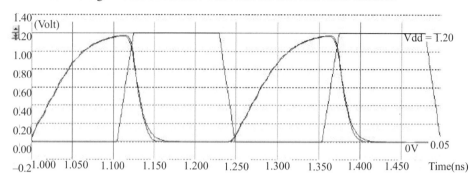

Fig. 5.64   Simulation of the signal propagation within a 500 µm interconnect with using the RC and RLC models (InductanceInv.MSK)

In order to observe the inductance effect, we change the configuration and now drive the same RLC line with strong buffers, as shown in Figure 5.65. Notice the ringing effect due to the combination of inductance and capacitance at the far end of the interconnect (Figure 5.66). In reality, the resistance, capacitance and inductance are distributed along the wire, which tends to limit the amplitude of the ringing effect, and fastens the damping of the oscillation to a stable logic value.

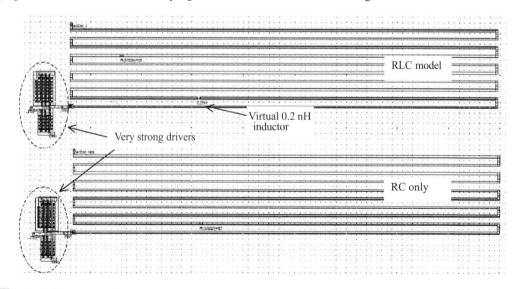

Fig. 5.65   Layout of the interconnect with and without serial parasitic inductance (Inductance.MSK)

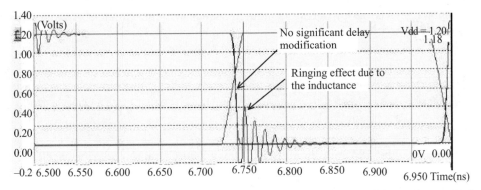

Fig. 5.66 Simulation of the signal propagation with and without inductance. This result was obtained with a simulation time step decreased to 0.02 ps (inductance.MSK)

Although the ringing effect is very important in this new simulation (Figure 5.66), the switching delay is not altered in a significant way. Consequently, the inductance effect may be neglected in a first-order approximation in the analysis of signal propagation. This assumption is valid if the buffer strength is not too large, and the interconnect not too short. Inductance effect will play an important role in supply networks, which are considered in Part II of the book.

5.13 Conclusion

In this chapter, we have described some layout techniques for designing metal interconnects between devices. We have given some information regarding the design rules and the electrical parameters for metal interconnects, such as the resistance, capacitance, and later the inductance. The signal propagation has been analyzed from the point of view of RC delay, technology scale down, and parasitic crosstalk effect. We have also made a rapid investigation of the role of the parasitic inductance in the signal transport.

References

[Weste] N. Weste, K. Eshraghian "Principles of CMOS VLSI design", Addison Wesley, ISBN 0-201-53376-6, 1993.

[Baker] R.J. Baker, H. W. Li, D.E. Boyce "CMOS circuit design, layout and simulation", IEEE Press, ISBN 0-7803-3416-7, 1998.

[Hastings] Alan Hastings "The Art of Analog Layout", Prentice-Hall, ISBN 0-13-087061-7.

[Sakurai] Sakurai T. " Closed-form expressions for interconnection delay, coupling and crosstalk in VLSIs", IEEE Transactions on Electron Devices, Vol. 40, No.1, pp 118-124, January 1993.

[Delorme] Delorme N., Belleville M., Chilo J. "Inductance and capacitance analytic formulas for VLSI interconnects" Electronic letters, Vol .32, No. 11, pp 996-997, May 1996.

[Lee] Thomas H. Lee "The design of CMOS radio frequency integrated circuits", Cambridge University Press, ISBN 0-521-63922-0.

[Itrs]

[Bendhia] Sonia Delmas-Bendhia, Fabrice Caignet, Etienne Sicard "A new method for measuring signal integrity in CMOS Ics" Paper in Microelectronics International, Vol. 17, No.1, January 2000, pp 17–21.

[ST] www.st.com

[Smith] John M., Smith S. "Application-Specific Integrated Circuits", Addison-Wesley, 1977, ISBN 0-201-50022-1, Chapter 7

EXERCISES

1. We consider two coupled lines in metal 1, with a 5 mm length, in 0.12 μm technology (Figure 5.67). What is the equivalent model of the line, in a first order approximation? Calculate the total serial resistance R, ground capacitance $Csub$ and crosstalk capacitance $C12$, using the command **Analysis → interconnect analysis with FEM** in Microwind. What is the best technological option for interconnect/oxide material as far as signal integrity is concerned (we suppose Low $K = 2$)?

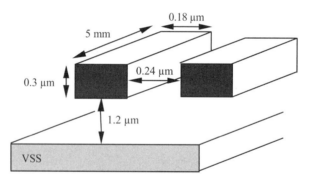

Case	R	C_{12}	C_{sub}
No 1: Al, SiO$_2$			
No 2: Al, Low K			
No 3: Cu, SiO$_2$			
No. 4: Cu, Low K			

Fig. 5.67 Two coupled lines in metal 1, 10 mm length

2. An experimental measurement of the RC effect has been realized in 0.18 μm technology (Figure 5.68). Details of the real case configuration are also provided. Create the corresponding layout for this configuration and compare the simulation with measurements.

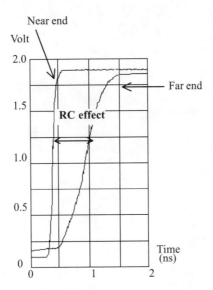

Technology	CMOS 0.18 µm
RUL file	CMOS 018.RUL
Interconnect length	10 mm
Interconnect width	4 lambda (0.4 µm)
Metal layer	Metal 5
nMOS buffer size	W = 32 µm, L = 0.18 µm
pMOS buffer size	W = 54 µm, L = 0.18 µm

Fig. 5.68   Parameters of the RC propagation test-case in 0.18 µm technology

Basic Gates

The basic logic gates are described in this chapter. The principles for building combinational logic circuits are developed. Details of layout implementation are also provided.

6.1   Introduction

Table 6.1 gives the corresponding symbol for each basic gate as it appears in the logic editor window as well as the logic description. In this description, the symbol & refers to the logical AND, | to Or, ~to INVERT, and ^ to XOR.

Table 6.1   The list of basic gates

Name	Logic symbol	Logic equation
INVERTER	⊳o	Out = ~in;
AND	⊃	Out = a&b;
NAND	⊃o	Out = ~(a&b);
OR	⊐⊃	Out = (a|b);
NOR	⊐⊃o	Out = ~(a|b);
XOR	⊐⊃⊃	Out = a^b;
XNOR	⊐⊃⊃o	Out = ~(a^b);

6.2   Combinational Logic

The construction of logic gates is based on MOS devices connected in series and in parallel. If two n-channel MOS switches are connected in series (Figure 6.1), the resulting switch connects ports $C1$ and $C2$ if both gates A and B are set to '1'. This yields an AND operator, represented by the symbol '&' in Equation 6.1, where C12 is the logical variable which represents the connection between C1 and C2.

$$C_{12} = A \& B \qquad \text{(Equation 6.1)}$$

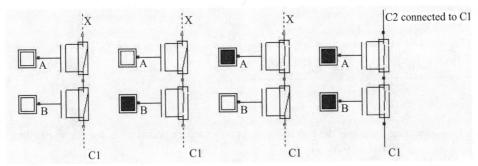

Fig. 6.1   Connecting n-channel devices in series creates a path between C1 and C2 when *A* and *B* are set to '1' (BaseCmos.SCH)

When two nMOS switches are connected in parallel (Figure 6.2), the resulting switch is ON if either gates *A* and *B* are set to '1'. This yields an OR operator (described as '|' in Equation 6.2).

$$C_{12} = A \mid B$$ (Equation 6.2)

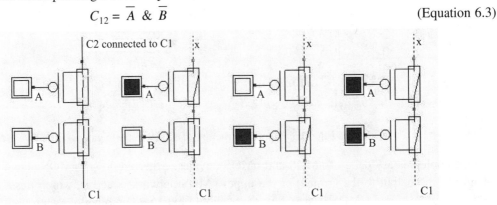

Fig. 6.2   Connecting n-channel devices in parallel creates a path between *C1* and *C2* when either *A* or *B* are set to '1' (BaseCmos.SCH)

Considering p-channel devices, we observe that two pMOS switches connected in series (Figure 6.3) behave as an AND between negative logic values: the resulting switch is ON if both gates *A* and *B* are set to '0'. The corresponding Boolean operator is as follows:

$$C_{12} = \overline{A} \ \& \ \overline{B}$$ (Equation 6.3)

Fig. 6.3   Connecting p-channel devices in series creates a path between *C1* and *C2* when *A* and *B* are set to '0' (BaseCmos.SCH)

When two pMOS switches are connected in parallel (Figure 6.4), the resulting switch is ON if either gates *A* and *B* are set to '0'. The Boolean operator is described by Equation 6.4.

$$C_{12} = \overline{A} \mid \overline{B}$$ (Equation 6.4)

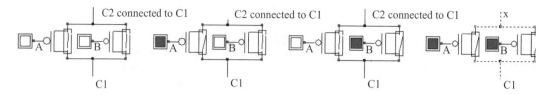

Fig. 6.4   Connecting p-channel devices in parallel creates a path between C1 and C2 when either *A* or *B* are set to '0' (BaseCmos.SCH)

6.3   CMOS Logic Gate Concept

The structure of a CMOS logic gate is based on complementary networks of n-channel and p-channel MOS circuits. Remember that the pMOS switch is good at passing logic signal '1', while nMOS switches are good at passing logic signal '0'. The operation of the gate has two main configurations:

- the nMOS switch network is ON, the output s = 0 (Figure 6.5 left)

- the pMOS switch network is ON, the output s = 1 (Figure 6.5 right)

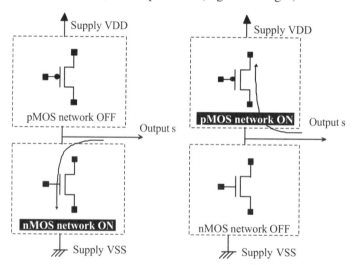

Fig. 6.5   General structure of a CMOS basic gate

Using complementary pairs of nMOS and pMOS devices, either the lower nMOS network is active, which ties the output to the ground, or the upper pMOS network is active, which ties the output to VDD. In conventional CMOS basic gates, there should exist no combination when both nMOS and pMOS networks are ON. If this case happens, a resistive path would be created between VDD and VSS supply rails. The situation where neither nMOS nor pMOS networks are OFF should also be avoided, because the output would be undetermined. These illegal situations are illustrated in Figure 6.6.

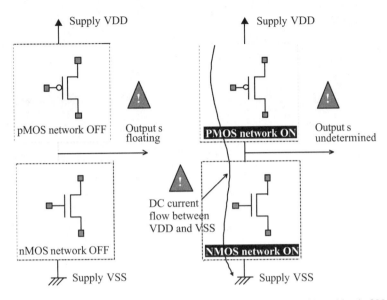

Fig. 6.6 Network configurations to be avoided: both OFF (left) and both ON (right)

6.4 The NAND Gate

6.4.1 Truth Table

The truth table and logic symbol of the NAND gate with two inputs are shown below. The truth tables use 0 for logic level zero (usually 0 V), 1 for logic level 1, (also called VDD, equal to 1.2 V in 0.12 μm technology) and X for unknown value. Some books [Uyemura] also include the z state, that is the high impedance value. We shall make no difference between the unknown and high impedance state in this book.

A	B	Out
0	0	1
0	1	1
1	0	1
1	1	0
X	0	1
X	1	X
0	X	1
1	X	X

NAND

Fig. 6.7 The truth table and symbol of the NAND gate

In DSCH, the NAND gate is part of the symbol palette, which appears at initialization on the right side of the main window. Select the NAND symbol in the palette, keep the mouse pressed and drag the shape to the editing window at the desired location. Also add two buttons and one lamp as shown in Figure 6.8. Add interconnects, if necessary, to link the button and lamps to the cell pins. The icon **Add a Line** is available for this purpose. Notice that the right click with the mouse enables the editing of an interconnect too. You may verify the logic behaviour of the cell by a click on **Simulate → Start Simulation** or the icon **Run Simulation**. The logic level '0' corresponds to a white colour, or to dot lines, and the logic level '1' is drawn in black, or with solid lines.

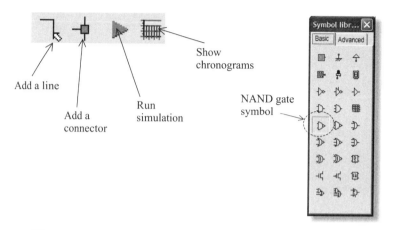

Fig. 6.8   Editing commands for simulating the NAND gate

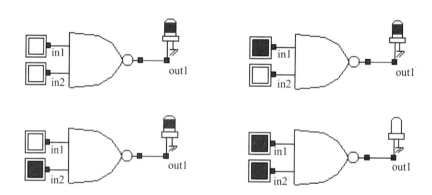

Fig. 6.9   The logic simulation of the NAND gate verifies the truth table (NandTruthTable.SCH)

6.4.2   Logic Design of the CMOS NAND Gate

In CMOS design, the NAND gate consists of two nMOS in series connected to two pMOS in parallel. The schematic diagram of the CMOS NAND cell is reported below. The nMOS devices in series tie the output to the ground for one single combination $A=1$ and $B=1$. For the three other combinations, the nMOS path is cut, but at least one pMOS ties the output to the supply VDD. Notice that both nMOS and pMOS devices are used in their best regime: the nMOS devices let '0' pass, the pMOS let '1' pass.

A B	nMOS	pMOS	Nand 2
0 0	off	on	1
0 1	off	on	1
1 0	off	on	1
1 1	on	off	0

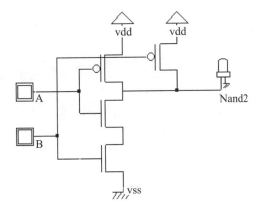

Fig. 6.10   Schematic diagram of the CMOS NAND gate (NandCmos.SCH)

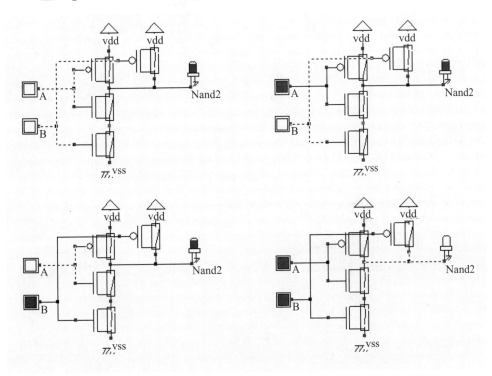

Fig. 6.11   The logic simulation of the NAND gate (Nand2Cmos.SCH)

Furthermore, the circuit of Figure 6.11 eliminates the static power consumption when *A* and *B* are steady by ensuring that the situation 'pMOS ON', 'nMOS ON' never happens.

6.4.3   Automatic Generation of the NAND Layout

MICROWIND features a built-in cell compiler that can generate the NAND gate automatically. In MICROWIND, click on **Compile → Compile One Line**. Either select the line corresponding to the

2-input NAND description (Figure 6.12) or type the logical expression that describes the link between one output and two inputs. The '~' operator represents the NOT operator, the '&' symbol represents the AND operator.

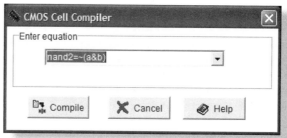

Fig. 6.12   The CMOS cell compiler is used to generate a NAND gate

When you click "Compile", the layout of the NAND gate appears in the screen, as drawn in Figure 6.13. The compiler has fixed the position of the VDD power supply in the upper part of the layout, near the p-channel MOS devices. The compiler has also fixed the ground VSS in the lower part of the window, near the n-channel MOS devices. The texts *A*, *B*, are placed in the layout, on the gates, and the text *nand2* is fixed on the output, at the upper location, near the contact. The layout generation is driven by the default design rules, corresponding in this case to a CMOS 0.12 µm technology. Depending on the design rules and the technology generation, the layout may look a little different. The implantation of the four devices is detailed in the schematic diagram at the right side of Figure 6.13. The MOS devices have been arranged in such a way that the diffusion regions are merged, both for the two nMOS in series and the two pMOS in parallel. Sharing a common diffusion always leads to more compact designs, which saves silicon area and minimizes parasitic capacitance.

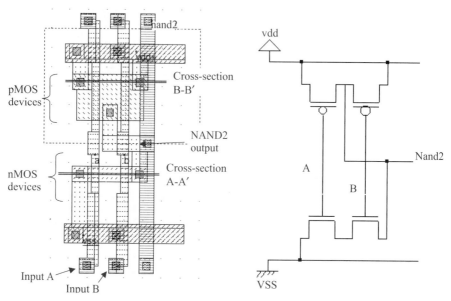

Fig. 6.13   The layout of the NAND gate generated by the CMOS cell compiler (Nand2.MSK)

6.4.4 Inside the NAND Gate

The 2D-process viewer is a useful tool to display the two nMOS in series and the two pMOS in parallel. Select the corresponding icon and draw a horizontal line in the layout in the middle of the nMOS channels, at location A-A′ shown in Figure 6.13. The figure below appears. The path from *VSS* to the output *nand2* goes through two transistors connected in series, one controlled by A, the other controlled by B. In Figure 6.14, the output *nand2* may be tied to VDD either through the pMOS device controlled by A or through the pMOS device controlled by B. Notice the n-well under the pMOS devices, polarized to VDD.

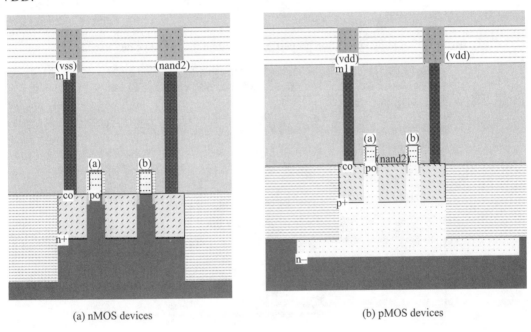

(a) nMOS devices   (b) pMOS devices

Fig. 6.14 The nMOS devices in series and the pMOS devices in parallel (Nand2.MSK)

6.4.5 Adding Simulation Properties

The simulation icons add properties to the nodes. Properties are applied to the electric nodes of the circuit in order to serve as simulation guides. The list of properties required to perform the analog simulation of the NAND gate are shown in Figure 6.15. Firstly, the NAND gate should be supplied by a 0 V (VSS) and 1.2 V (VDD). The VDD and VSS properties are already placed in the layout by the CMOS cell compiler.

Secondly, clocks should be assigned to the input gates *A* and *B*. The clock icon is the fourth icon from the right in the palette menu. Simply activate the clock icon, and click in the letter a in the layout window. The following screen appears (Figure 6.16):

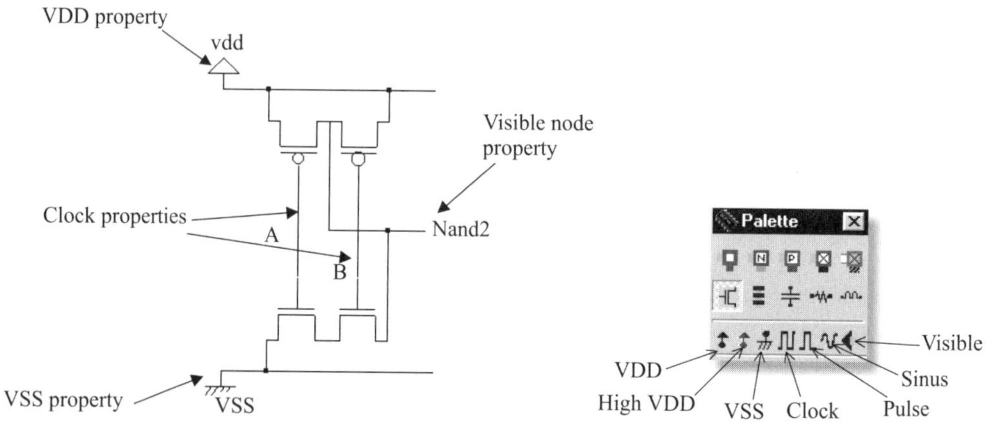

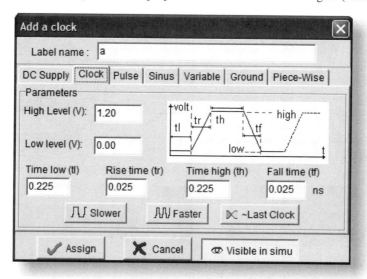

Fig. 6.15   Adding simulation properties to simulate the NAND gate (Nand2.MSK)

Fig. 6.16   Clock property added to node A (Nand2.MSK)

The parameters of the clock are divided as follows: time at low level (*tl*), rise time (*tr*), time at high level (*th*) and fall time (*tf*). All values are expressed in nanosecond (ns). Clock **Assign** to assign a clock to label *a*. Click again the clock icon, and this time click on label *b* in the layout. As you ask for a second clock, the period is automatically multiplied by two.

- You may alter level 0 and level 1 by entering a new value with the keyboard.

- To generate a clock which works in opposite phase, click ~**Last Clock**.

- Use **Slower** to multiply the clock period by two.

- Use **Faster** to divide the clock period by two.

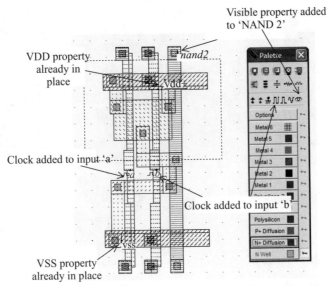

Fig. 6.17 Simulation properties added to the NAND gate (Nand2.MSK)

Finally, click on the "eye" in the palette, and click on the text *nand2* in the layout to make the chronograms of the node appear. Initially, all nodes are invisible. However, the nodes with clocks, impulse and sinus properties are subsequently made visible.

6.4.6 Electrical Structure of the NAND Gate

The icon above is useful for getting an insight into the electrical node structure of the layout. Select the icon and simply click inside the layout at the desired location. Information corresponding to the parasitic capacitance, resistance, as well as simulation properties is also displayed in a separate window, called navigator. In Figure 6.18, the ground node and supply node are illustrated. In the Navigator window, the electrical properties of the node are displayed.

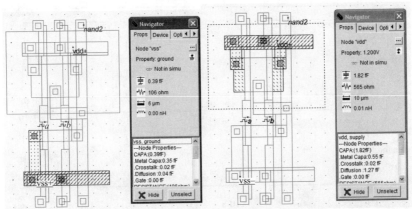

Fig. 6.18 The VSS node and VDD node with their associated properties (Nand2.MSK)

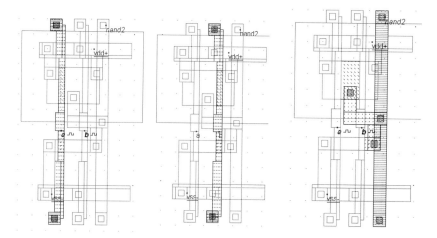

Fig. 6.19 The structure of the input nodes *A*, *B* and the output node *nand2* (Nand2.MSK)

The electrical connection of the inputs *A*, *B* and the output *nand2* is revealed in Figure 6.19. The inputs correspond to the nMOS gates, to a small portion of polysilicon between the pMOS and nMOS areas, and to the pMOS gates. The contacts situated on the lower and upper parts of the cell serve as a simple connection point to the upper metal layers. The output node is quite complex. It consists of a drain area in the n-channel MOS region, connected to the central part of the pMOS diffusion area. The output is also connected to routing contacts up and down, thanks to a vertical bar in metal2.

6.4.7 Analog Simulation of the NAND Gate

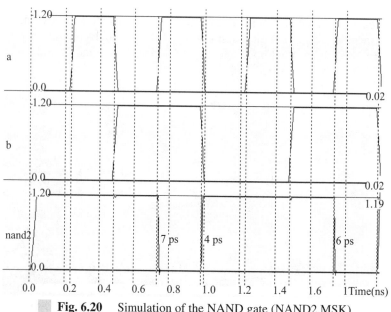

Fig. 6.20 Simulation of the NAND gate (NAND2.MSK)

The simulation of Figure 6.20 is obtained by the command **Simulate → Run simulation**, or the above icon. We verify that *nand2* is equal to 0 when both $A = 1$ and $B = 1$, according to the truth table, otherwise the output is at 1. The rise time and fall time are computed according to the following scenario. The simulator starts computing the delay when the selected signal, chosen here as *a*, crosses VDD/2. The simulator stops when the output *nand2* also crosses VDD/2 (Figure 6.21). We should not emphasize the extremely small switching delay too much (7 ps for the fall edge, 4 ps for the rise edge). There are two reasons for such a short delay: one is the delay computation mode, based on VDD/2 which is always optimistic as compared to a 10 per cent-90 per cent delay evaluation, the second is the absence of any output load, which gives a best-case estimation of the switching delay.

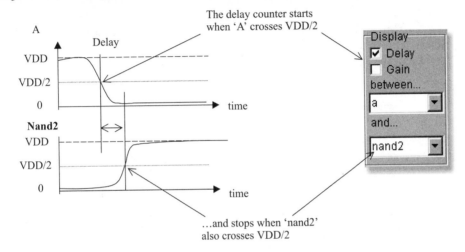

Fig. 6.21 Delay computation in the simulation menu

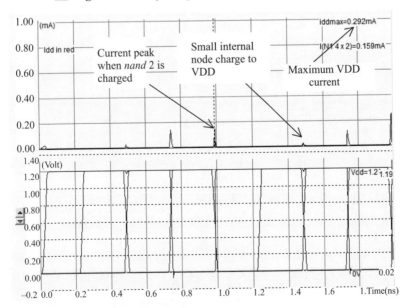

Fig. 6.22 Current consumption on VDD supply line vs. time (Nand2.MSK)

Let us now consider another simulation mode, Voltage and Currents, accessible through the main menu via the command **Simulate → Run Simulation → Current, Voltage vs. Time**. The simulator displays all voltages in the lower window, and a selection of currents in the upper window (the IDD current, which is the sum of currents flowing from the supply VDD, and the current of one selected MOS device). The ISS currents can also be displayed. A transient current peak appears on IDD when the output node *nand2* is charged, as shown in Figure 6.22. The current consumption is important only during a very short period corresponding to the charge and discharge of the output node.

Without any switching activity, the current is very small and cannot be seen accurately in linear scale. When looking at the same diagram in logarithmic scale (assert **Scale I in log** in the simulator parameter window), we observe that the NAND gate consumes around 1 nA of standby current. The default simulation model is Model 3, which does not account accurately for leakage currents. The BSIM4 model, accessible by the command **Simulate → Using Model → BSIM4**, configures the simulation with BSIM4 model parameters.

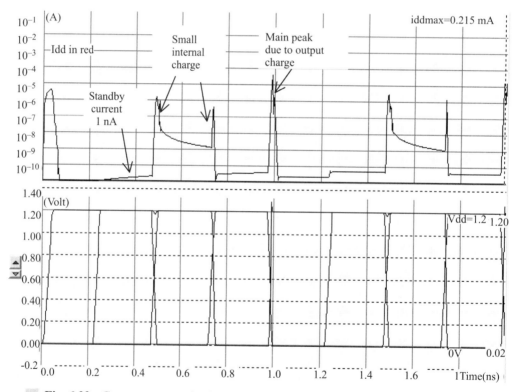

Fig. 6.23   Current consumption in log scale showing the standby current (Nand2.MSK)

At time 0.5 ns, the internal node situated in the n-channel MOS area between gates *A* and gate *B* is charged with a peak of current around 1 µA. The area consists of a small n+ diffusion which is shown in Figure 6.24. This diffusion region creates a N+/p-substrate junction, polarized in invert, which may be considered as a parasitic capacitance. This capacitance is charged and discharged under certain conditions. For example, at time 0.75 ns, the short-circuit current resulting from a temporary situation where both n-channel MOS and p-channel MOS devices are ON, produces a current consumption also around 1 µA.

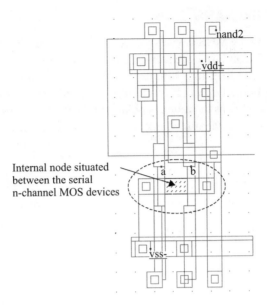

Internal node situated
between the serial
n-channel MOS devices

Fig. 6.24   At time 0.5ns, an internal node is charged and induces a 1 μA current on VDD supply line
(Nand2.MSK)

6.4.8   MOS Sizing

In the time-domain chronograms of Figure 6.20, we observed a 7 ps fall time and a 4 ps rise time. This
is due to the non-symmetrical structure of the NAND gate regarding low-to-high and high-to-low output
switching. A fall edge of *nand2* is provoked by a discharge current IVSS through the two n-channel
MOS in series, meaning an equivalent of 2xRN, where RN is the nMOS resistance when the channel is
ON. In the case where both *a* and *b* are set to 0, the rise edge of *nand2* is provoked by the charge current
IVDD through the two n-channel MOS in parallel. In this case, the equivalent resistance of the path is
RP/2. Consequently, there is a switching speed difference due to the asymmetry inherent in the NAND
gate structure. This asymmetry can be improved by resizing the nMOS or pMOS devices.

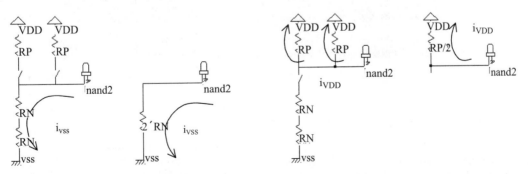

Fig. 6.25   The unsymmetrical charge and discharge paths for the currents IVDD and IVSS lead to a faster
rise time of the output

6.4.9   Optimization of the NAND Surface

The main advantage in joining the MOS devices diffusion whenever possible, rather than implementing all devices separately, is the silicon surface reduction, and consequently cost savings. A second advantage is the speed improvement. Joint diffusions lead to smaller areas, meaning lower parasitic capacitance, and thus shorter charge/discharge delays. The origin of the parasitic capacitance is mainly the N+/P-substrate junction capacitance due as the diode is polarized the other way round (P at low voltage VSS, N at higher voltage). Furthermore, the direct link between diffusions leaves space for metal routing.

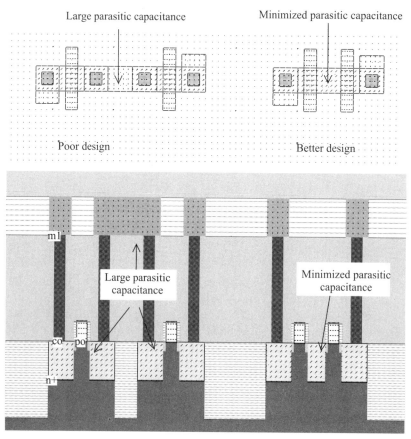

Fig. 6.26   Joined diffusions lead to compact designs and speed improvements (NandComp.MSK)

6.4.10   Optimum pMOS Placement

There are two solutions in implementing the pMOS devices, according to the schematic diagram of Figure 6.27. One solution consists in placing the pMOS with two connections to VDD (left circuit), the second one with two connections to the output (right circuit). Both solutions work fine. However, from the simulation of both structures, it can be seen that the left structure with minimum diffusion and metal connected to the output switches a little faster than the right structure which includes two diffusion areas.

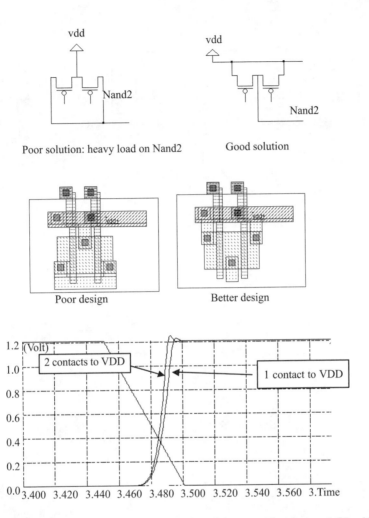

Fig. 6.27   A minimum path for the output is preferred for an optimum speed (NandComp.MSK)

6.4.11   Submicron vs. Deep-submicron Technology

In 0.8 μm technology, the design rules concerning the design of a link between polysilicon and metal 2 layers lead to the complicate and area consuming layout displayed on the left side of Figure 6.28. The cross-section of this poly/metal 2 contact is shown in Figure 6.29. Notice that the via plug between metal 1 and metal 2 is slightly larger (7×7 lambda) than the contact from polysilicon to metal (6×6 lambda). In 0.12 μm technology, the contacts have the same small dimensions (4×4 lambda), and can be stacked on top of each other. Consequently, the routing can be more dense and the cell design can be compacted.

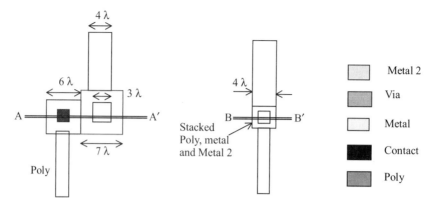

Fig. 6.28 Submicron via (left) and deep-submicron stacked vias (right)

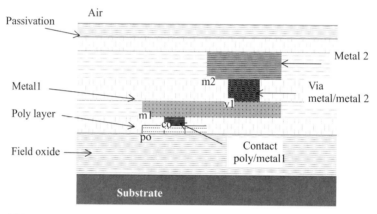

Fig. 6.29 2D cross-section of the poly/metal 12 contact in 0.8 µm (A-A′ of Figure 6.28)

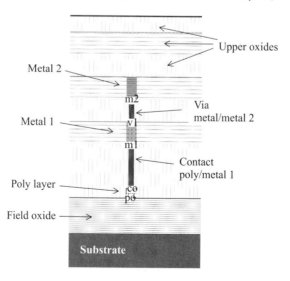

Fig. 6.30 2D cross-section of the poly/metal 12 contact in 0.12 µm (B-B′ of Figure 6.28)

The benefits of deep-submicron technology for a complete logic gate are illustrated in Figure 6.31. On the left side, the NAND gate compiled with the 0.8 μm technology parameters (cmos08.rul) is drawn. On the right side, the same NAND gate compiled with the 0.12 μm technology parameters (cmos012.rul) is reported. Although both layout designs are in lambda scale, the gain in surface is obvious.

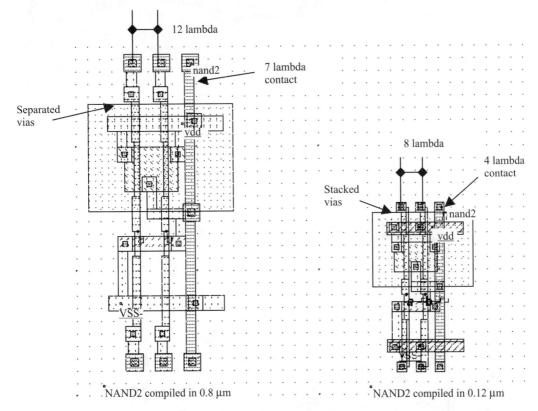

Fig. 6.31 Silicon surface improvement in deep-submicron technology (NandCompo.MSK)

The major reasons for this improvement are:

- A smaller routing pitch: 12 lambda in 0.8 μm, 10 lambda in 0.35 μm, 8 lambda in 0.12 μm thanks to more aggressive rules, specifically for the size of contacts.

- The possibility to stack via: two routing pitches are required in 0.8 μm to connect polysilicon to metal 2, one single pitch is required for starting 0.35 μm technology.

6.4.12 Three-input NAND Gate

The schematic diagram of the n-input NAND gate is derived from the architecture of the NAND2 gate, with n-channel MOS devices in series and p-channel MOS devices in parallel. Using the built-in cell compiler, you can generate the 3- or 4-input NAND gate. The command is **Compile → Compile One Line**. Enter the text nand3 = ~(a & b & c) and click **Compile**.

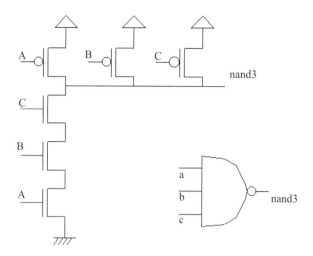

Fig. 6.32 Symbol and implementation of the 3-input NAND gate (nand3Cmos.SCH)

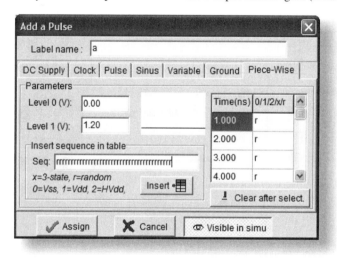

Fig. 6.33 Editing the pulse properties to generate a random logic signal (Nand3Rand.MSK)

A convenient way to observe the switching performances of the NAND3 gate is to stimulate the inputs using random patterns and to observe the simulation results in **eye diagram** mode. The way to proceed consists in changing clock properties into pulse properties, and to assign the value 'r' in place of '1' or '0', which is understood as a random logic value. The procedure is as follows: enter an "rrrrrrrrrrrrrrrr" sequence and click **insert** to transfer these values into the table. Then click **Assign**. The result is a series of random logic values each 1 ns, as shown in the window reported in Figure 6.34. The inputs (Figure 6.34) consist of a random series of logic values. Click **Reset** and a new and uncorrelated series of samples will be generated. The output change may be seen in this particular example at time 3 ns.

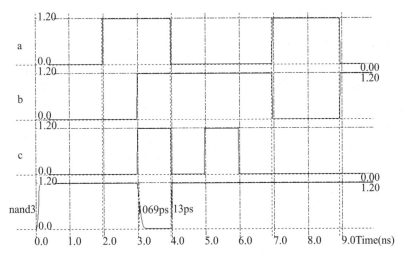

Fig. 6.34 Random stimulation of the 3-input NAND gate (nand3Rand.MSK)

6.4.13 Eye Diagram Simulation

The eye diagram is created by a press on the "**eye diagram**" in the choice bar situated in the lower menu of the simulation window. Each time the input data changes, the simulation chronograms of the output are replaced in the left corner of the simulation window. This makes it possible to compare the switching delays. What we observe is a constant fall delay when discharging a 10 fF load. Depending on the value of A, B and C, the rise delay may change significantly, as one, two or three p-channel MOS devices may be put in parallel to charge the 10 fF load. Notice that the case A, B, C "on" simultaneously is a very rare event that has not been observed in Figure 6.35.

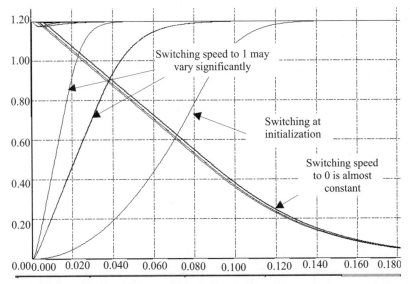

Fig. 6.35 Eye diagram of the random simulation showing significant variations of the rise delay in the 3-input NAND gate (nand3Rand.SCH)

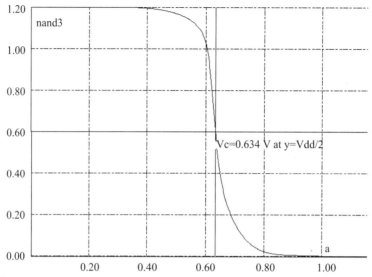

Fig. 6.36 Static characteristics of the 3-input NAND gate (Nand3.MSK)

The switching characteristics of the 3-input NAND gate may be improved by reducing the width of the pMOS devices and increasing the width of nMOS devices. However, the commutation point of the cell is not significantly different from VDD/2. As seen in Figure 6.36, the transfer characteristics simulated using BSIM4 between input a and the output *nand3* exhibit a commutation point Vc near 0.63 V which is close to VDD/2.

6.5 The AND Gate

The truth table of the AND gate is shown in Figure 6.37. In CMOS design, the AND gate is the sum of a NAND gate and an inverter. More generally, the negative gates (NAND, NOR, INV) are simpler to implement in CMOS technology, than the non-negative gates (AND, OR, Buffer). In the logic simulation at switch level (Figure 6.38), the NAND output serves as the input of the inverter output stage, and verifies the truth table.

A	B	AND2
0	0	0
0	1	0
1	0	0
1	1	1
X	0	0
X	1	X
0	X	0
1	X	X

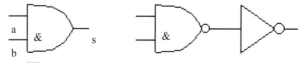

Fig. 6.37 Truth table of the AND gate

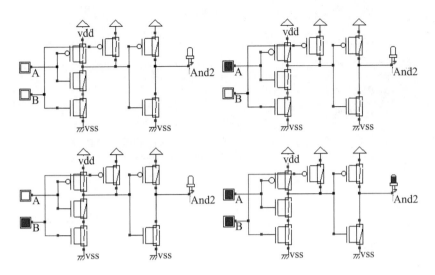

Fig. 6.38 The switching details of the 2-input AND gate (And2Cmos.SCH)

The layout of the AND cell may be compiled by using the command **Compile → Compile One Line**. Notice that a more compact layout of the AND cell may be found by joining the diffusions of the NAND and the inverter cells. This leads to several design rule errors that must be corrected by moving contacts, decreasing polysilicon width, etc. The final arrangement (Figure 6.39 right) saves one horizontal routing pitch, that is 8 lambda.

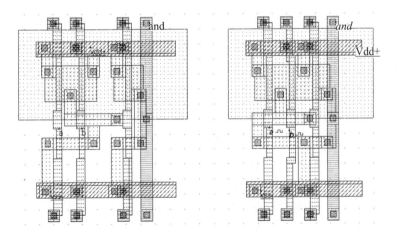

Fig. 6.39 Compiled layout of the AND gate (left) and manual arrangement (right) to achieve a more compact cell (And2.MSK)

The simulation described in Figure 6.40 concerns the 2-input AND gate with a 10 fF load on the output *and*. We also observe the internal node named *nand2*. We confirm that the *and2* output reacts with a certain delay, due to a cascade of two logic cells. However, the major cause of delay is the 10 fF load on the output.

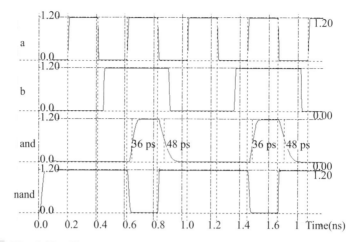

Fig. 6.40 Simulation of the AND gate with a 10 fF load (And2.MSK)

6.6 The NOR Gate

A	B	Out
0	0	1
0	1	0
1	0	0
1	1	0
X	0	X
X	1	0
0	X	X
1	X	0

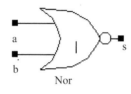

Fig. 6.41 The truth table and symbol of the NOR gate

In CMOS design, the NOR gate consists of two nMOS in parallel connected to two pMOS in series. The schematic diagram of the CMOS NOR cell is reported below. The nMOS in parallel ties the output to the ground if either A or B is at 1. When both A and B are at 0, the nMOS path is cut, but the two pMOS devices in series tie the output to the supply VDD.

A	B	nMOS	pMOS	Out
0	0	off	on	1
0	1	on	off	0
1	0	on	off	0
1	1	on	off	0

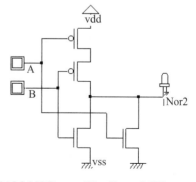

Fig. 6.42 Schematic diagram of the CMOS NOR gate (NorCmos.SCH)

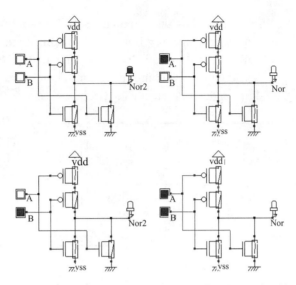

Fig. 6.43 The logic simulation of the NOR gate verifies the truth table (NorCmos.SCH)

Using the cell compiler, we generate the NOR gate from its VERILOG description by entering the equation Nor2=~(a|b). The '~' operator represents the NOT operator, the 'l' symbol represents the OR operator. The compiled layout of the NOR gate appears on the screen, as drawn in Figure 6.44. The compiler has fixed the position of the VDD power supply in the upper part of the layout, near the p-channel MOS devices. The compiler has also fixed the ground VSS in the lower part of the window, near the n-channel MOS devices. The texts *A*, *B*, and *out* have been fixed to the layout as for the NAND gate.

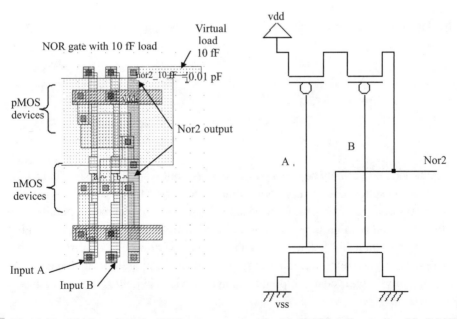

Fig. 6.44 The layout of the NOR gate generated by the CMOS cell compiler (Nor2.MSK)

The simulation of Figure 6.45 is obtained by the command **Simulate → Run simulation**. We verify that *nor2* output signal is equal to 0 when either $a = 1$ or $b = 1$, according to the truth table. The rise time and fall time are shown for a NOR2 gate without any load connected to its output or with a virtual 10 fF load. We see that the 10 fF slows down the switching speed considerably. Furthermore, the rise time is significantly larger than the fall time, due to p-channel MOS in series.

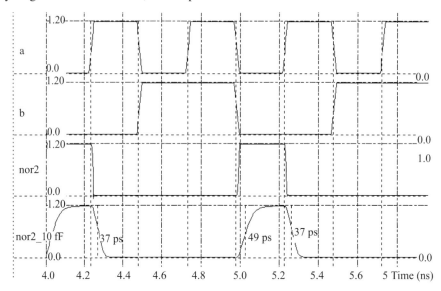

Fig. 6.45   Simulation of the NOR gate with and without loading capacitance (Nor2.MSK)

6.6.1   The NOR3 Gate

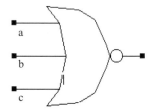

The 3-input NOR gate circuit may be deduced from the 2-input NOR gate design, but a problem arises. The pMOS devices in series lead to very poor output node charge to VDD, and consequently poor rise time performances, as shown in Figure 6.46. Meanwhile, the nMOS devices in parallel provoke a very efficient path for discharge to ground, meaning a short fall time. This non-symmetrical behaviour can be tolerated up to a certain limit. This is why the n-input NOR gates (with n=3,4.) are rarely used. NAND gate-based designs are preferred because of the pMOS in parallel which naturally compensate the poor hole mobility of their channel, producing symmetrical switching characteristics.

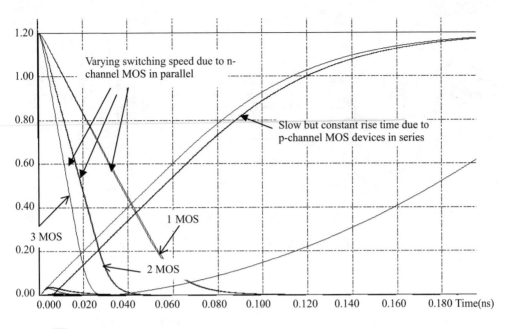

Fig. 6.46   The non-symmetrical behaviour of the NOR3 gate (Nor3.MSK)

6.7   The OR Gate

a	b	S
0	0	0
0	1	1
1	0	1
1	1	1
X	0	X
X	1	1
0	X	X
1	X	1

OR2 Gate

Fig. 6.47   The truth table and symbol of the OR gate

Just like the AND gate, the OR gate is the sum of a NOR gate and an inverter. The implementation of the OR2 gate in CMOS layout requires six transistors. An arrangement may be found to obtain continuous diffusions on n-MOS regions and pMOS regions, as illustrated in Figure 6.48.

OR2 Gate   OR2 made from NOR2 and Inverter

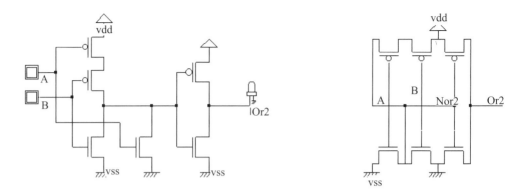

Fig. 6.48 The schematic diagram of the OR gate and its CMOS structure (Or2.SCH)

The layout generated by the Verilog compiler (or2 = a|b) is shown in Figure 6.49. A more compact version of the OR2 gate is also provided (Figure 6.50 right) where the supply pins VSS and VDD are shared by the NOR2 and inverter cells.

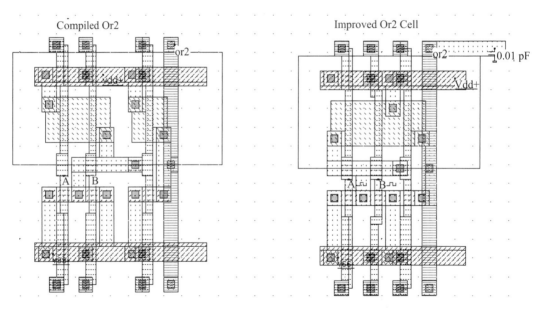

Fig. 6.49 Layout of the compiled OR2 cell (left) and a more area-efficient manual arrangement (right) (OR2.MSK)

6.8 The XOR Gate

The truth table and the usual symbol of the CMOS XOR gate are shown in Figure 6.50. There exist many possibilities of implementing the XOR function into CMOS, which are presented in the following paragraphs.

a	b	Xor2
0	0	0
0	1	1
1	0	1
1	1	0
X	0	X
X	1	X
0	X	X
1	X	X

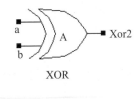

Fig. 6.50 Truth table and symbol of the XOR gate

6.8.1 A Poor Design

The least efficient design, but the most forward, consists in building the XOR logic circuit from its Boolean equation given in Equation 6.4. The direct translation into primitives leads to a very complicated circuit shown in Figure 6.51. The circuit requires two inverters, two AND gates and one OR gate, that is, a total of 22 devices. Furthermore, the XOR gate has a critical path based on six stages of CMOS gates, which slows down the XOR switching response considerably.

$$Xor2 = A\overline{B} + B\overline{A} \qquad\qquad \text{(Equation 6.4)}$$

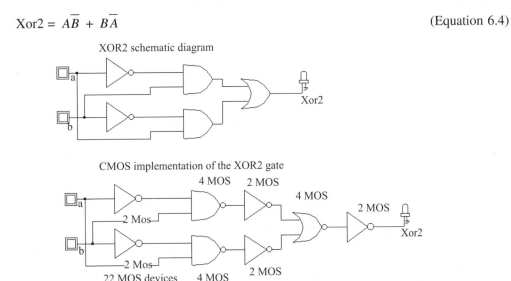

Fig. 6.51 The direct translation of the XOR function into CMOS primitives leads to a 22-transistor cell with five delay stages (Xor2.SCH)

Keeping in mind that negative logic is preferred in CMOS, we can re-arrange the expression of the **xor** cell in a less complex circuit, by replacing the AND/OR stage by NAND gates, following the Equation 6.5. The total number of transistors is decreased to 16.

$$Xor2 = \overline{\overline{A\overline{B}} + \overline{B\overline{A}}} = \overline{\overline{A\overline{B}} \cdot \overline{B\overline{A}}} \qquad\qquad \text{(Equation 6.5)}$$

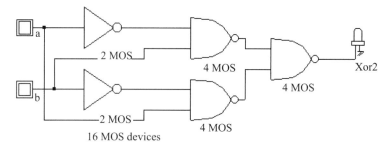

Fig. 6.52   An XOR function based on NAND gates leading to 16 transistors and 3 delay stages (XOR2_16.SCH)

6.8.2   Compiling a Schematic Diagram

An alternative to the manual design consists in describing the logic circuit of the XOR gate in DSCH, and then in compiling the schematic diagram into layout using MICROWIND. The design flow is detailed in Figure 6.53. The XOR circuit is created according to the schematic diagram of Figure 6.52. The circuit is saved under the name **Xor2_16.SCH**, for example. Next, we create the Verilog description corresponding to the circuit, using the command **File → Make Verilog File**. The text file **Xor2_16.TXT**, which includes the Verilog description, serves as the input of MICROWIND, to drive the automatic compilation of the circuit into layout. In MICROWIND, the command is **Compile → Make Verilog File**.

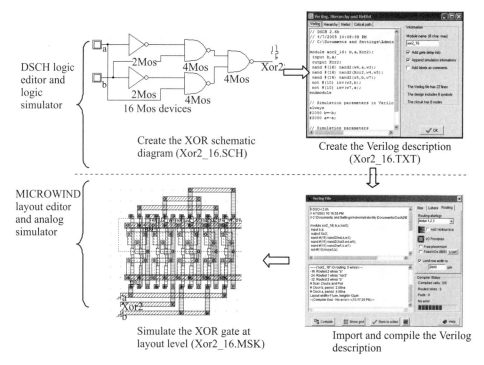

Fig. 6.53   Compiling the XOR circuit from its logic description, through the Verilog format (Xor2_16.MSK)

The result is a layout dynamically created by Microwind which corresponds to the initial circuit defined at logic level. The simulation works fine, as shown in Figure 6.54. Notice that no simulation property is required as it is inherited from the logic circuit simulation. The Verilog text file not only includes the structural description of the circuit but also the list of inputs and the associated simulation parameters. Still, the layout is quite large, and the switching performances suffer from the three delay stages.

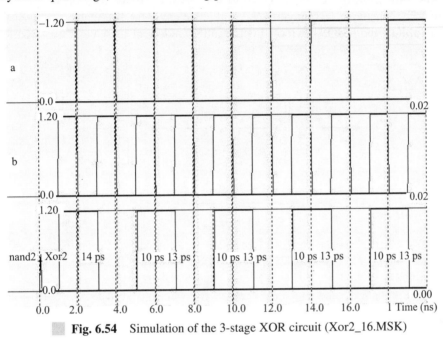

Fig. 6.54   Simulation of the 3-stage XOR circuit (Xor2_16.MSK)

6.8.3   A Better Design

An interesting design is shown in Figure 6.55. The XOR function is built using AND/OR inverted logic (AOI logic). The function created by the n-channel MOS network is equivalent to (A|~B) and (~A|B). The p-channel MOS network gives the function where all AND functions are transformed into OR, and vice versa. In other words, the pMOS network realizes the function (A & ~B)|(~A & B).

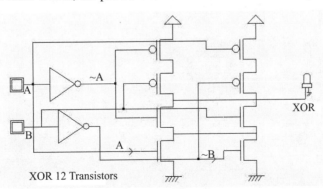

Fig. 6.55   A 12-transistor XOR gate (XorAoi.SCH)

6.8.4 An Efficient Design

The most efficient solution consists of two inverters and two pass transistors, that is only six MOS. The cell is not a pure CMOS cell as two MOS devices are used as transmission gates. The truth table of the XOR can be read as follows: If $B = 0$, $OUT = A$, IF $B = 1$, $OUT = \sim A$. The principle of the circuit presented below is to enable the A signal to flow to node $N1$ if $B = 1$ and to enable the $\sim A$ signal to flow to node $N1$ if $B = 0$. The node OUT inverts $N1$, so that we cover the truth table of the XOR operator. Notice that the nMOS and pMOS devices situated in the middle of the gate serve as pass transistors (Figure 6.56).

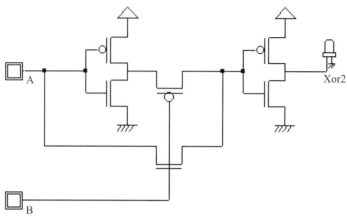

Fig. 6.56 The schematic diagram of the XOR gate with only six MOS devices (Xor2.SCH)

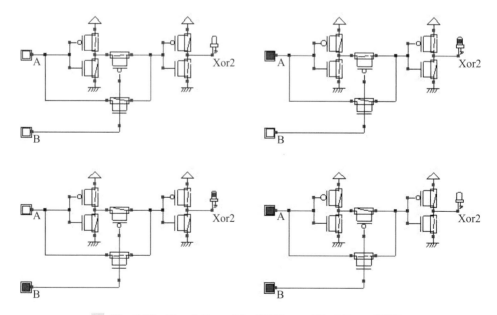

Fig. 6.57 Simulation of the XOR gate (Xor2Cmos.SCH)

The Verilog description of the XOR gate in Microwind is xor = a^b. The layout compiler produces an XOR cell layout as shown in Figure 6.58.

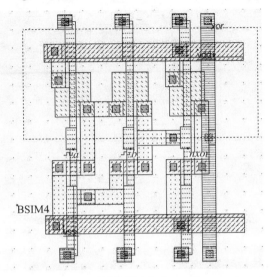

Fig. 6.58 Layout of the XOR gate. The BSIM4 label configures the simulator with BSIM4 model (XOR2.MSK)

When you add a visible property to the intermediate node **xnor** which serves as an input of the output stage inverter, see how the signal is altered by *Vtn* (when the nMOS is ON) and *Vtp* (when the pMOS is ON).

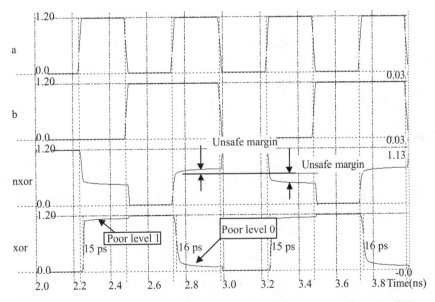

Fig. 6.59 Simulation of the XOR gate using BSIM4 model (XOR2.MSK)

The inverter regenerates the signal, but fails to produce clean 0 and 1 levels. This is because the MOS devices used as pass transistor reduce the voltage amplitude, resulting in dangerous voltage levels, close to the switching point of the logic. The alternative is to abandon the single pass-gate and to use a complete pair of n-channel MOS and p-channel MOS devices, as shown in Figure 6.60.

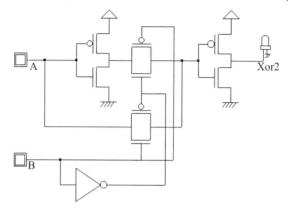

Fig. 6.60   Schematic diagram of the XOR gate with complete transmission gates (XOR2Full.MSK)

6.8.5   About the XNOR Gate

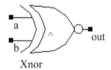

The XNOR gate symbol is shown above. The XNOR circuit is usually an exact copy of the XOR gate, except that the role of the B and ~B signals are opposite in the transmission gate structures. Removing the last inverter is a poor alternative as the output signal is no longer amplified. Adding a supplementary inverter would increase the propagation delay of one stage.

6.9   Complex Gates

6.9.1   Principles

The complex gate design technique applies for any combination of operators AND and OR. The AND operation is represented in logical equations by the symbol "&". The OR operation is represented by "|" (Table 6.2). The complex gate technique described below produces compact cells with higher performances, in terms of spacing and speed, than conventional logic circuits.

Table 6.2   Logic operator used to compile complex gates

Bit operation	Verilog symbol used in complex gates
Not	~
And	&
Or		

In order to illustrate the concept of complex gates, let us take the example of the following Boolean expression (Equation 6.6). Its truth table is reported in Table 6.3.

$$F = A \,\&\, (B \mid C)$$ (Equation 6.6)

Table 6.4 Truth table of the function F = A|(B&C))

A	B	C	A&(B|C)	F
0	0	0	0	0
0	0	1	0	0
0	1	0	0	0
0	1	1	1	0
1	0	0	1	0
1	0	1	1	1
1	1	0	1	1
1	1	1	1	1

A schematic diagram corresponding in a straightforward manner to its equation and truth table is reported below. The circuit is built using a 2-input OR and a 2-input AND cell, that is 12 transistors and four delay stages.

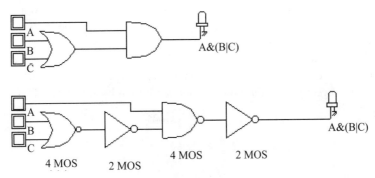

Fig. 6.61 The conventional schematic diagram of the function F (ComplexGate.SCH)

A much more compact design exists in this case (Figure 6.62), consisting of the following steps:

1. For the nMOS network, translate the AND operator '&' into nMOS in series , and the OR operator 'I' into nMOS in parallel.

$$Fn = A \text{ serie s } (B \text{ parallel } C)$$ (Equation 6.7)

2. For the pMOS network, translate the AND operator '&' into pMOS in parallel, and the OR operator 'I' into pMOS in series.

$$Fp = A \text{ parallel } (B \text{ series } C)$$ (Equation 6.8)

3. If the function is non-inverting, as for F, the inverter is mandatory.

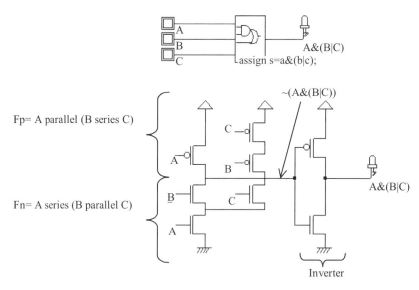

Fp= A parallel (B series C)

~(A&(B|C))

Fn= A series (B parallel C)

A&(B|C)

Inverter

Fig. 6.62   The complex gate implementation of the function F = A&(B|C) (ComplexGate.SCH)

6.9.2 Complex Gates in Dsch

Specific symbols exist to handle complex gate description in DSCH. The location of these symbols (3-input complex gate and 5-input complex gate) is shown in Figure 6.63. At a double click inside the symbol, the menu shown in Figure 6.64 appears. You must describe the logical function that links the output s to the inputs a, b, c. The syntax corresponds to the examples proposed in the previous paragraph (~for NOT, & for AND and I for OR). By using this behavioural description approach instead of building the function with basic cells, the switching performances of the gate are improved. Furthermore, the complex gate can be directly compiled into a compact layout by using MICROWIND.

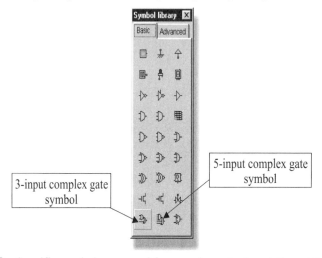

3-input complex gate symbol

5-input complex gate symbol

Fig. 6.63   Specific symbols proposed for complex gate description at the logic level

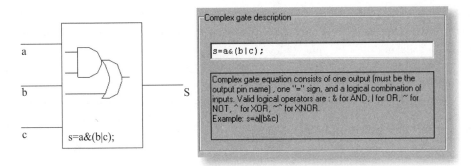

Fig. 6.64   The complex gate symbol and its logic description

6.9.3   Complex Gates in Microwind

MICROWIND2 is able to generate the CMOS layout corresponding to any description based on the operators **AND** and **OR**, using the command **Compile → Compile one line**. Using the keyboard, enter the cell equation, or modify the items proposed in the list of examples. In Equation 6.9, the first parameter *AndOr* is the output name. The sign '=' is obligatory, and follows the output name. The '~' sign corresponds to the operation NOT and can be *used only* right after the '=' sign. The parenthesis '( )' are used to build the function, where '&' is the AND operator and '|' is the OR operator.

$$\text{Andor} = A \& (B \mid C) \qquad\qquad (\text{Equation 6.9})$$

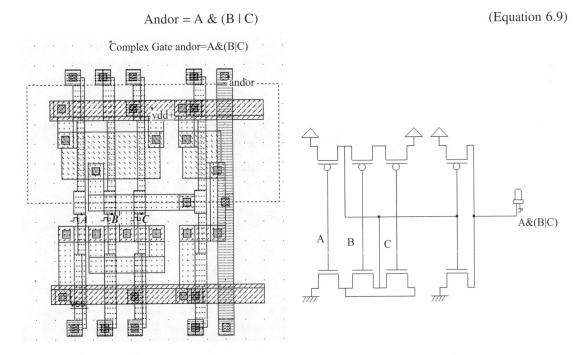

Fig. 6.65   Compiled complex gate andor=A&(B|C) (ComplexGate.MSK)

The MOS arrangement proposed by Microwind consists of a function ~A & (B|C) and one inverter. Notice that the cell could be re-arranged to avoid the diffusion gap by a horizontal flip of the left structure and the sharing of VDD and VSS contacts.

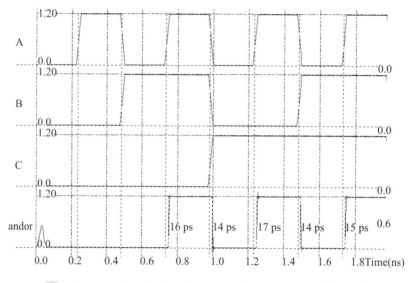

Fig. 6.66 Simulation of the complex gate andor=A&(B|C)

6.10 Multiplexor

Generally speaking, a multiplexor is used to transmit a large amount of information through a smaller number of connections. A digital multiplexor is a circuit that selects binary information from one of many input logic signals and directs it to a single input line. A behavioural description of the multiplexor is the case statement:

```
Case (sel)
0 : f = in0;
1 : f = in1
endcase
```

Table 6.3 The multiplexor truth table

sel	in0	in1	f
0	X	0	0
0	X	1	1
1	0	X	0
1	1	X	1

The usual symbol for the multiplexor is given in Figure 6.67. It consists of the two multiplexed inputs *in0* and *in1* on the left side, the command *sel* at the bottom of the symbol, and the output *f* on the right.

The simulation of the multiplexor proposed in Figure 6.68 uses two different clocks *clk*1 and *clk*2. When *sel*=0, *f* copies *clk*1. When *sel*=1, *f* copies *clk*2.

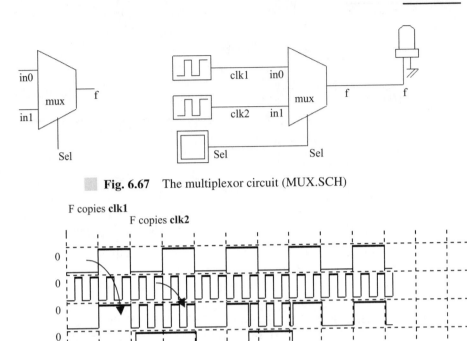

Fig. 6.67 The multiplexor circuit (MUX.SCH)

Fig. 6.68 Logic simulation of the multiplexor circuit (MUX.SCH)

6.10.1 Design of the Multiplexor

The most simple schematic diagram of the 2-input multiplexor is based on two MOS devices. The main problem of using single pass devices is the degradation of voltage levels. In 0.12 μm, the default MOS devices ("low leakage") have a high threshold voltage, meaning that *f* does not reach clean high and low voltages in many cases. Furthermore, the output is not buffered, so *f* cannot drive significant loads. Moreover, inputs in0 and in1 are connected to the diffusions of n-channel and p-channel transistors. Depending on the value of *sel*, the load may vary, so the cell delay may vary accordingly. This circuit, appearing as structure (1) in Figure 6.69 is rarely found in CMOS design as the signal *f* is weak and very sensitive to noise.

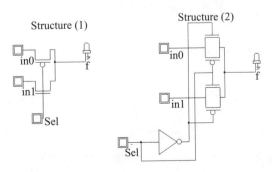

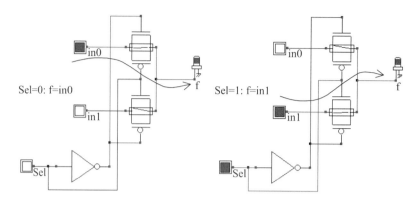

Fig. 6.69 Two-transistor and 6-transistor implementation of the multiplexor (MUX.SCH)

A better circuit, proposed in Figure 6.69 as structure (2) is commonly used for low power operations. The threshold voltage degradation is eliminated by the use of transmission gates. However, the signal *f* is not amplified. Two implementations of the multiplexor are proposed in Figure 6.70. The layout on the left has been generated automatically by placing one inverter and four single MOS devices one after the other. A manual arrangement of the MOS devices (Figure 6.70 right) leads to a much more compact circuit. When the output is loaded by 10 fF, the circuit behaves quite well in terms of delay. As predicted, no parasitic threshold effect may be seen.

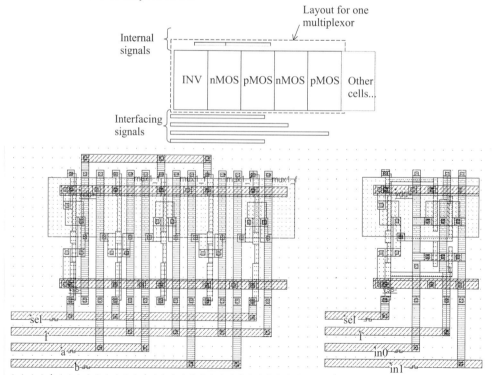

Fig. 6.70 Compiled and manual implementation of the 6-transistor multiplexor (MUX.MSK)

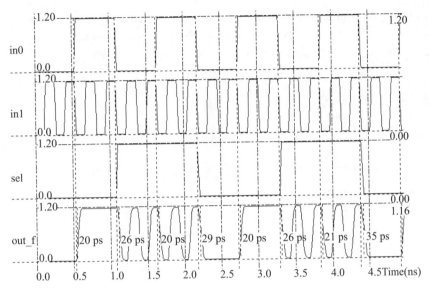

Fig. 6.71 Simulation of the 6-transistor multiplexor with a 10 fF load (MUX.MSK)

Still, inputs in0 and in1 are connected to the diffusions of n-channel and p-channel pass transistors, which may lead to varying switching delays. Adding an output buffer and providing isolation from input signals to drain/sources lead to the multiplexor structure (3), presented in Figure 6.72. This design is safe, easy to modellize at the logic level, but requires many transistors and consumes a lot of power. When a multiplexor with large strength is required, the output inverter can be modified by enlarging the output buffer width. The structure (4) also implements the multiplexor function, using complex gates for the combined OR/AND function, instead of transmission gates.

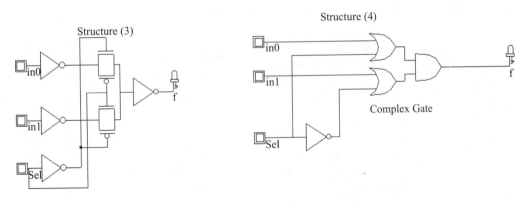

Fig. 6.72 Multiplexor implemented with transmission gates or with complex gates (MUX.SCH)

6.10.2 n-to-1 Multiplexer

The n-to-1 multiplexor performs the selection of a particular input line among n input lines. The selection is controlled by a set of lines which represent a number *sel*. Normally, there are 2^m input lines and m selection lines whose bit combinations determine which input is selected. Figure 6.73 shows one possible implementation of the 8-to-1 multiplexor, based on an elementary multiplexor as described in paragraph 6.69. A behavioural description of the n-to-1 multiplexor is given below:

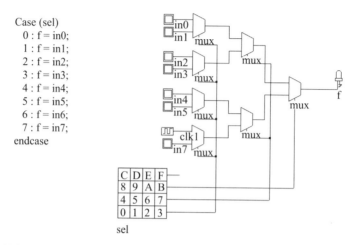

```
Case (sel)
   0 : f = in0;
   1 : f = in1;
   2 : f = in2;
   3 : f = in3;
   4 : f = in4;
   5 : f = in5;
   6 : f = in6;
   7 : f = in7;
endcase
```

Fig. 6.73   8-to-1 multiplexing based on elementary multiplexor cells (Mux8to1.sch)

6.10.3   Timing Analysis

One important characteristic of the multiplexor cell is the propagation delay between the change of the input and the corresponding change of the output. The propagation delay is usually named t_{PD}. Another important delay, named t_{SD} is measured between the change of the selection and the effective set-up of the corresponding channel. These delays are illustrated in Figure 6.74.

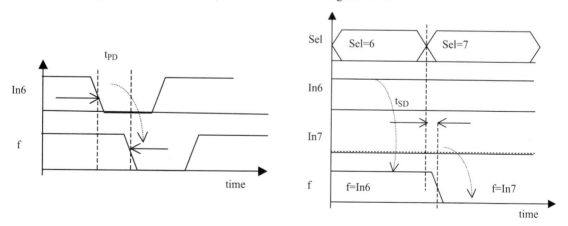

Fig. 6.74   Definition of the propagation delay t_{PD} and set-up delay t_{SD}

The main drawback of the multiplexor design proposed in Figure 6.73 is the use of local inverters at each elementary multiplexor gate, that lead to important power consumption and set-up delay.

Figure 6.75 shows the direct transmission gate implementation of the 8-to-1 multiplexor. The result is simpler than for the multiplexor implementation, it works faster and requires fewer devices.

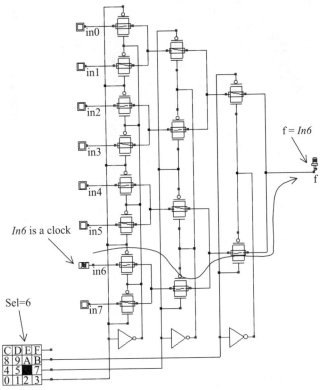

Fig. 6.75 8-to-1 multiplexor based on transmission gates (Mux8to1Tgate.SCH)

We have implemented the 8-to-1 transmission gate multiplexor by compiling its Verilog description, as shown in Figure 6.76. The layout properties have been changed to consider only the selection of *In0* and *In6* alternatively. *In0* is assigned a fast clock while *In6* is assigned a slow clock. The simulation scenario consists in first multiplexing the input *In0* to the output f (Named *pmos_f* in the layout), with *Sel1* = 0, *Sel2* = 0, and then in multiplexing the input *In1* with *Sel1*=1, *Sel2* = 1, at time 2.0 ns in Figure 6.77. The output copies successively *In0* and *In1*, as expected. The eye diagram shows a homogeneous switching delay performance.

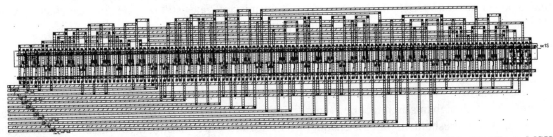

Fig. 6.76 Automatic generation of the 8-to-1 multiplexor based on transmission gates (Mux8to1Tgate.MSK)

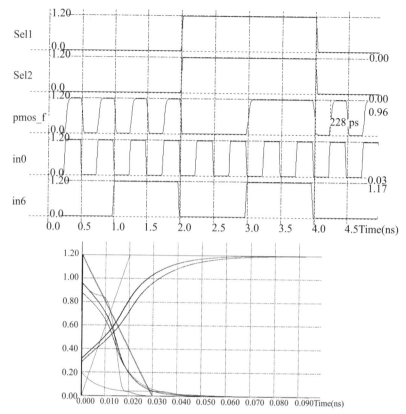

Fig. 6.77 Analog simulation and eye diagram of the 8-to-1 multiplexor (Mux8to1Tgate.MSK)

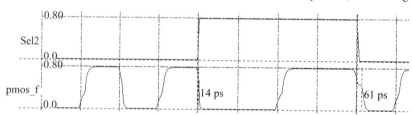

Fig. 6.78 Low voltage operation of the 8-to-1 multiplexor (Mux8to1Tgate.MSK)

If we try to perform the same operation with a voltage supply significantly lower than VDD (here 0.8 V, that is 66 per cent of VDD), the circuit based on transmission gates still operates in a satisfactory way, though significant delays are observed (Figure 6.78).

6.10.4 All nMOS Multiplexor

Some authors have proposed multiplexor designs based only on n-channel MOS devices, as shown in Figure 6.79. Although this architecture is easy to implement in a regular way, the waveform is degraded by the parasitic threshold effect of n-channel MOS devices when passing high signals. Therefore, the output should later be refreshed thanks to a buffer. Although the gain in silicon area is evident, no improvement in switching speed should be expected, as the cascaded threshold loss at each pass transistor leads to very slow rise times.

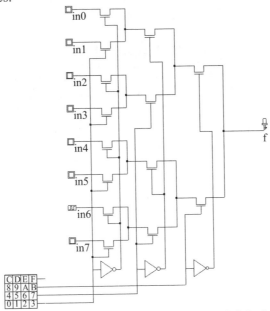

Fig. 6.79 8-to-1 multiplexor based on n-channel MOS (Mux8to1Nmos.SCH)

We have implemented the 8-to-1 n-channel MOS multiplexor by compiling its Verilog description, as shown in Figure 6.80. The layout properties have been changed in a similar way as for transmission gate design. The layout is much more compact as expected. Unfortunately, the switching performances are very poor, which is especially visible in the eye diagram shown in Figure 6.81. When trying to pass the "one", the series of n-channel MOS device degrades the levels considerably. The signal reaches VDD/2 only after 100 ps, that is three times slower than the transmission gate design.

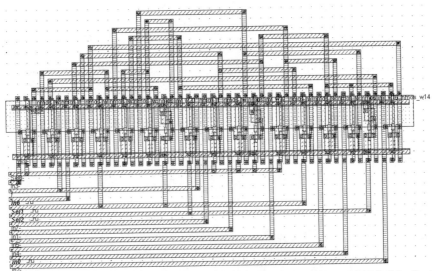

Fig. 6.80 Automatic generation of the 8-to-1 multiplexor based on n-channel MOS (Mux8to1Nmos.MSK)

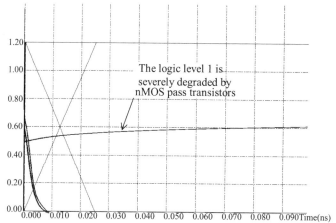

Fig. 6.81 Eye diagram simulation of the 8-to-1 multiplexor based on n-channel MOS (Mux8to1Nmos.MSK)

When we try to operate the circuit at low supply voltage (0.8 V), the circuit does not work any more as the final voltage achieved by the output *f* is as low as 0.3 V (Figure 6.82), still below the switching point of most logic gates.

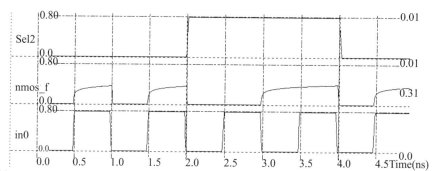

Fig. 6.82 The n-channel multiplexor does not operate at low voltage (here 0.8 V) (Mux8to1Nmos.MSK)

Demultiplexor

Probably one of the most well-known demultiplexors is the 74LS138 circuit, which is presented in Figure 6.83. If the enable circuit is active (G1 must be at 1), all outputs O0..O7 are at 1 except the one issued from the binary decoding of the input A,B and C, according to the statements given below.

```
Case (address)
  0 : O1=0;
  1 : O2=0;
  2 : O3=0;
  3 : O4=0;
  4 : O5=0;
  5 : O5=0;
  6 : O6=0;
  7 : O7=0;
endcase
```

Note that the multiplexor can be implemented using NAND4 cells, to avoid the use of AND logic circuits which require NAND and inverter circuits.

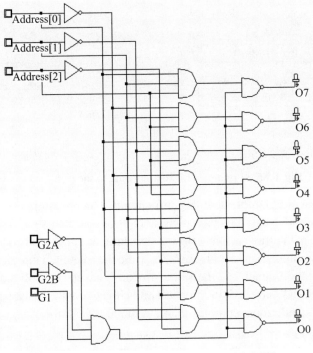

Fig. 6.83 The demultiplexor (74ls138.SCH)

6.11 Shifters

Shifters are very important circuits found in virtually all processor cores. Shifters are able to manipulate data and shift the bits to the right or to the left. Taking the example of an 8-bit input data A, initially set to 0xB3 in hexadecimal (10110011 in binary), the result of shifting A two bits to the right is 0x3c (00101100 binary), as illustrated in Figure 6.84. The corresponding symbol is >>. Now, the result of shifting A three bits to the left would be 0x98 (10011000 binary). The corresponding symbol is <<.

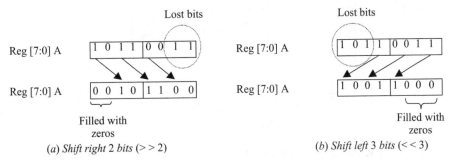

Fig. 6.84 Principles for shifting data to the right and to the left

The rotate circuit is based on the shift mechanism, but has the property to re-inject the lost bits in the place left for zeros. An illustration of the rotate structure is given in Figure 6.85.

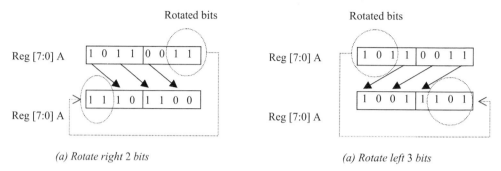

(a) Rotate right 2 bits   *(a) Rotate left 3 bits*

Fig. 6.85   Principles for rotating data to the right and to the left

There are mainly two ways to implement the shift circuits. One [Weste] is based on multiplexor cells, while the other [Uyemura] is based on pass transistors. Pass transistors yield simpler and more regular layout structures. The main drawback is a slower switching and less predictable timing characteristic. The 4-bit shift right circuit is shown in Figure 6.86, while the rotate left circuit is shown in Figure 6.87. The shifter based on multiplexor has the same structure, except that all single pass transistors are replaced by a pair of nMOS and pMOS in parallel. Each gate control must be inverted, which makes the final circuit significantly more complex than the single MOS shifter.

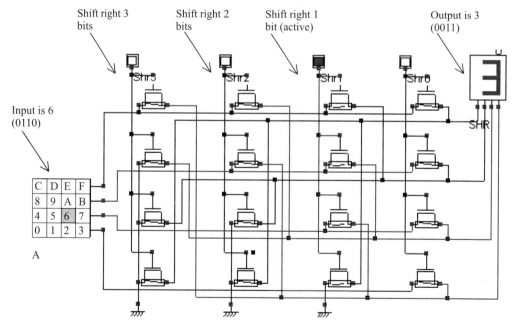

Fig. 6.86   Simulation of the shift right circuit (ShiftRotate4b.SCH)

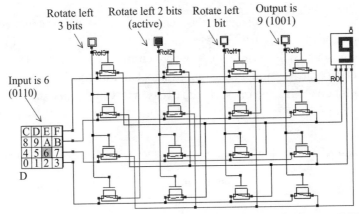

Fig. 6.87   Simulation of the rotate left circuit (ShiftRotate4b.SCH)

6.12   Description of Basic Gates in Verilog

We give in this paragraph a short introduction to Verilog Hardware Description Language (HDL). This language is used by MICROWIND and DSCH tools to describe digital logic networks. The description is a text file, which includes a header, a list of primitive keywords for each elementary gate, and node names specifying how the gates are connected together. Let us study how the logic network corresponding to "BASE.SCH" is translated into a Verilog text.

```
// DSCH 2.5d                                                              1
// 14/04/02 16:52:45                                                      2
// C:\Dsch2\Book on CMOS\Base.sch                                         3
                                                                          4
module Base (Enable,A,B,s_BUF,s_NOT,s_NOTIF1,s_AND,s_NAND,               5
s_NOR,s_OR,s_XOR,s_XNOR,s_PMOS,s_NMOS);                                   6
input Enable,A,B;                                                         7
output s_BUF,s_NOT,s_NOTIF1,s_AND,s_NAND,s_NOR,s_OR,s_XOR;               8
output s_XNOR,s_PMOS,s_NMOS;                                              9
and #(16) and2(s_AND,A,B);                                               10
notif1 #(13) notif1(s_NOTIF1,A,Enable);                                  11
not #(10) not(s_NOT,A);                                                  12
buf #(13) buf1(s_BUF,A);                                                 13
nand #(10) nand2(s_NAND,A,B);                                            14
nor #(10) nor2(s_NOR,B,A);                                               15
or #(16) or2(s_OR,B,A);                                                  16
xor #(16) xor2(s_XOR,B,A);                                               17
xnor #(16) xnor2(s_XNOR,B,A);                                            18
nmos #(10) nmos(s_NMOS,A,Enable); // 1.0u 0.12u                          19
pmos #(10) pmos(s_PMOS,A,Enable); // 2.0u 0.12u                          20
endmodule                                                                21
                                                                         22
// Simulation parameters                                                 23
// Enable CLK 10 10                                                      24
// A CLK 20 20                                                           25
// B CLK 30 30                                                          26
```

Fig. 6.88   The Verilog description of the circuit BASE.SCH (Base.TXT)

The Verilog description is a text file. Double slash characters are used for comments. The rest of the line is ignored. Consequently, lines 1 to 3 are comments. In line 5, the keyword **module** defines the start of the description, which will end after the keyword **endmodule** (line 21). The name of the module is **base**. Its parameters are the list of input and output signals. From line 7 to line 9, signals are declared as **input**, **output**, or internal **wire**. No internal wire exists in this file. Three input and 11 output signals are listed. Then, from line 10 to 20, the schematic diagram is described as a list of primitives.

Table 6.4   The Verilog primitives supported by DSCH and MICROWIND

Name	Logic symbol	Verilog primitive
INVERTER		Not <name>(out,in);
BUFFER		Buf <name>(out,in);
TRI-STATE INVERTER		Notif1 <name>(out,in,enable);
TRI-STATE BUFFER		Bufif1 <name>(out,in,enable);
AND		And <name>(out,in1,in2,..);
NAND		Nand <name>(out,in1,in2,..);
OR		Or <name>(out,in1,in2,..);
NOR		Nor <name>(out,in1,in2,..);
XOR		Xor <name>(out,in1,in2,..);
XNOR		Xnor <name>(out,in1,in2,..);
NMOS		Nmos <name>(out,source,gate);
PMOS		Pmos <name>(out,source,gate);

The gate delay is described for the accurate simulation of time delays in large structures. The logic delay through each gate is specified in integer units. One unit corresponds to a single logic simulation cycle. The correspondence between the elementary cycle and the real time varies depending on the technology. In 0.12 µm, which is the default directory used by DSCH, the elementary unit is 0.01 ns, that is 10 ps. Consequently, the AND gate described in line 10 has a switching delay 16×0.01 ns = 0.160 ns, or 160 ps. This delay is observed between each active transition of A, B and s_AND.

In the simulation chronograms, when we zoom strongly on the time scale in order to see each delay step (0.01ns), we observe the discrete aspect of the delay estimation, as shown in Figure 6.89. A series of 10 elementary delay is used for the NOT cell, and 16 for the OR cell, according to the Verilog description made in Figure 6.88.

The gate delay description is optional. If no delay information has been specified, the logic simulator assigns a default gate delay, which is a basic parameter of the technology. In 0.12 µm, the default gate delay is 0.03 ns, and each wire connected to the gate adds a supplementary 0.07 ns delay. A more detailed description of the Verilog language may be found in [Uyemura].

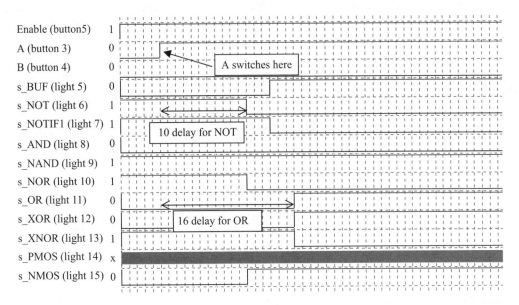

Fig. 6.89 Illustration of the elementary logic delay in the simulation chronograms (Base.SCH)

6.13 Conclusion

In this chapter, the design of basic cells has been reviewed. We have described in details the NAND, AND, OR and NOR gates, and the asymmetrical switching problems of multiple-input NOR circuits. The specific design techniques to design a compact XOR gate have then been reviewed. The techniques for translating AND/OR combinations into an optimized have also been studied. Finally, we have described the structure of multiplexors, shifters, and given some information about the Verilog format used by DSCH and MICROWIND to exchange structural descriptions of logic circuits.

References

[Weste] Neil Weste, K. Eshraghian "Principles of CMOS VLSI design", Addison Wesley, ISBN 0-201-53376-6, 1993.

[Baker] R.J. Baker, H. W. Li, D.E. Boyce "CMOS circuit design, layout and simulation", IEEE Press, ISBN 0-7803-3416-7, 1998.

[Uyemura] John.P. Uyemura "Introduction to VLSI Circuits and Systems", Wiley, 2002, ISBN 0-471-12704-3.

EXERCISES

1. Give the switching point voltage of the gate shown in Figure 6.90. Find its logic function.

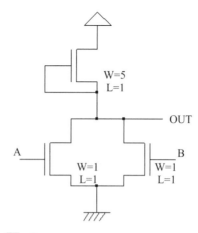

Fig. 6.90   A logic gate case study

2. Design a CMOS 3-input XOR using CMOS And-OR-Invert logic.

3. Design the following function using CMOS And-OR-Invert logic and try to find continuous diffusions.

$$F = \sim(A \ \& \ B|c \ \& \ (A|B))$$

4. Using MICROWIND, compare the switching point voltage of a 3-input NOR gate (minimum size MOS) to the switching point voltage of a 3-input NAND gate (minimum size MOS). Which one is closer to the ideal and why?

5. What is the functionality of the circuit shown in Figure 6.91? Justify with a chronogram using DSCH.

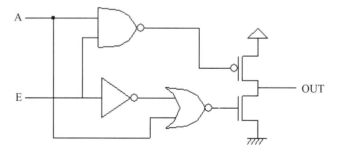

Fig. 6.91   Logic circuit case study

6. What is the functionality of the circuit shown in Figure 6.92? Justify with a chronogram using DSCH. Implement this circuit into a layout. What are the conditions to match the logic behaviour?

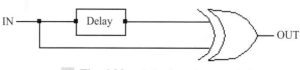

Fig. 6.92   A logic gate case study

7

Arithmetics

This chapter introduces basic concepts concerning the design of arithmetic gates. Firstly, we illustrate data formats. Secondly, the adder circuit is presented, with its corresponding layout created manually and automatically. Then the comparator, multiplier and the arithmetic and logic unit are also discussed. This chapter also includes details of a student project concerning the design of binary-to-decimal addition and display.

7.1 Data Formats

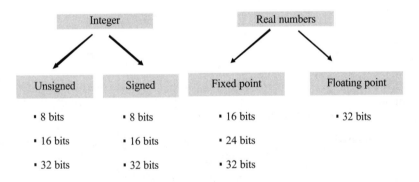

Fig. 7.1 Most common data formats used in ASIC designs

The two classes of data formats are the integer and real numbers (Figure 7.1). The integer type is separated into two formats: unsigned format and signed format. The real numbers are also sub-divided into fixed point and floating point descriptions. Each data is coded in 8, 16 or 32 bits. In particular cases, other formats are used, like the exotic 24-bit in some optimized applications, such as in application-specific digital signal processors.

7.1.1 Integer Format

Table 7.1 Size and range of usual integer formats

Type	Size (bit)	Usual name	Range
Unsigned integer	8	Byte	0..255
	16	Word	0..65535
	32	Long word	0..4294967295
Signed integer	8	Short integer	−128..+127
	16	Integer	−32768..+ 32767
	32	Long integer	−2147483648..+ 2147483647

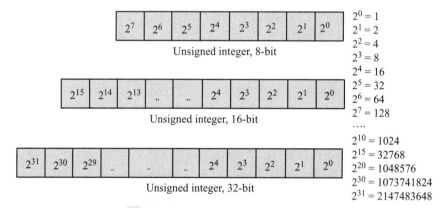

Fig. 7.2 Unsigned integer format

A summary of integer formats is given in Table 7.1. The signification of each bit is given in Figure 7.2. The unsigned integer format is simply the series of power of 2. As an example, the number 01101011 corresponds to 107, as detailed in Equation 7.1.

$$01101011 = 2^6 + 2^5 + 2^3 + 2^1 + 2^0 = 64 + 32 + 8 + 2 + 1 = 107 \qquad \text{(Equation 7.1)}$$

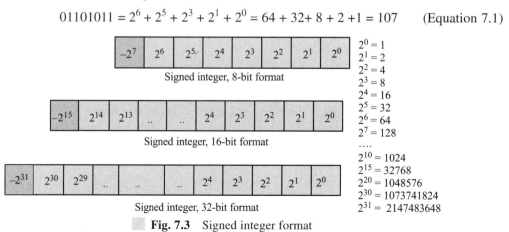

Fig. 7.3 Signed integer format

The signed integer format uses the leftmost bit for the sign. The coding of the data works as for the unsigned integer, except that the leftmost bit accounts for a negative number. In a 16-bit format, a 1 in the sign bit equals to –32768. As an example, the 8-bit signed number 11101011 is detailed in Equation 7.2. The sign bit appears in the sum as –128, with all the other bits remaining positive.

$$11101011 = -2^7 + 2^6 + 2^5 + 2^3 + 2^1 + 2^0 = -128 + 64 + 32 + 8 + 2 + 1 = -21 \qquad \text{(Equation 7.2)}$$

7.1.2 Real Format

A second important class of numbers is the real format. In digital signal processing, real numbers are often implemented as fixed point numbers. The key idea is to restrict the real numbers within the range [–1.0..+1.0] and to use arithmetic hardware that is compatible with integer hardware. More general real numbers are coded in a 32-bit format. A summary of real formats is given in Table 7.2.

Table 7.2 Size and range of usual real formats

Type	Size (bit)	Usual name	Range
Fixed point	16	Fixed	–1.0..+1.0 (Minimum 0.00003)
	32	Double fixed	–1.0..+1.0 (Minimum 0.00000000046)
Floating point	32	Real	–3.4^{e}38..3.4^{e}38 (Minimum 5.8^e-39)

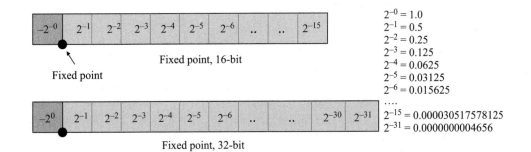

Fixed point, 16-bit

Fixed point, 32-bit

Fig. 7.4 Fixed point numbers in 16-and 32- bit formats

In the case of fixed point arithmetic, we read bits as fractions in negative power of 2 (example of Equation 7.3). When the leftmost bit is set to 1, it accounts for –1.0 in the addition (Equation 7.4). The main limitation of this format is its limited range from –1.0 to 1.0. Its main advantage is a hardware compatibility with integer circuits, leading to low power computing, a particularly attractive feature for embedded electronics. For example, most digital signal processing of mobile phones work in fixed point arithmetic.

$$01100100 = 2^{-1} + 2^{-2} + 2^{-5} = 0.5 + 0.25 + 0.03125 = 0.78125 \qquad \text{(Equation 7.3)}$$

$$11100100 = -2^0 + 2^{-1} + 2^{-2} + 2^{-5} = -1.0 + 0.5 + 0.25 + 0.03125 = -0.21875 \qquad \text{(Equation 7.4)}$$

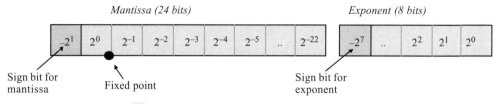

Mantissa (24 bits)   *Exponent (8 bits)*

Sign bit for mantissa

Fixed point

Sign bit for exponent

Fig. 7.5   Floating point arithmetic format

Finally, floating point data is coded using a mantissa multiplied by an exponent (Figure 7.5). The general formulation of the real number format is as follows:

$$data = mantissa.2^{exponent}$$

(Equation 7.5)

The illustration of this format is given in Equation 7.6. Numbers may range from -3.4×10^{38} to 3.4×10^{38}.

$$(0110100...)(0..101) = (2^0 + 2^{-1} + 2^{-3}) \times (2^2 + 2^0)$$

$$= (1.0 + 0.5 + 0.125) \times (4 + 1)$$

$$= 1.625 \times 2^5$$

$$= 52.0$$

(Equation 7.6)

7.2   The Adder Circuit

Let's consider two 8-bit unsigned integers A and B. These numbers range from 0 to 255. The arithmetic addition of the two numbers is given in Equation 7.7.

$$S = A + B$$

(Equation 7.7)

In order to obtain the arithmetic addition of these numbers, we build elementary circuits which realize the bit-level arithmetic addition, as described below. Remember that if $a0=1$ and $b0=1$, a carry bit is generated in the next column. We need to take into account the carry bit ci for each column, by creating a circuit with three logic inputs ai, bi and $ci-1$ to produce si and ci. This circuit is called "full adder". The adder in column 0 is a more simple circuit, which realizes the addition of two logic data $a0$ and $b0$, which is called "half adder".

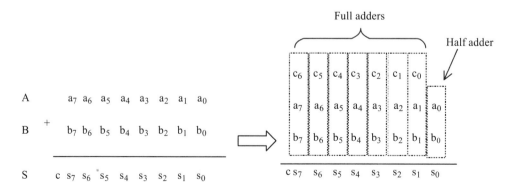

Fig. 7.6   8-bit unsigned integer addition principles

7.2.1   From Logic Design to Layout

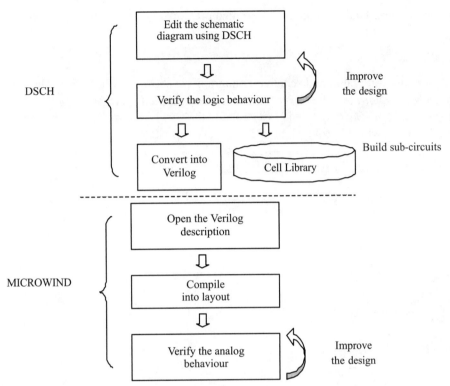

Fig. 7.7   Design flow from logic design to layout implementation

When the design starts to be complex, the manual layout is very difficult to conduct, and an automatic approach is preferred, at the price of a less compact design and more rigid implementation methodology. The steps from logic design to layout validation are listed in Figure 7.7. The schematic diagram constructed using DSCH is validated first at the logic level. At that level, both timing analysis as well as power consumption estimation are available. The accuracy of these predictions is quite fair as accurate layout information is still missing. The designer iterates on his circuit until the specified performances are achieved. At that point, it is of key importance to build a sub-circuit and to feed a cell library. These sub-circuits may be re-used in other designs, if the user provides sufficient information about the performances of the cells.

A specific command in DSCH creates the Verilog description of the logic design, including the list of primitives and some stimulation information. This Verilog text file is understood by MICROWIND to construct the corresponding layout, with respect to the desired design rules. This means that the layout result will significantly change irrespective of whether we use 0.8 μm or 0.12 μm design rules. The supply properties and most stimulation information are added to the layout automatically, according to the logic simulation. Finally, the analog simulation permits us to validate the initial design and to verify its switching and power consumption performances.

7.3 Adder Cell Design

In this section, we use the design method presented previously to build an 8-bit adder.

7.3.1 Half Adder

The half adder gate truth table and schematic diagram are shown in Figure 7.7. The Sum function is made with an XOR gate, while the *Carry* function is a simple AND gate. When both *A* and *B* are at 1 (in black in Figure 7.8), the *Carry* output is asserted, and the *sum* is at zero. The combination of *Carry* and *Sum* is equal to (10), which is equivalent to 2 in binary format.

A	*B*	*Carry*	*Sum*	*Result*
0	0	0	0	0
0	1	0	1	1
1	0	0	1	1
1	1	1	0	2

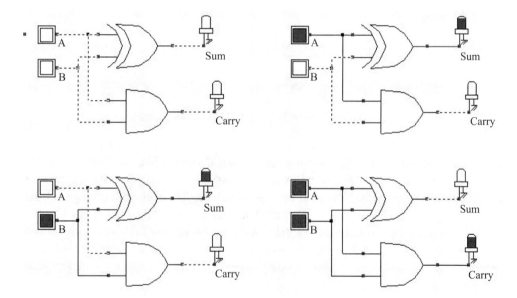

Fig. 7.8 Truth table and schematic diagram of the half-adder gate (halfAdderTest.SCH)

The correct behaviour of the adder is proven by the values of the display, which show a result in accordance with the truth table. The command **File → Schema to New Symbol** in DSCH enables us to create a sub-circuit. This technique is useful to construct hierarchical designs, by embedding a complete circuit into a single component. The command menu is shown in Figure 7.9. By a click of the OK button, a symbol **HaldAdder.SYM** is created. This symbol includes the Verilog description of the circuit, that is the XOR and AND gate, as seen in the left part of the window. Once created, the symbol may be instantiated in any logic design.

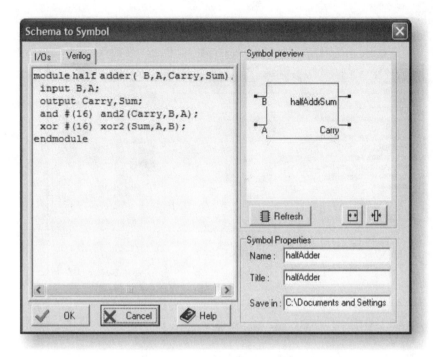

Fig. 7.9 Creating a half-adder symbol (HalfAdder.SYM)

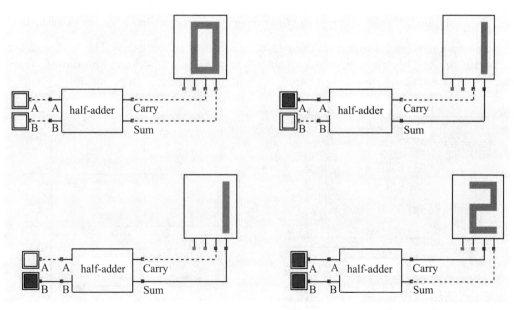

Fig. 7.10 Logic validation of the half-adder symbol (halfAdderTest.SCH).

The logic validation of the half adder is provided in the simulation of Figure 7.10. When the symbol was created, a Verilog text was also generated, namely **HalfAdder.TXT**, In MICROWIND, the same

Verilog text file is used to construct the corresponding layout automatically. Just invoke the command **Compile → Compile Verilog File**, select the corresponding text file and click **Compile**.

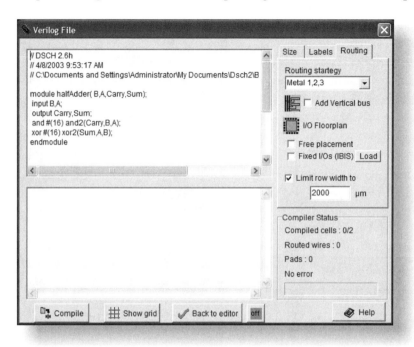

Fig. 7.11   Using the half adder Verilog description to pilot the layout compiling in Microwind (halfAdder.TXT)

When the compiling is complete, the resulting layout appears as shown below. The XOR gate is routed on the left and the AND gate is routed on the right. Click on **Simulate → Start Simulation**. The timing diagrams of Figure 7.12 appear where the truth table of the half adder can be verified.

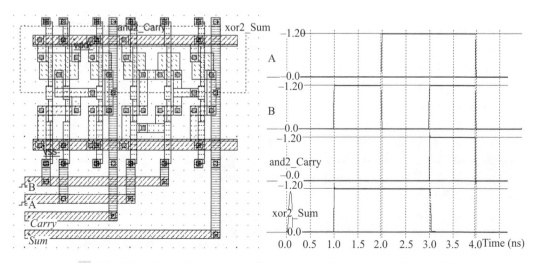

Fig. 7.12   Compiling and simulation of the half adder gate (Hadd.MSK)

7.3.2 Full Adder Gate

The truth table and schematic diagram for the full adder are shown in Figure 7.13. The most straightforward implementation of the CARRY cell is to use a combination of AND, OR gates. Using negative logic equivalence, the AND, OR combination becomes a series of NAND gates, as suggested by Equation 7.8. Concerning the sum, one common approach consists in using a 3-input XOR gate. This 3-input XOR is usually constructed from two stages of 2-input XOR.

A	B	C	Carry	Sum	Result
0	0	0	0	0	0
0	0	1	0	1	1
0	1	0	0	1	1
0	1	1	1	0	2
1	0	0	0	1	1
1	0	1	1	0	2
1	1	0	1	0	2
1	1	1	1	1	3

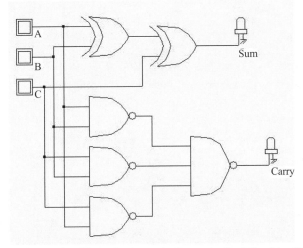

Fig. 7.13   The truth table and schematic diagram of a full-adder (FADD.SCH)

$$sum = A \wedge B \wedge C$$

$$carry = (A \ \& \ B) \ | \ (A \ \& \ C) \ | \ (B \ \& \ C)$$

$$carry = \overline{\overline{(A \ \& \ B)} \ \& \ \overline{(A \ \& \ C)} \ \& \ \overline{(B \ \& \ C)}}$$

(Equation 7.8)

7.3.3 Full Adder Using Complex Gate

A more efficient circuit involves the use of complex gates [Uyemura]. We apply this technique for the *carry* cell: rather than assembling NAND gates (Total $3 \times 4 + 6 = 18$ transistors), we assemble MOS devices in AND/OR combinations. The carry function built by using MOS in parallel or in series is presented in Figure 7.14. A tiny re-arrangement of the Boolean expression leads to a 12-transistor complex gate (Equation 7.9) rather than a 14-transistor one, if we include the two MOS devices of the final stage inverter. The carry circuit shown in Figure 7.14 is strictly equivalent to the carry circuit of Figure 7.13. We avoid the needs for cascaded NAND gates, and so the number of transistors is lower.

$$carry = (A \& B) \mid (C \& (A \mid B)) \qquad \text{(Equation 7.9)}$$

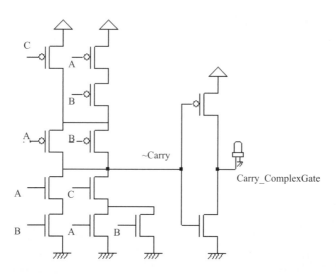

Fig. 7.14   The carry cell based on a complex gate requires less transistors (Fadd.SCH)

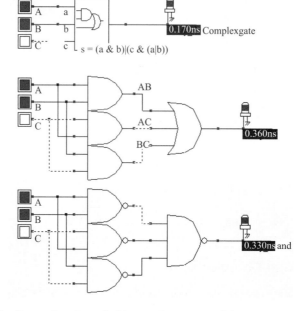

Fig. 7.15   Comparing the switching performances of the carry cells (Carry.SCH)

A switching speed comparison is proposed in Figure 7.15: the upper cell is the fastest, as the number of stages is restricted to the complex gate itself, with short internal connections and small diffusion areas. The cell in the middle is the slowest due to the four internal stages (the AND is a NAND plus an inverter). The lower circuit is a little faster, but not so. The reason is that DSCH automatically assigns a typical delay (around 70 ps) to each interconnect. At this stage, DSCH does not make the difference between a short and a long interconnect. In all probability, the interconnect between the NAND2 output and the NAND3 input will be very short and the delay is severely over-estimated. But it is possible that the routing is quite large, in which case the supplementary delay is justified.

7.3.4 Full Adder Using Complex Gate

The procedure for generating the symbol of the full adder from its schematic diagram is the same as that for the half adder. When invoking **File → Schema to new symbol**, the screen of Figure 7.16 appears. Simply click **OK**. The symbol of the full-adder is created, with the name *FullAdder.sym* in the current directory. Meanwhile, the Verilog file **fullAdder.txt** is generated, the contents of which are given in the left part of the window (Item **Verilog**).

We see that the XOR gates are declared as primitives while the complex gate is declared by using the **Assign** command, as a combination of AND (&)and OR (|) operators. If we were to use AND and OR primitives instead, the layout compiler would implement the function in a series of AND and OR CMOS gates, losing the benefits of the complex gate approach in terms of cell density and switching speed.

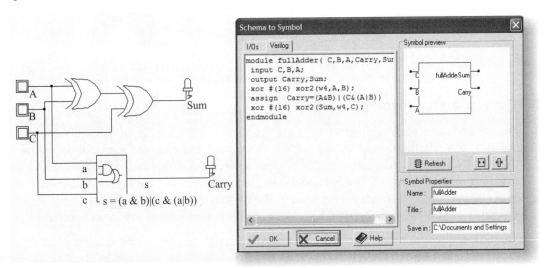

Fig. 7.16 Verilog description of the full adder (FullAdder.SYM)

Use the command **Insert → User Symbol** to include this symbol into a new circuit. For example, the circuit **FaddTest** includes the hierarchical symbol and verifies its behaviour. Three clocks with 20, 40 and 80 ns, respectively, are declared as inputs. Using such clocks is of particular importance during the scanning of all possible combination of inputs, by following the truth table line by line. What we observe in the chronograms of Figure 7.17 is the addition of three numbers, which follows the requested result given in the initial truth table. The addition of *A*, *B* and *C* appears in the output *sum*. For example, at time t = 70 ns, $A = B = C = 1$, the output is 3 after a little delay.

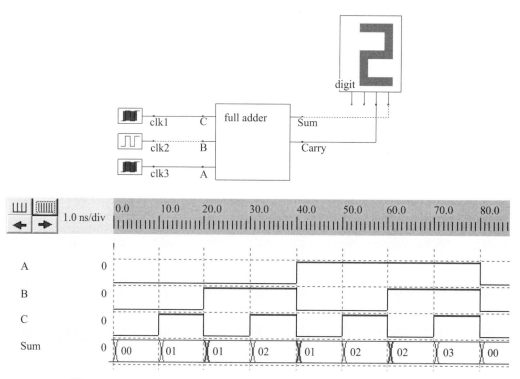

Fig. 7.17   Testing the new adder symbol using clocks (FaddTest.SCH)

7.3.5   The Full Adder Layout

You may create the layout of the full adder by using the Verilog cell compiler of MICROWIND. It is recommended to compile the full adder symbol which includes the complex gate, for an optimum result. MICROWIND handles complex gate descriptions and has the ability to convert AND/OR logic combinations into a compact layout. The full adder compiled layout is shown in Figure 7.18. By default, all signals are routed to the left side of the logic cells, for clarity. In real case designs, the routing creates connections in all directions, depending on the position of the signal inputs and outputs.

The complex gate implementation includes some interesting design techniques for achieving a compact layout. The layout and schematic diagram are detailed in Figure 7.19. The cell compiler has organized the contacts to ground and metal bridge in such a way that the n-diffusion and p-diffusion areas are continuous. The MOS arrangement is electrical and equivalent to the schematic diagram of Figure 7.14. Notice that the diffusion has been stretched to build an internal connection, in the nMOS area. Diffusion is rarely used as an interconnect due to its huge parasitic resistance. In that case, however, the role of this diffusion connection is not significant.

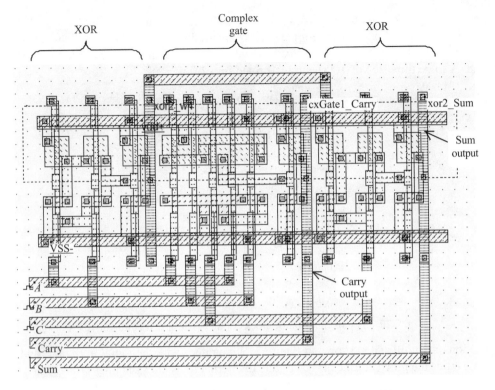

Fig. 7.18   The compiled full adder (FullAdder.MSK)

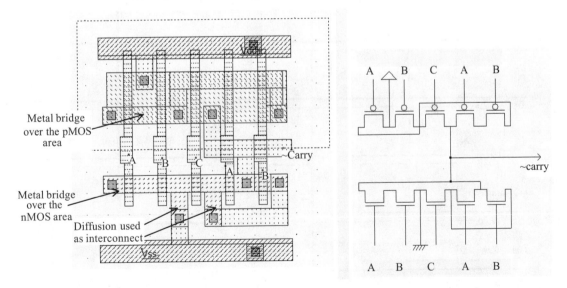

Fig. 7.19   The carry cell layout and associated structure (CarryCell.MSK)

There is no need to add the clock properties in the layout as the clock signals properties have been extracted from the Verilog text, and automatically placed in the compiled layout. The clock description in Verilog format appears after the keyword "Always". Three clocks are declared: A, B and C. The text signifies that for a given time step, the logic signal is inverted. For example clock C switches from 0 to 1 at time 1000, from 1 to 0 at time 2000, etc. There exists a time scale conversion between DSCH and Microwind. A logic time step of 1000 is transformed into 1.0 ns.

```
module fullAdder (C,B,A,Carry,Sum);
    input C,B,A;
    output Carry,Sum;
    xor #(16) xor2(w4,A,B);
    assign  Carry=(A&B)|(C&(A|B))
    xor #(16) xor2(Sum,w4,C);
endmodule


// Simulation parameters in Verilog Format
always
#1000 C=~C;
#2000 B=~B;
#4000 A=~A;
```

Fig. 7.20   The declaration of clocks in the Verilog text (FullAdder.TXT)

The simulation of the full adder, shown in Figure 7.21, complies with the initial truth table and the logic simulation. Notice the glitch at time t = 6.0 ns due to internal transient switching of the logic circuit.

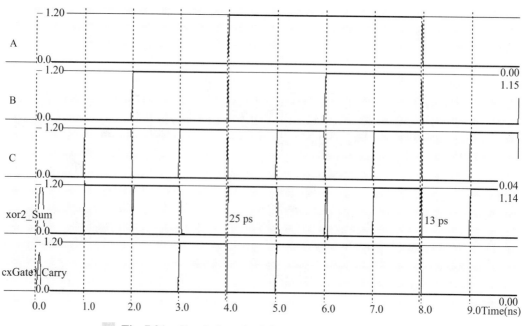

Fig. 7.21   Simulation of a full adder (File fullAdder.MSK)

7.4 Ripple-carry Adder

In this section, the design of the adder is addressed in its simplest approach, wherein the carry of one stage is propagated as an input for the next stage. This type of circuit based on a carry chain is called "ripple carry" adder.

7.4.1 Structure of the Ripple-carry Adder

The 4-bit ripple-carry adder circuit includes one half adder and three full-adders in serial, according to the elementary addition shown in Figure 7.22 [Belaouar]. Each stage produces the Boolean addition of three logical information. For example, the Boolean output $s[1]$ is the addition of input signals $a[1]$ and $b[1]$, together with the internal carry $c[1]$. Some examples of 4-bit addition are given in Table 7.3. We give the output s in hexadecimal and decimal format.

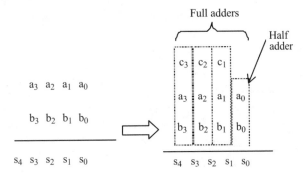

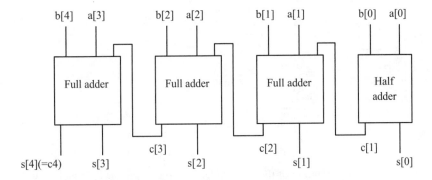

Fig. 7.22 Structure of the 4-bit ripple-carry adder

The logic circuit shown in Figure 7.4 allows a four-bit addition between two numbers $a[0..3]$ and $b[0..3]$. The user symbols 'Hadd.sym' and 'Fadd.sym' are added to the design using the command **Insert → User Symbol**. In DSCH, the numbers a and b are generated by keyboard symbols, which produces at the press of a desired key, a 4-bit logic value. In the example shown in Figure 7.22, the value of a is 1 (0001 in binary), and the value of b is F (1111 in binary form, or 15 in decimal). The result s, which combines s[0], s[1], s[2], s[3] and the last carry bit, is equal to 0×10 in hexadecimal, or 16 in decimal.

Table 7.3   Adder result examples

a[0..3]	b[0..3]	s[0..4] (hexa)	s[0..4] (Decimal)
0	0	00	0
0	F	0F	15
9	7	10	16
6	C	12	18
F	F	1E	30

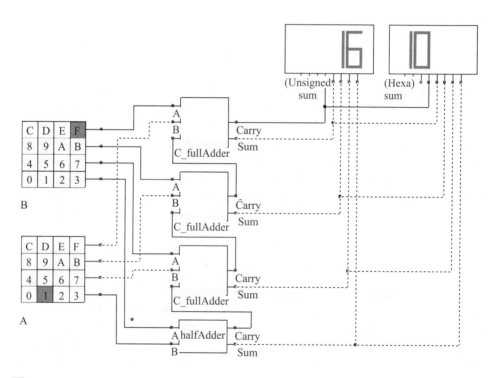

Fig. 7.23   Schematic diagram of the 4-bit adder and some examples of results (Add4.SCH)

The two displays are connected to the identical data, but are configured in different mode: hexadecimal format for the rightmost display, and integer mode for the leftmost display. In order to change the display mode, double click inside the symbol, and change the format in the symbol property window, as shown in Figure 7.24. The default option is the hexadecimal format. The unsigned integer format is used in the schematic diagram of Figure 7.24 for the left display. The integer and fixed point display formats are used for arithmetic circuits. The ASCII character corresponding to the 8-bit input data may also be displayed.

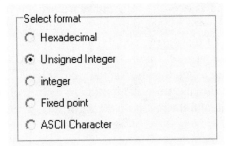

Fig. 7.24   The display symbol has five format options (Add4.SCH)

7.4.2   Critical Path

The worst case delay of the circuit is calculated in DSCH by the command **Simulate → Find critical Path → In the Diagram**. In the ripple-carry circuit, the critical path passes through the carry chain. The predicted worst case delay is 0.8 ns.

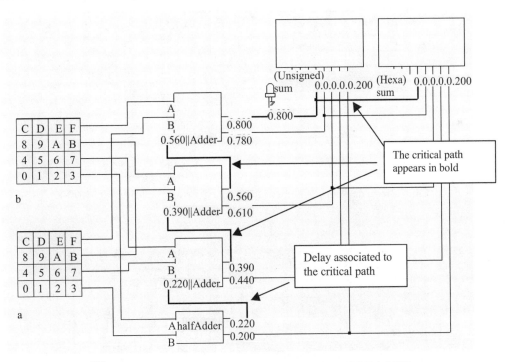

Fig. 7.25   The critical path of the 4-bit adder (ADD4.SCH)

7.4.3   A Manual Design of the 4-bit Ripple-carry Adder

We described in this paragraph an example of manual design of a 4-bit ripple-carry adder. The elementary adder circuit **FullAdder.MSK** from Figure 7.18 is re-used. We tried to compact the interconnect network by re-routing the input interconnects over the cell, and shortened the carry wires. The modified layout is **fadd.MSK**. A proposed strategy for connecting adders together in order to build a ripple carry adder is detailed in Figure 7.26. The cell placement, supply and I/O routing positioning ease the connection between blocks, leading to a compact design and short connections. Each adder is supplied by VDD and VSS rails all of which are connected together on the right side of the bloc. The carry propagation is realized by a short interconnect flowing from the bottom to the top of each cell. The *a* and *b* data are routed on the left side of the cell, and the result on the right side.

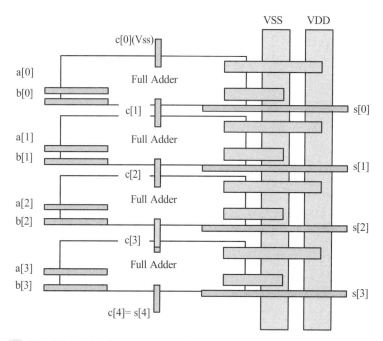

Fig. 7.26 The floor planning of the 4-bit ripple-carry adder (ADD4.MSK)

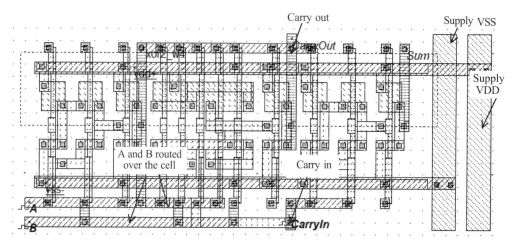

Fig. 7.27 Arranging the full adder to create a compact layout and to ease further connection (fadd.MSK)

Figure 7.28 details the 4-bit adder layout based on the manual cell design of the adder. In MICROWIND, the command **Edit → Duplicate X,Y** has been used to duplicate the full adder layout vertically. The input and output names use brackets with an index ($A[0]..A[3]$, $B[0]..B[3]$ and $Sum[0].. Sum [4]$), so that MICROWIND automatically displays the logic value corresponding to A, B and Sum. For example, at time $t = 1.5$ ns (Figure 7.29), $A = 3$, $B = 1$, $Sum = 4$.

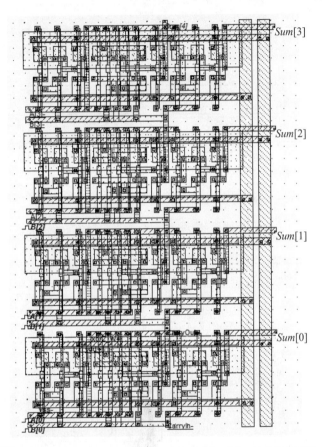

Fig. 7.28 Implementation of the 4-bit adder (Add4.MSK)

A convenient method for characterizing the worst case circuit delay is to use the **Eye Diagram** simulation. This simulation mode superimposes output signals at any rise or fall edges of inputs $A0..A3$ or $B0..B3$. The cumulative drawing of the outputs is called the eye diagram. Figure 7.30 gives the eye diagram of the 4-bit ripple-carry adder. It can be seen that the worst case delay is around 80 ps. Remember that the critical delay was 0.8 ns in the logic simulation. How do we explain that difference? Firstly, the layout level simulation does not investigate all input configurations, specifically the ones where the critical delay is involved. The inputs $A2$, $A3$, and $B3$ are inactive, and set to 0. Secondly, the logic simulation assigned a default 70 ps delay per interconnect. This average delay is over-estimated in our case, as most interconnects are routed very short.

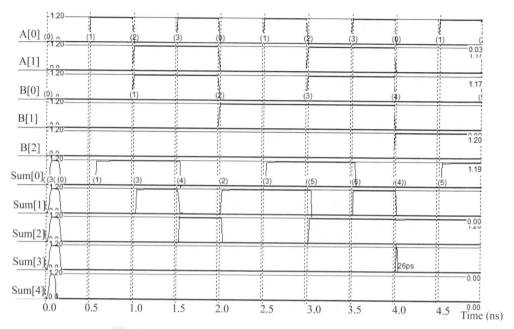

Fig. 7.29   Simulation of the 4-bit adder (Add4.MSK)

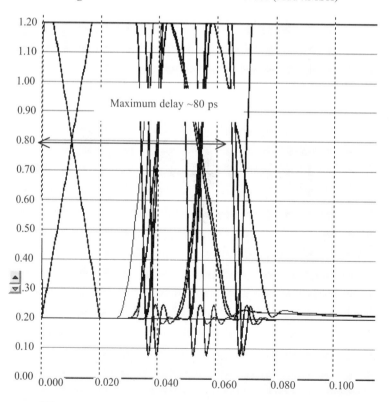

Fig. 7.30   Eye diagram of the 4-bit adder (Add4.MSK)

7.5   Signed Adder

The signed integer format for a 4-bit input data is specified in Table 7.4. The correspondence between unsigned and signed data for a 4-bit information is also reported. A specific symbol, namely **KbdSigned.SYM**, is available to support the generation of 4-bit signed input. The symbol may be loaded using the command **Insert → User Symbol**. Select the symbol **KbdSigned.SYM** in the list, as shown in Figure 7.31. The display symbol on the left displays the 4-bit input data in unsigned integer form. The other symbol displays the same input information as a signed integer.

$$2^0 = 1$$
$$2^1 = 2$$

-2^3	2^2	2^1	2^0

$2^2 = 4$

4-bit signed integer $2^3 = 8$

Table 7.4   Correspondence between signed and unsigned 4-bit data

I[3]	I[2]	I[1]	I[0]	*Unsigned integer*	*Signed integer*
0	0	0	0	0	0
0	0	0	1	1	1
0	0	1	0	2	2
0	0	1	1	3	3
0	1	0	0	4	4
0	1	0	1	5	5
0	1	1	0	6	6
0	1	1	1	7	7
1	0	0	0	8	-8
1	0	0	1	9	-7
1	0	1	0	10	-6
1	0	1	1	11	-5
1	1	0	0	12	-4
1	1	0	1	13	-3
1	1	1	0	14	-2
1	1	1	1	15	-1

It can be seen from Figure 7.32 that the unsigned adder circuit can be re-used without any hardware modification for the addition of signed integers. The addition of $a = -2$ with $b = -4$ gives *sum* $= -6$.

Fig. 7.31   Inserting the specific keyboard symbol which supports signed input (KbdSigned.SYM)

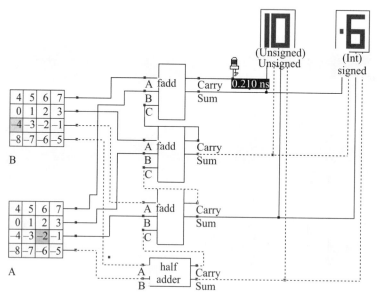

Fig. 7.32   The signed addition uses the same hardware as for the unsigned addition (ADD4Signed.SCH)

7.6   Fast Adder Circuits

The main drawback of ripple-carry adders is the very large computational delay due to the carry chain built in series. Several techniques have been proposed to speed up the addition, at the cost of more complex and power-consuming circuits [Weste] [John]. We construct in this paragraph one type of high-speed adder and compare its performances with that of the ripple-carry adder.

7.6.1   Carry Look Ahead Adder

The carry look-ahead adder computes the c_2 and c_3 carry functions by an optimized hardware, based on a complex gates. The start circuit is the elementary half adder which produces p_i and g_i at each stage

(Equations 7.10 and 7.11). The sign "&" accounts for logical AND operator, "|" for OR and "^" for XOR. Then, the complex gate is built for each carry function, according to the Boolean expressions given in Equations 7.12.

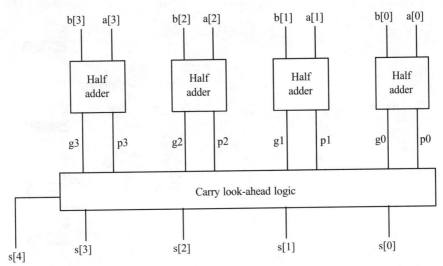

Fig. 7.33 The carry look-ahead logic circuit principles

$$g0 = a[0] \& b[0]$$
$$g1 = a[1] \& b[1]$$
$$g2 = a[2] \& b[2] \quad \text{(Equation 7.10)}$$
$$g3 = a[3] \& b[3]$$

$$p0 = a[0] \wedge b[0]$$
$$p1 = a[1] \wedge b[1]$$
$$p2 = a[2] \wedge b[2] \quad \text{(Equation 7.11)}$$
$$p3 = a[3] \wedge b[3]$$

$$s[0] = p0 \quad \text{(Equation 7.12)}$$
$$c1 = g0$$
$$s[1] = c1 \wedge p1$$
$$c2 = g1 \, (p1 \& g0)$$
$$s[2] = c2 \wedge p2$$
$$c3 = g2 | (p2 \& (g1|(p1 \& g0)))$$
$$s[3] = c3 \wedge p3$$
$$c4 = g3 | (p3 \& (g2|p2 \& (g1| (p1 \& g0))))$$
$$s[4] = c4$$

Equation 7.12 can be translated into layout by using XOR gates and complex gates. The schematic diagram of the carry look-ahead adder is given in Figure 7.33. Each half adder circuit produces the gi and pi data. Complex gates produce the internal carry ci, while XOR gates generate the outputs $s[i]$.

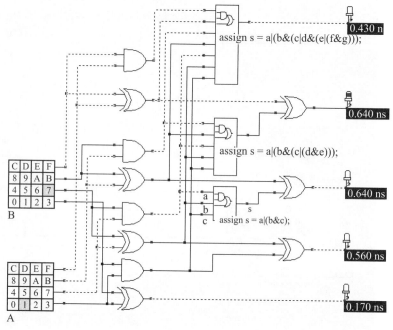

Fig. 7.34   The logic implementation of the carry look-ahead adder (Add4LookAhead.SCH)

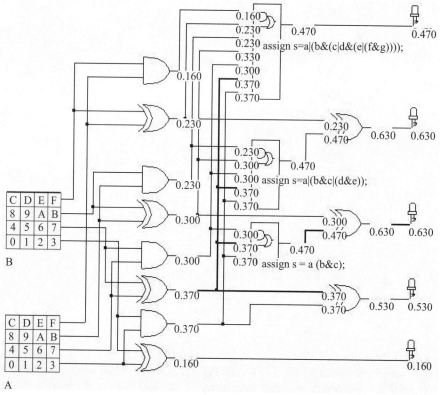

Fig. 7.35   Evaluation of the critical delay in the carry look-ahead adder (Add4LookAhead.SCH)

The key question is to know to which extend the critical delay has been reduced. The answer is given in Figure 7.34. We notice a reduction of the critical delay from 0.81 ns (Previous ripple-carry circuit) down to 0.63 ns (this look-ahead circuit). A remarkable point is the homogeneous switching delay for most outputs, as compared to the ripple-carry result which increases with the stage number.

7.6.2 Complex Gate Implementation

Most logic cells involved in the construction of the carry look-ahead adder are conventional AND and XOR gates. However, the internal carry signals *C2*, *C3* and *C4* are combinations of AND and OR operators that are perfect candidates for complex gate implementation.

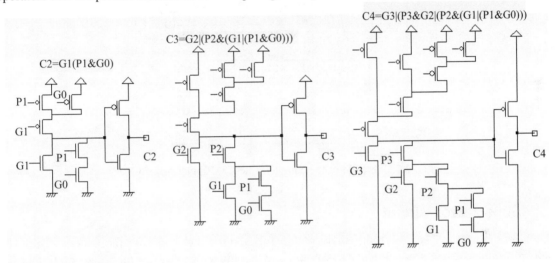

Fig. 7.36 The complex gates used in the carry look-ahead adder (Add4LookAheadCmos.SCH)

The circuit *c3* can be generated by the CMOS cell compiler, through the command **Compile → Compile One Line**. The complex gate description is shown in Figure 7.37. The layout generated from this description appears in Figure 7.38. It is interesting to notice that the cell compiler could not implement the MOS devices using a continuous diffusion. Three separate p-diffusion areas have been used for implementing the pMOS devices.

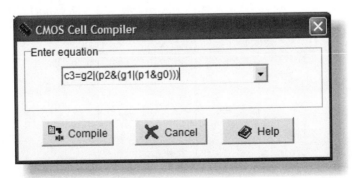

Fig. 7.37 Layout of the complex gate C3 (RippleCarryC3.MSK)

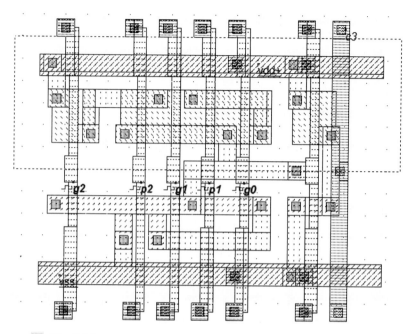

Fig. 7.38   Layout of the complex gate C3 (RippleCarryC3.MSK)

7.7   Substractor Circuit

The substractor circuit can be built easily with a full adder structure as for the adder circuit. The main difference is the needs for a 2's complement circuit which inverts the value of *b*, and the replacement of the half adder by a full adder, as the initial carry must be 1 (Figure 7.38). The logic circuit corresponding to the 4-bit substractor is shown in Figure 7.39. Some examples of substractor results are also listed.

Table 7.5   Substractor result examples

a[0..3]	b[0..3]	s[0..4] = a − b (hexa)	s[0..4] (Decimal)
0	0	00	0
0	1	0F	− 1
1	0	01	1
7	9	0E	− 2
C	6	06	6
F	F	00	0

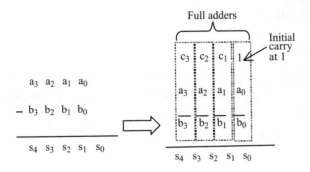

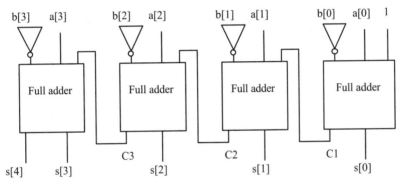

Fig. 7.39 Structure of the 4-bit substractor

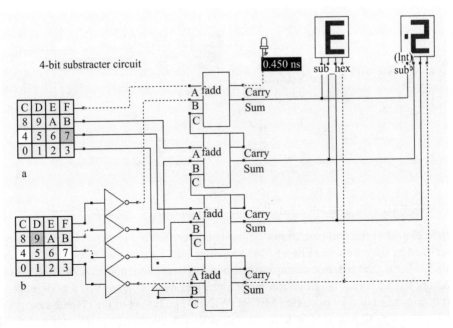

Fig. 7.40 Logic simulation of the 4-bit substractor (Sub4.SCH)

7.8   Comparator Circuit

7.8.1   One-bit Comparator

The truth table and the schematic diagram of the comparator are given in Figure 7.41. The $A = B$ equality is built by using an XNOR gate, and $A>B$, $A<B$ are operators obtained by using inverters and AND gates.

A	B	A>B	A<B	A=B
0	0	0	0	1
0	1	0	1	0
1	0	1	0	0
1	1	0	0	1

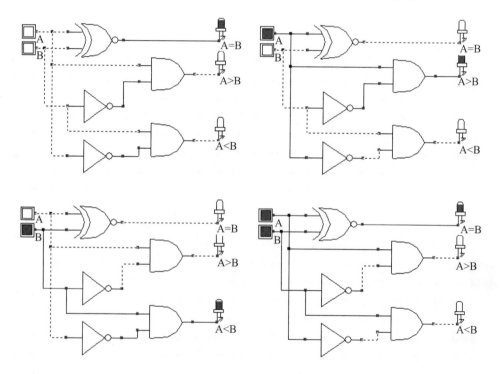

Fig. 7.41   The truth table and schematic diagram of the comparator (CompTest.SCH)

Once the logic circuit of the comparator is designed and verified at the logic level, the conversion into Verilog is realized by the command **File → Make verilog File**. During the conversion, the node name of some signals has been changed. For example, the node $A<B$ has been modified into AiB. This is because some characters have another signification in Verilog, and cannot be part of a node name. Then, the Verilog text is converted into layout using MICROWIND. The layout of the comparator circuit is given in Figure 7.42. The XNOR gate is located on the left side of the design. The inverter and NOR gates are on the right side.

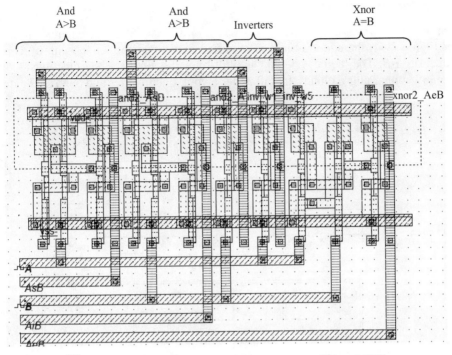

Fig. 7.42 Implementation of the comparator (Comp.MSK)

The simulation of the comparator is given in Figure 7.43. After the initialization, $A = B$ rises to 1. The clocks A and B produce the combinations 00, 01, 10 and 11.

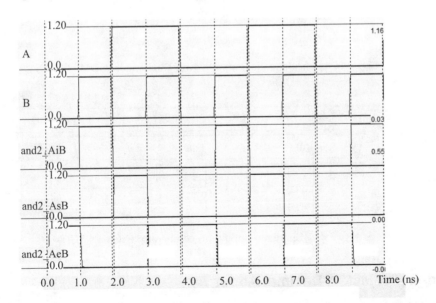

Fig. 7.43 Simulation of the comparator (COMP.MSK file)

7.8.2 n-bit Comparator

An efficient technique for n-bit comparison is based on the use of adder circuits. In the schematic diagram of Figure 7.44, the adder system is modified into a comparison system, by computing the Boolean function *A-B*. The 'equal' operator is built by using simple AND functions. Consequently, each elementary comparator includes the full adder layout, one inverter and one AND gate.

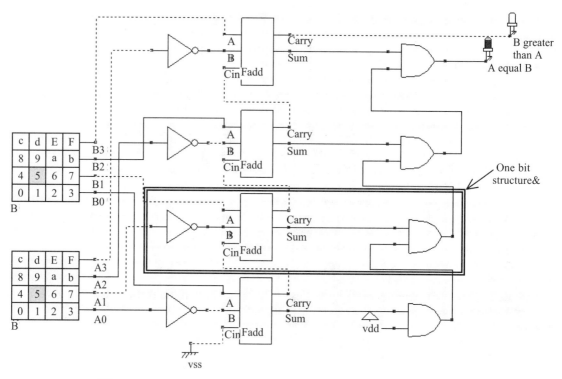

Fig. 7.44   Schematic diagram of a 4-bit comparator (COMP4.SCH)

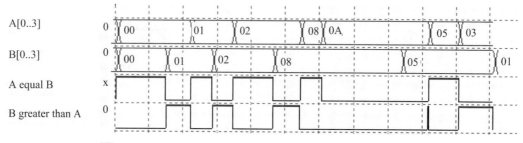

Fig. 7.45   Simulation of a 4-bit comparator (COMP4.SCH)

7.9   Student Project: A Decimal Adder

This paragraph details a student project that performs the addition of two Binary Decimal Coded (BCD) numbers, X and Y. The numbers range from 0 to 9, and the result (between 0 and 18) is visualized on hexadecimal displays. The specification of this project is described in the schematic diagram of Figure 7.46.

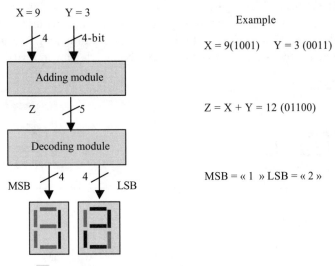

Example

X = 9(1001)   Y = 3 (0011)

Z = X + Y = 12 (01100)

MSB = « 1 » LSB = « 2 »

Fig. 7.46   Structure of the decimal adder project

7.9.1   4-BIT ADDER

Four 1-bit adders linked in cascade construct the 4-bit adder. You may use the adder symbol created previously (Fadd.SYM). The 4-bit adder from Figure 7.21 may be re-grouped into a new user symbol Add4.SYM, which will be re-used in the project.

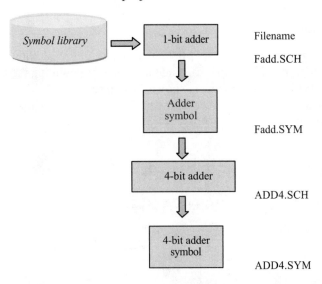

Fig. 7.46   Hierarchical construction of the 4-bit adder symbol

Use the command **File → Schema to New Symbol** to transfer the 4-bit adder schema into a user symbol. The schematic diagram is analyzed and a symbol is proposed by re-grouping all declared inputs and outputs. The button, clock, pulse and keyboard symbols are considered as inputs. The led and displays are considered as output. The user symbol with a name corresponding to the schematic diagram name, with the '.SYM' appendix, is stored in the current directory. In Figure 7.48, the symbol ADD4.SYM is connected to two keyboards and one hexadecimal display to verify the correct behaviour of the 4-bit adder.

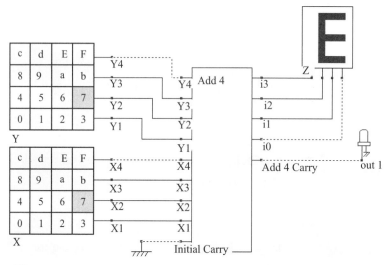

Fig. 7.48   Validation of the 4-bit adder symbol (ADD4Test.SCH)

7.9.2   Decoder Module

The objective of the decoding module is to split the binary result of the addition into two BCD codes, one representing the tenth bit ranging from 0 to 1, and the other representing the unit bit ranging from 0 to 9. The principle of the decoding circuit is shown in Figure 7.49. Firstly, the result $Z = X+Y$ is passed through a comparator. If $Z<10$, the result is sent directly to the visualizing module. If not, the result is adjusted by subtracting 10.

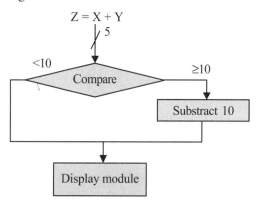

Z = X + Y (Hex Binary)		SupOrEqualTo10	Less Than 10
0	00000	0	1
1	00001	0	1
2	00010	0	1
3	00011	0	1
4	00100	0	1
5	00101	0	1
6	00110	0	1
7	00111	0	1
8	01000	0	1
9	01001	0	1
A	01010	1	0
B	01011	1	0
	...		
1E	11110	1	0

Fig. 7.49   Principles and truth table of the decoder module

7.9.3   Compare To 10

The function *SupOrEqualTo10* may be written by using the following Boolean equation. One possible implementation is reported in Figure 7.50. The positive logic was used for clarity, though negative logic is a better choice from the delay and power consumption points of view.

$$SupOrEqualTo10 = Z4|(Z3 \& Z2)|(Z3 \& Z1) \qquad \text{(Equation 7.13)}$$

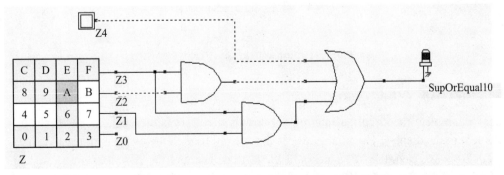

Fig. 7.50   A possible implementation of the function *SupOrEqualto10* (SupEqu10.SCH)

7.9.4   Substract 10

The substractor module is enabled when the result $Z = X + Y$ is greater then or equal to 10. In that case, a substract-by-10 operation is performed. The substractor circuit is simply an adder with inverted input B, and an input carry set to 1. Consequently, the substractor by 10 can be derived from the 4-bit substractor circuit, as shown in Figure 7.51. A **sub10.SYM** symbol is then created, with Z as inputs and LSB (least significant bit) as outputs.

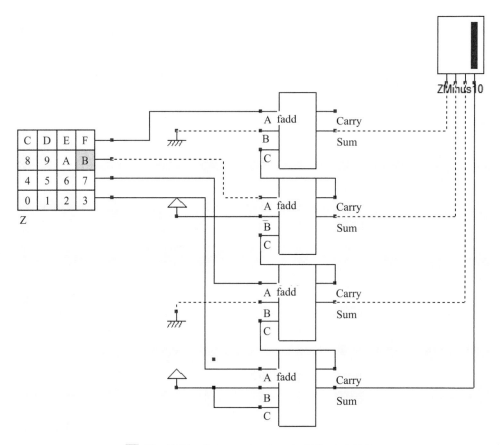

Fig. 7.51   Substraction by 10 (SUB10.SCH)

7.9.5   Final BCD Adder

In order to complete the circuit, a multiplexor circuit decides whether the Z result or the *ZMinus10* result is sent to the display. The multiplexor is made from four elementary multiplexor devices. When *SupOrEqu10* is asserted, the "1" appears in the left display, and the *ZMinus10* result appears in the right display. When *SupOrEqu10* is 0, a "0" appears in the left display and Z appears in the right display. The final circuit is shown in Figure 7.52.

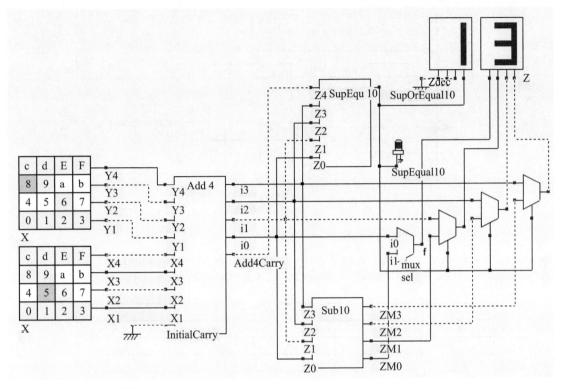

Fig. 7.52 Final circuit of the BCD adder (AdderBcd.SCH)

7.10 Multiplier

Let us illustrate the multiplication process through a small example based on a 4-bit unsigned integer. The basic mechanism is a product $A_i \& B_j$, combined with the addition which involves a carry. The result propagates down to the result.

$$
\begin{array}{llr}
A & 0110 & (6) \\
B & 0111 & (7) \\
\hline
  & 0110 & \\
  & 0110 & \\
  & 0110 & \\
  & 0000 & \cdot \\
\hline
  & 00101010 & (42) \\
\end{array}
$$

The multiplication of integer numbers A and B can be implemented in a parallel way by using elementary binary multiplication circuits [Bellaouar]. Within each multiplication cell, the key idea is to compute the product $P = Ai \& Bi$, and to add the previous sum and previous carry. The next sum is propagated to the bottom, while the next carry is connected to the multiply cell situated on the left side (Figure 7.53).

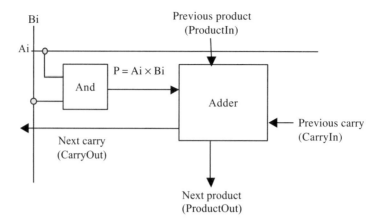

Fig. 7.53   Principles for the elementary multiplication

The elementary multiplication cell should verify the truth table given below. The cell can be made up of a full-adder cell and an AND gate, as shown in the schematic diagram given in Figure 7.54.

Multiplier				
$Ai \times Bi$	ProductIn	CarryIn	CarryOut	ProductOut
0	0	0	0	0
0	0	1	0	1
0	1	0	0	1
0	1	1	1	0
1	0	0	0	1
1	0	1	1	0
1	1	0	1	0
1	1	1	1	1

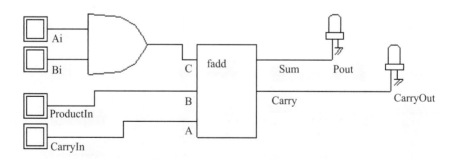

Fig. 7.54   Principles for the elementary multiplication (MUL1.SCH)

A 4×4 bit multiplication is proposed in Figure 7.55. The circuit multiplies input *A* (upper keyboard) with input *B* (lower keyboard) which produces an 8-bit result, *P*. In the logic simulation, the 8-bit display is configured in decimal mode to ease the interpretation of the result.

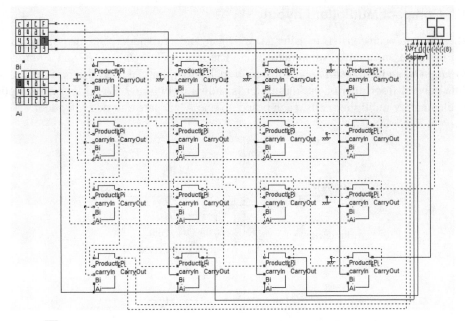

Fig. 7.55   Schematic diagram of the 4 × 4 bit multiplier (Mul44.SCH)

7.10.1   Compiled Multiplier

The 4 × 4 bit multiplier can be translated into layout thanks to the Verilog compiler. By default, the circuit is constructed by placing all basic gates on a single row and then performing the routing. The input and output signals are routed below the active area, and the internal wire routing is performed on the upper part of the active area. Notice that the routing is performed by using metal 3 for horizontal connections and metal 2 for vertical connections. Industrial routing tools take advantage of the other metal layers (metal 4, 5, 6, etc.) to create a more layout-effective implementation.

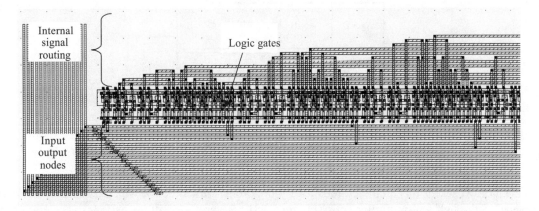

Fig. 7.56   Linear implementation of the 4 × 4 bit multiplier (Mul44.MSK)

7.10.2   A Compact Multiplier Layout

A more efficient approach consists in using an AND gate and a full adder, and placing the interconnects in order to facilitate an iterative implementation. The inputs *CarryIn*, *Ai*, *Bi* and *ProductIn* and the outputs *ProductOut* and *CarryOut* are organized in such a way that the array can be extended to a larger format. The general floor plan of the multiplier is shown in Figure 7.57. An $n \times n$ multiplier would require n^2 elementary multiplier cells, each based on one AND gate and one full adder.

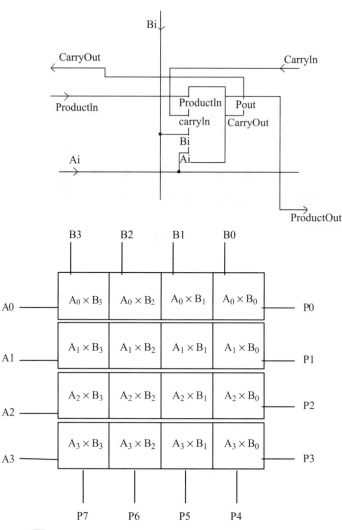

Fig. 7.57   Floor planning of the 4×4 bit multiplier

The main drawback of this architecture is a very important critical delay, which consists of the carry propagation through ten multiplier cells (Figure 7.58). Several algorithms exist for accelerating the multiplication, more or less by a factor of two. Another problem is the design time needed to implement this array structure, and the severe risk of error due to manual arrangement of the cells and interconnects.

In ultra-deep submicron technologies, compact 8×8 bit or 16×16 bit multipliers are proposed as intellectual property (IP) blocks.

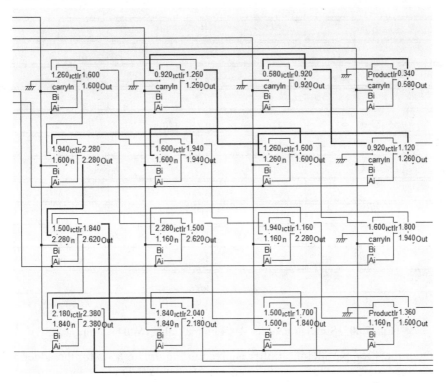

Fig. 7.58 Estimation of the critical delay path (Mul44.SCH)

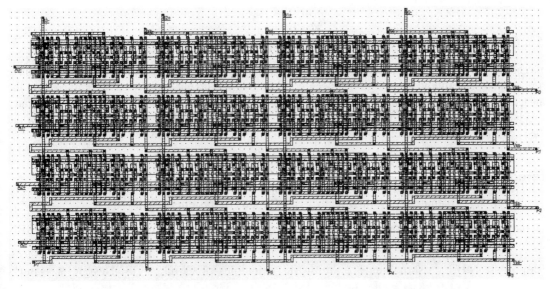

Fig. 7.59 Design of the 4×4 bit multiplier (Mul44Compact.MSK)

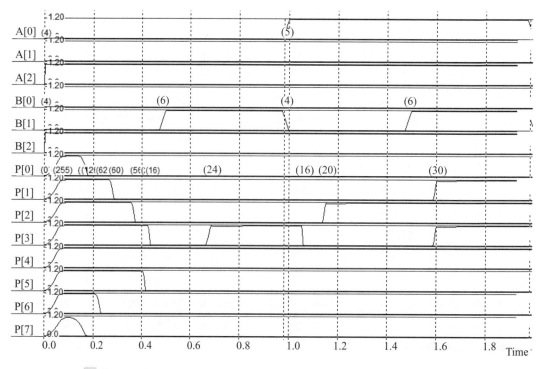

Fig. 7.60 Simulation of the 4 × 4 bit multiplier (Mul44Compact.MSK)

The simulation of the 4 × 4 parallel multiplier is given in Figure 7.60. As for the adder, we use signal names with brackets, starting with index 0, to enable MICROWIND to compute the equivalent logic value of the *A*, *B* and *P* signals. Notice that the initialization phase is almost equal to 0.5 ns. At time *t* = 1.0 ns, *A* = 5, *B* = 4, and *P* = 20 after 150 ps delay.

7.11   Conclusion

In this chapter, the design of basic arithmetic gates has been presented. The half adder and full adder have been presented. The full adder design has been conducted at the layout level, with an emphasis on the advantages of manual design against automatic design regarding the silicon area efficiency. The comparator circuits have also been described. A student project concerning the addition in binary-to-decimal format has been described in detail, and an illustration of multiplier circuits has also been given.

References

[Uyemura] John P. Uyemura "Introduction to VLSI Circuits & Systems", Wiley, 2002.

[Belaouar] Elmasry M.I. Bellaouar I. "Low-Power Digital Vlsi Design" Kluwer Academic Pub, ISBN 0-792-39587-5.

[Weste] Weste N.H.E., Eshraghian K., Smith M. "Principles of CMOS VLSI Design: A systems Perspective with VERILOG/VHDL Manual", ISBN 0-201-73389-7, Addison-Wesley, 2000.

[John] Michael John, Sebastien Smith, "Application Specific Integrate Circuits", Addison-Wesley, 1997, ISBN 0-201-50022-1.

EXERCISES

1. What is the 16 bit-data equivalent to –3? What is the 32-bit data equivalent to 10^6?

2. Design a half adder by using the AOI technique described in Chapter 6.

3. Implement the carry look-ahead adder of Figure 7.33, by using the CMOS cell compiler of MICROWIND. Do you confirm the switching speed gains forecast by logic simulation?

4. Compile into layout the binary-decimal-coded adder described in section 7.10. What is the average power consumption per MHz? From the analysis of the logic circuitry, which part could be re-designed in order to reduce the dynamic power consumption?

5. Are there alternative circuits to reduce the standby current of the binary-decimal-coded adder described in Section 7.10?

Sequential Cell Design

This chapter details the structure and behaviour of latches. The RS latch, the D latch and the edge-sensitive register are presented. The counters are also introduced, with an application concerning a 24-hour clock.

8.1 The Elementary Latch

Combinational circuits are able to add, multiply, shift, etc. However, combinational circuits cannot store and retain values. The basis for storing an elementary binary value is called a latch. The simplest CMOS circuit is made from two inverters.

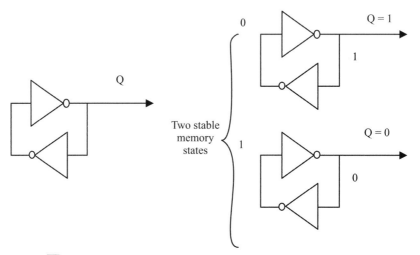

Fig. 8.1 Elementary memory cell based on an inverter loop

Several techniques exist to modify the data inside the loop. One solution consists in adding a pass transistor, and making sure that the information coming from D overcomes the information retained by the feedback inverter. Another solution is based on the NAND gate, which is extensively used in RS and D latch structures. The *Set* and *Reset* signals are active low. A similar design, based on NOR gates, is also shown in Figure 8.2. The *Set* and *Reset* signals are active high in that circuit.

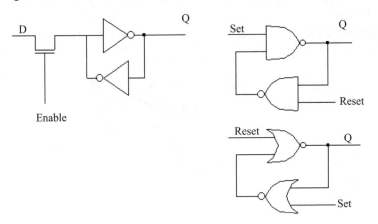

Fig. 8.2   Methods to change the contents of the loop (latches.SCH)

8.2 RS Latch

The RS latch, also called Reset-Set Flip Flop (RS FF), transforms a pulse into a continuous state. The RS latch can be made up of two interconnected NAND gates, as shown in Figure 8.3. In the truth table, we see that the *Reset* and *Set* inputs are active low. The memory state corresponds to *Reset* = *Set* = 1 and is written "Q" to express the fact that the last value stored in the latch is still at the output. The combination *Reset* = *Set* = 0 should not be used, as it means that Q should be reset and set at the same time. In this case, $Q = nQ = 1$.

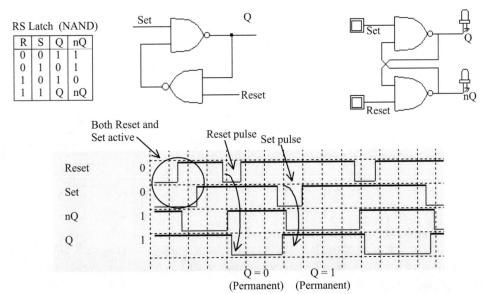

RS Latch (NAND)

R	S	Q	nQ
0	0	1	1
0	1	0	1
1	0	1	0
1	1	Q	nQ

Fig. 8.3   The RS-NAND latch and its typical simulation waveform (RSNand.SCH)

The simultaneous change from *Reset* = *Set* = 0 to *Reset* = *Set* = 1 (Figure 8.4) provokes what is called the meta-stable state, that corresponds to a high frequency oscillation. The meta-stability appears in the chronograms of Figure 8.4 at the simultaneous rise of *Set* and *Reset*. At the layout level, the parasitic oscillation does not really exist, but a parasitic delay effect may be observed, up to 2 or 3 times the nominal gate delay.

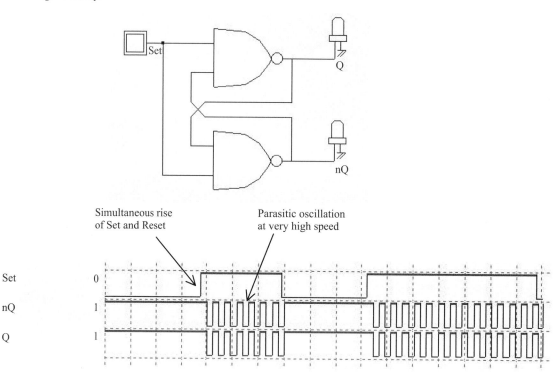

Fig. 8.4 Illustration of meta-stability at the logical level (LatchMetaStable.SCH)

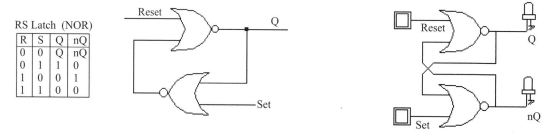

Fig. 8.5 The truth table and schematic diagram of a RS-NOR latch (RsNor.SCH)

An alternative implementation of the RS latch is made from NOR gates (Figure 8.5). In that case, the *Reset* and *Set* inputs are active high. The cell transforms positive pulses into continuous states, as seen in the chronograms of Figure 8.6.

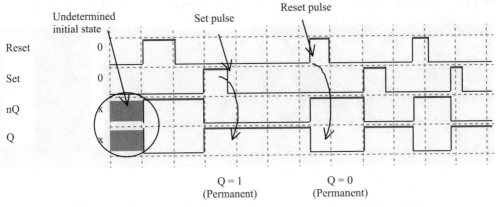

Undetermined
initial state

Set pulse

Reset pulse

Reset

Set

nQ

Q

Q = 1
(Permanent)

Q = 0
(Permanent)

Fig. 8.6   The typical simulation of a RS-NOR latch (RsNor.SCH)

8.2.1 RS Latch Layout

You may create the layout of a RS latch manually. The two NAND gates may share the VDD and VSS supply achieving continuous diffusions. The internal routing may also save routing area, leading to the layout shown in Figure 8.7.

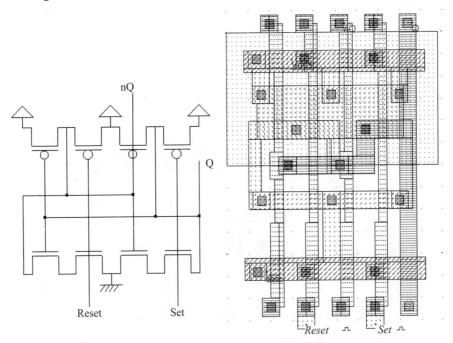

Fig. 8.7   Manual design of the RS-NAND (RsNandManual.MSK)

An alternative approach is to compile the layout directly from the schematic diagram. Using DSCH, create the RS NAND logic circuit, and generate the Verilog text by using the command **File → Make**

Verilog File. In MICROWIND, click on the command **Compile → Compile Verilog File**. Select the Verilog text file corresponding to the RS-NAND gate.

```
module RSNand( Reset,Set,Q,nQ);
 input Reset,Set;
 output Q,nQ;
 nand #(23) nand2(Q,Set,nQ);
 nand #(23) nand2(nQ,Q,Reset);
endmodule

// Simulation parameters
always
#100 Set=~Set;
#200 Reset=~Set;
```

Fig. 8.8 The Verilog description of the RS-NAND gate

Click on **Compile**. When the compiling is complete, the resulting layout appears as shown in Figure 8.8. The NAND implementation of the RS gate is completed. As compared to the layout shown in Figure 8.7, the compiled version is a little larger, due to the fact that the layout conversion tool did not merge the supply connections.

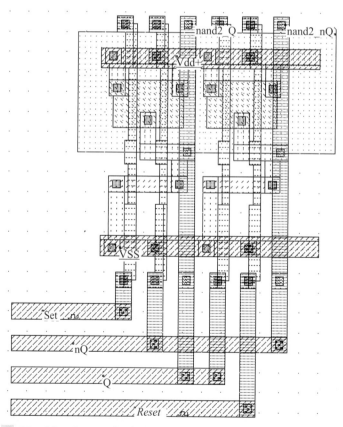

Fig. 8.9 Automatic design of the RS-NAND (RsNand.MSK)

8.2.2   Control of the RS-Latch

If we keep the clock properties assigned by default to *Set* and *Reset*, we get chronograms which are inappropriate to validate the RS latch behaviour, particularly the memory effect. A better approach consists in declaring pulse properties. For example, an active pulse is created on *Reset* followed by an active pulse on *Set*. The pulse parameters are shown in Figure 8.10. In order to obtain a negative pulse, click the small icon placed in the middle of the window ($\prod$).

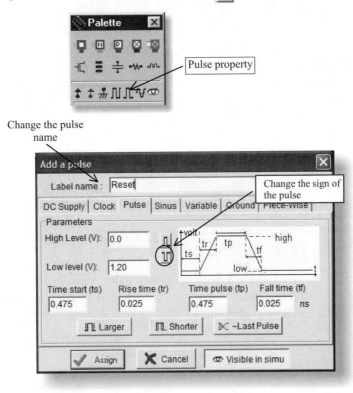

Fig. 8.10   The pulse property in Microwind (RSNand.MSK)

The steps to add a pulse property are as follows. Select the Pulse icon in the palette. Click on the layout of the desired node. The screen shown in Figure 8.10 appears. Change the label name, fix the sign of the pulse (positive by default), and click **Assign**.

Repeat the same procedure to assign a pulse property to node *Set*. The active level of the pulse is automatically delayed. Finally, click on **Simulate → Start Simulation**. The timing diagrams of Figure 8.11 appear. A negative pulse on *Reset* turns *Q* to a stable low state. Notice that when *Reset* goes back to 1, *Q* remains at 0, which demonstrates the ability of the latch to transform a transient pulse into a permanent state. When a negative pulse occurs on *Set*, *Q* goes high, *nQ* goes low. The combination *Reset* = *Set* = 1 corresponds to the memory state.

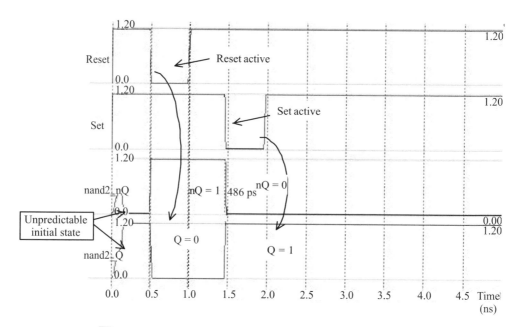

Fig. 8.11 Simulation of the RS-NAND latch (RSNand.MSK)

8.2.3 Minimum Pulse Width

One important characteristic of the RS cell is the minimum pulse width that provokes the *Set* or *Reset* effect. The pulse width is usually named t_w. One possible procedure consists in increasing the pulse width until the output Q is set to its proper value.

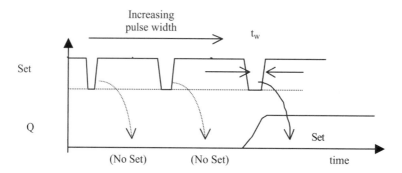

Fig. 8.12 Definition of the minimum pulse width t_w that sets the latch correctly

In order to conduct this analysis, we program the input *Set* as a Piece-Wise-Linear signal (PWL). This property is derived from the pulse property, but not limited to one single shot. The signal is described using pairs of (time, voltage) information, listed in an array. In order to access the PWL property, click the Pulse first, then change the selection in the property window.

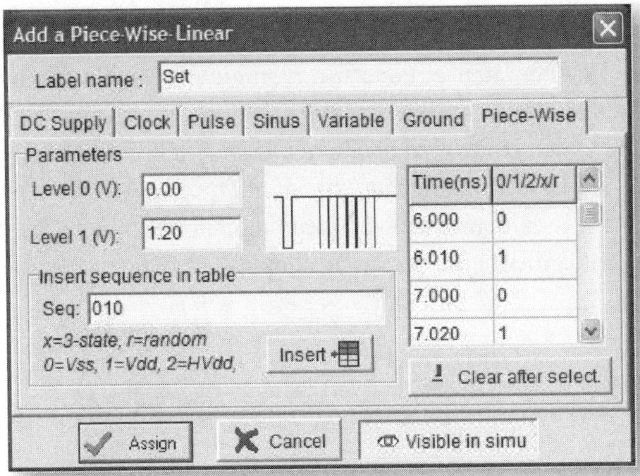

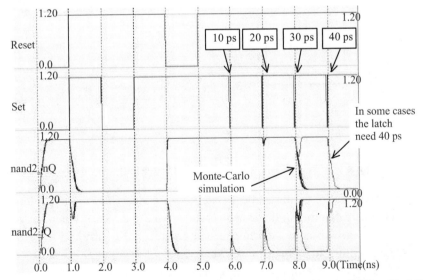

Fig. 8.13   Setting a series of negative pulses with increased width, step 10 ps on the Set signal using a Piece-Wise-Linear waveform (RSNandTw.MSK)

For this study, the latch outputs should be connected to a load in order to conduct the simulations in a realistic environment. A 10 fF virtual capacitor is connected to Q and nQ. In Figure 8.14, the pulse width which leads to a correct setting of the output Q is approximately 40 ps. We used the Monte-Carlo simulation mode (**Simulation → Simulation Parameters**) to use a random set of technological parameters which accounts for the process variations. Depending on the process parameters, the result is slightly changed. The parameter tw is usually specified in the data sheet of the RS latch, and more generally, in that of all latches. Notice that this value is strongly dependent on the loading conditions. We use the BSIM4 model for an accurate analog simulation. In that case, however, the results are similar with model Level 3.

Fig. 8.14   Finding the minimum pulse width of the RSNand latch (RSNandTw.MSK)

8.3 D Latch

The vast majority of CMOS integrated circuits use a single clock to synchronize the sequences of operations. This technique is called synchronous design, and is very popular as it allows automated and safe design. A well known *D* latch controlled by a clock is shown in Figure 8.15. When the clock input is high, the latch output *Q* follows the changes of the input *D*. The latch is transparent. Now, when the clock input goes low, the latch is in memory mode, meaning that it produces the value stored in the loop at the output *Q*. The latch holds the memory state as long as the two inverters are supplied.

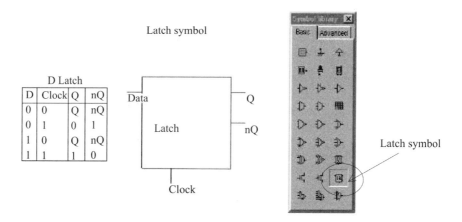

Latch symbol

D Latch			
D	Clock	Q	nQ
0	0	Q	nQ
0	1	0	1
1	0	Q	nQ
1	1	1	0

Fig. 8.15 The truth-table and symbol of the D latch

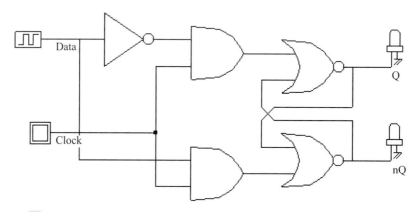

Fig. 8.16 The schematic diagram of a D latch (File DLATCH.SCH)

The truth table and the symbol of the static D latch, also called Static D-Flip-Flop, are shown in Figure 8.15. The schematic diagram of the D latch is shown in Figure 8.16. When performing the logic simulation, *Q* and *nQ* start with an undetermined state (appearing in gray in Figure 8.17). Once the clock is active, *Q* and *nQ* turn to a deterministic state, as the data input *D* is transferred to *Q*, and its opposite to *nQ*. When the clock returns to level 0, the latch keeps its last value.

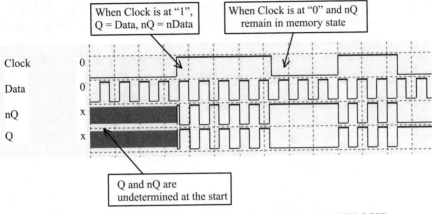

When Clock is at "1", Q = Data, nQ = nData

When Clock is at "0" and nQ remain in memory state

Clock 0

Data 0

nQ x

Q x

Q and nQ are undetermined at the start

Fig. 8.17 Logic simulation of the D latch (File DLATCH.SCH)

8.3.1 A Complex Gate Implementation of the D Latch

In order to obtain a compact design, complex gates should be used rather than discrete AND cells which require a NAND gate with an inverter. The two circuits are compared in Figure 8.18: the direct implementation uses 22 MOS devices, while the complex gate one uses only 14. The complex gate solution also leads to shorter propagation delay. Notice the description of the complex function using the expression:

$$S = \sim ((a \ \& \ b) \mid c) \qquad \text{(Equation 8.1)}$$

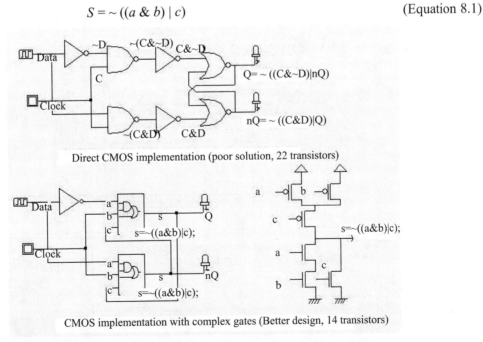

Direct CMOS implementation (poor solution, 22 transistors)

CMOS implementation with complex gates (Better design, 14 transistors)

Fig. 8.18 CMOS implementation of the D latch (Dlatch.SCH)

Generate the Verilog text by using the command **File → Make Verilog File**. In MICROWIND, click on the command **Compile → Compile Verilog File**. In the example given in Figure 8.19, the file 'DlatchCompile.TXT' contains the D latch structural description, where we recognize the two complex gates and the inverter. Also included in the Verilog text are the declarations of the control signals.

```
module DLatchCompile( Data,Clock,Q,nQ);
 input Data,Clock;
 output Q,nQ;
 assign      Q=~((w1&Clock)|nQ);
 assign      nQ=~((Data&Clock)|Q);
 not inv(w1,Data);
endmodule


// Simulation parameters in Verilog Format
always
#1000 Data=~Data;
#1000 Clock=~Clock;
```

Fig. 8.19   Generating a Verilog file corresponding to the D latch (File DLatchCompile.TXT)

When the compiling is complete, the resulting layout appears as shown in Figure 8.20. The layout of the D latch includes, from left to right, the two complex gates and the inverter. The internal wire is routed on the top of the cell, all I/Os being routed on the bottom.

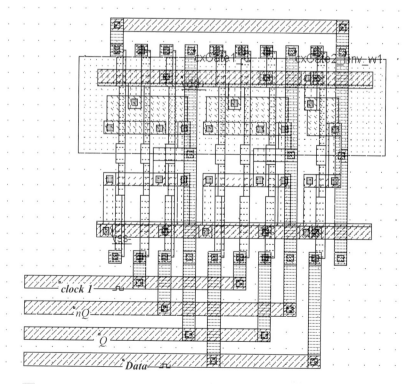

Fig. 8.20   Compiling of the D latch (File DLatchCompile.MSK)

The simulation is shown in Figure 8.21. The default clocks assigned during compilation have been modified. The parameters of the clock *Data* have been changed to avoid the perfect synchronization with *clock1*, in order to watch several situations in one single simulation. When *clock1* is asserted, the logic information contained in *Data* is transferred to *Q*, and its inverted value to *nQ*. When *clock1* falls down to 0, *Q* and *nQ* stay in memory state.

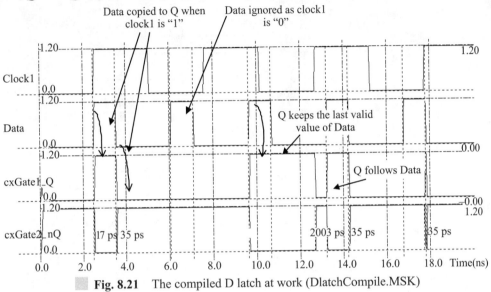

Fig. 8.21 The compiled D latch at work (DlatchCompile.MSK)

8.3.2 Timing Analysis

When the latch is transparent, the delay between the change of *D* and the change of *Q* is named t_{PD} (Figure 8.22). The outputs *Q* and *nQ* are usually loaded with a 10 fF capacitance to account for the connection to other logic gates through metal interconnects. The delay between the rise of *D* and the rise of *Q* may differ from the delay evaluated between the fall of *D* and the fall of *Q*.

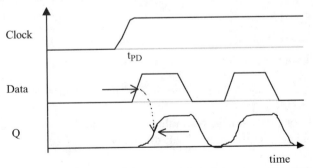

Fig. 8.22   Definition of the propagation delay t_{PD} within D latch

The minimum pulse width for the clock to pass the input *Data* to the output *Q* properly is labelled t_W. An illustration of the test set-up to characterize t_W is given in Figure 8.23. We change the clock property into a pulse property and increase the width of the clock pulse step-by-step, until the data passes to the output.

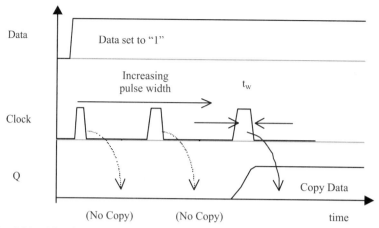

Fig. 8.23   Simulation setup to characterize the minimum clock pulse of the D latch

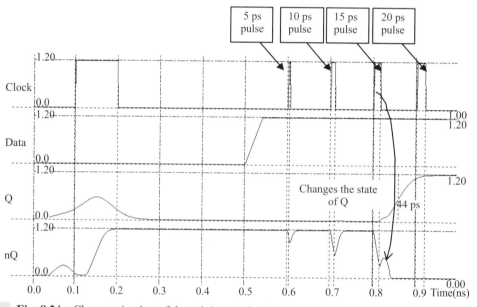

Fig. 8.24   Characterization of the minimum clock pulse of the D latch (DlatchLevel.MSK)

We find that a 15 ps pulse is required to enable the *Data* information to be transferred to *Q*. In this design, with a 10 fF parasitic load, a safe value for t_W would be 20 ps. However, the latch should be tested in extreme conditions (worst case technological parameters, worst case temperature), and the largest value of t_W should be kept. In Figure 8.24, the delay t_{PD} between the rise of *D* and the rise of *Q* is around 45 ps.

8.3.3   Compact D Latch Layout

A very compact design of the D latch is obtained by using the original two-inverter memory loop and by adding the minimum hardware to change its state: one n-channel pass transistor to inject the new

Data, and one p-channel to perform the feedback, or to suppress the loop effect when writing a new data. The resulting schematic diagram is shown in Figure 8.25. Only six MOS devices are required to construct the D latch. Recall that the complex gate implementation was based on 14 devices.

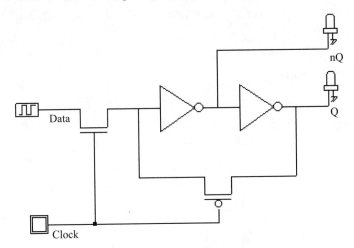

Fig. 8.25 The very compact D latch including two inverters and two pass transistors (Dlatch.SCH)

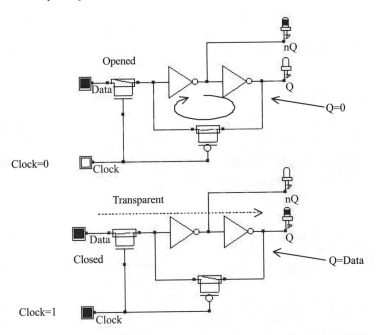

Fig. 8.26 Simulation of the very compact D latch (Dlatch.SCH)

An illustration of the latch at work is given in Figure 8.26. When the clock is at 0 (upper configuration), the loop is active thanks to the pMOS, and the *Data* information is isolated from the loop. When the clock is at 1 (lower configuration), the latch is transparent, and the loop is broken by the pMOS.

Usually, latches proposed in cell libraries propose designs based on transmission gates rather than on single pass transistors. The single pass transistor approach leads to more compact designs. However, the voltage amplitude is degraded, so the noise margin is reduced, and the switching speed is slower. Consequently, low power designs would prefer the pass transistor approach, while high speed circuits would use pass transistors.

8.3.4   D Latch Limitations

The main limitation of the D latch is its inadequacy to build shift registers or counters. On a positive level of the clock, the whole series of D latches is transparent to the input data.

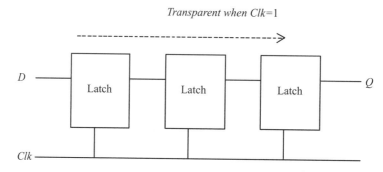

Fig. 8.27   Incorrect shift register design using level-sensitive D latch

8.4   Edge-trigged D Register

In order to overcome the limitation of latches when trying to build registers, we use edge-trigged latches, wherein the information flows from the input D to the output Q only at a rise edge of the clock (Figure 8.28). The latch is commonly known as D-Flip Flop (DFF) or D-register (Dreg) [Uyemura].

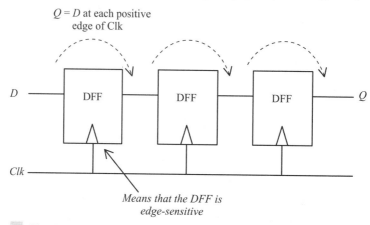

Fig. 8.28   Correct shift register design using edge-trigged flipflop

A well-known edge-trigged flip flop is the JK latch. However, the JK is rarely used due to its complexity. In practice, a more simple version that features the same function with one single input *D* is preferred. This simple type of edge-trigged latch is one of the most widely used cells in integrated circuit design.

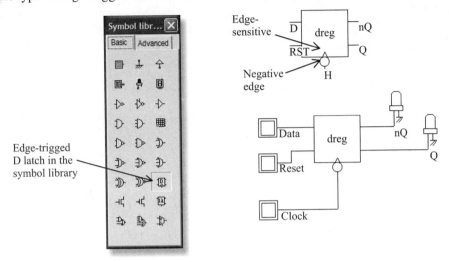

Fig. 8.29   The edge-trigged D-flip flop symbol

The symbol of the edge-trigged D-flip flop may be found in the symbol palette at the location shown in Figure 8.29. The symbol and its typical connection are also given in that figure. Notice the triangle at the clock input that recalls the sensitivity of outputs *Q* and *nQ* to the edge of the clock. The circle indicates that this Dreg cell is sensitive to a fall edge of the clock.

8.4.1   Behavioural Model of the DReg

In DSCH, the Dreg symbol is declared by a behavioural model shown in Figure 8.30. Notice the difference between previous Verilog descriptions, wherein the structure of the cell was explicit (AND, Inv, complex gates, XOR, etc.), and this type of declaration, which is purely a software description. At that stage, we do not know how the cell is constructed. The behavioural models are very attractive for fast simulations, but do not accurately account for the physical effects such as delay or consumption.

In the behavioural description of Table 8.1, the *q* and *nq* output signals are declared as registers. This specific variable is equivalent to a memory cell. In DSCH, a register *reg* is assigned an elementary memory, which may contain three possible values "1", "0" or "x". The line "always @(negedge clk)" is equivalent to a test on a fall edge of the signal *clk* (Figure 8.30). In that case, the instructions between the beginning and the end keywords are executed. The constant "default_delay" has a value which is updated depending on the loading conditions and the technology. Notice the asynchronous reset of the Dreg when *rst* = 1.

Table 8.1 Behavioural model of the edge-sensitive D-flip flop used in DSCH

```
MODULE DReg(d,clk,rst,q,nq);
input d,clk,rst;
output q,nq;
reg q,nq;

always @ (negedge clk)
begin
  #(default_delay) q = d;
  #(default_delay) nq = ~d;
end
if (rst)
begin
  q = 0;
  nq = 1;
end
endmodule
```

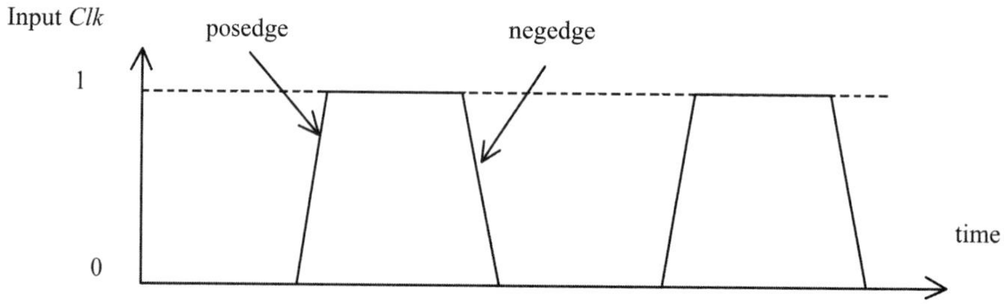

Fig. 8.30 Definition of clock edges in Verilog

8.4.2 Schematic Diagram of the DReg

One very compact implementation of the edge-trigged Dreg is reported below. For clarity, the *Reset* circuit has been removed. The schematic diagram is based on inverters and pass transistors. It is constructed from two memory loop circuits in series. The cell structure includes a master memory cell (left) and slave memory cell (right).

In Figure 8.32, *clock* is high, the master latch is updated to a new value of the input *D*. The slave latch produces the previous value of *D* on the output *Q*. When *clock* goes down, the master latch turns to memory state. The slave circuit is updated. The change of the clock from 1 to 0 is the active edge of the clock. This type of latch is a negative edge flip flop.

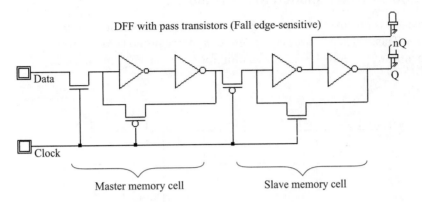

Fig. 8.31 A compact implementation of the edge-sensitive Dreg (Dreg.SCH)

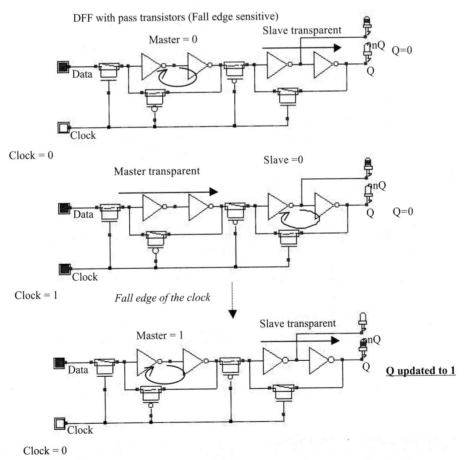

Fig. 8.32 The edge-trigged latch and its logic simulation (Dreg.MSK)

8.4.3 Adding a Reset Function to the DReg

The reset function is obtained by a direct ground connection of the master and slave memories, using nMOS devices. This added circuit is equivalent to an asynchronous *Reset*, which means that *Q* will be reset to 0 when *Reset* is set to 1, without waiting for an active edge of the clock.

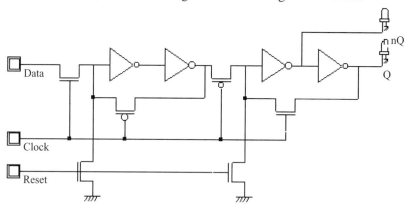

Fig. 8.33 Reset function added to the DReg (Dreg.MSK)

8.4.4 Timing Analysis

Three delay parameters are important in the case of Dreg cells (Figure 8.34): the set-up time t_{SU}, between a valid change of *D* and the active edge of *Clock*, the hold time t_H, between the active edge of *Clock* and a change of *Data*, and the propagation delay t_{PD}, between the active edge of *Clock* and an updated set-up of *Q*.

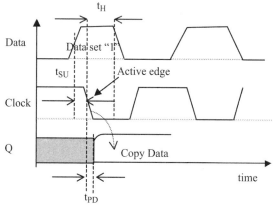

Fig. 8.34 Important timing parameters in the DReg cell

8.4.5 Implementation of the DReg

We create a schematic diagram including the register symbol proposed in the symbol palette of DSCH. Three buttons are added to the circuit to control the inputs *Data*, *Reset* and *Clock*, as shown in

Figure 8.35. Using the command **Make Verilog File**, we create the corresponding Verilog description. As can be seen, the register is built up from one single call to the primitive "dreg".

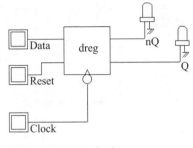

```
module dregCompile( Clock,Reset,Data,Q,nQ);
 input Clock,Reset,Data;
 output  Q,nQ;
 dreg #(19) dreg1(Q,nQ,Data,Reset,Clock);
endmodule

// Simulation parameters in Verilog Format
always
#1000 Clock=~Clock;
#2000 Reset=~Reset;
#3000 Data=~Data;
```

Fig. 8.35   The Verilog text corresponding to the Dreg circuit (DregCompile.SCH)

In Microwind, the Verilog text is converted into layout as shown in Figure 8.36. The *dreg* primitive is converted into a complex structure including the master and slave memories, each with two inverters and two pass transistors, as well as one pass transistor for the *Reset* function.

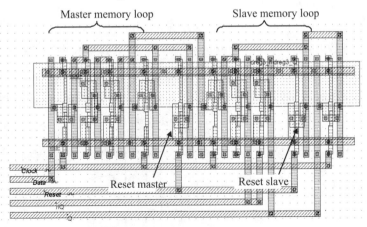

Fig. 8.36   Compiling the Dreg with Microwind (DregCompile.MSK)

For testing the Dreg, the *Reset* signal is activated twice, at the beginning and later, using a piece-wise-linear property. The *Clock* signal has a 2 ns period. The data *D* is not synchronized with *Clock*, in order to observe various behaviours of the register.

The simulation of the edge-trigged D-register is shown in Figure 8.37. The signals Q and nQ always act in opposite. When *Reset* is asserted, the output Q is 0, nQ is 1. When *Reset* is not active, Q takes the value of D at a fall edge of the clock. For all other cases, Q and nQ remain in memory state. The latch is thus sensitive to the fall edge of the clock.

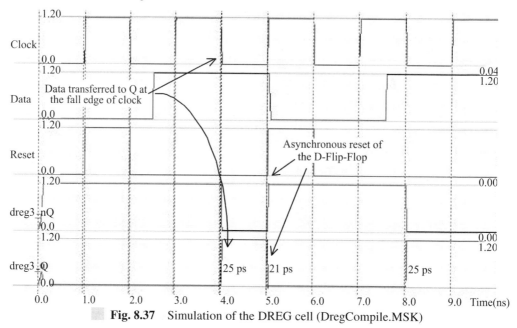

Fig. 8.37 Simulation of the DREG cell (DregCompile.MSK)

8.4.6 Improved Dreg Design

In cell libraries, most latch designs are based on transmission gates rather than single pass transistors. The single pass transistor approach leads to more compact designs, dissipating less power. However, the voltage amplitude is degraded, so the noise margin is reduced, and the switching speed is slower. The *Dreg* cell implemented with transmission gates requires one inverter to generate the *nClock* signal (Figure 8.38). Notice the NAND gate for the *Reset* function which is active on a low level of *nReset*.

DReg with transmission gates and Reset

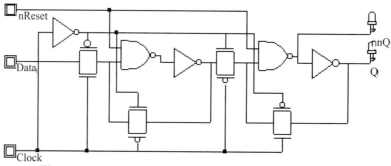

Fig. 8.38 Dreg with transmission gates (DregTgate.SCH)

In the case of long distance routing between the Dreg output and its cell destination, the output node Q is rarely linked to the storage node directly (Figure 8.39). This is crucial for reliability in case of coupling noise, as introduced in Chapter 5. Around one millimeter of coupled interconnects is sufficient to provoke crosstalk noise approaching the commutation point of the inverters. With a latch design with direct feedback from Q to the storage loop (Figure 8.39(a)), a strong noise injected by capacitance coupling in an adjacent line may change the state of the register. If the signal Q is isolated from the storage loop (Figure 8.39(b)), the noise vanishes and the memory state is not altered. A modified circuit with isolated outputs is proposed in Figure 8.40. Notice the supplementary inverters that increase the propagation delay of the circuit and increase the power consumption.

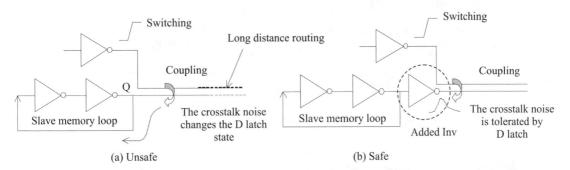

Fig. 8.39   The storage loop directly connected to the output node Q may be affected by a coupling noise.

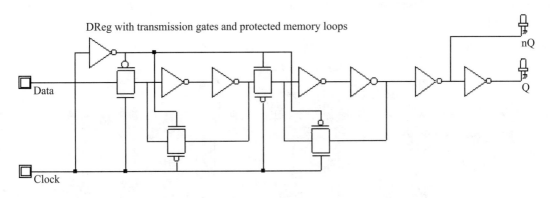

Fig. 8.40   The isolated Dreg circuit (DregTgate.SCH)

8.5   Clock Divider

The 1-bit counter is able to produce a signal featuring half the frequency of a clock. The most simple implementation consists of a *Dreg* where the output *nQ* is connected to *D*, as shown in Figure 8.41. In the logic simulation, the input *Clock* changes the state of *ClockDiv2* at each fall edge. The *Reset* signal is active high, and sticks the output to 0.

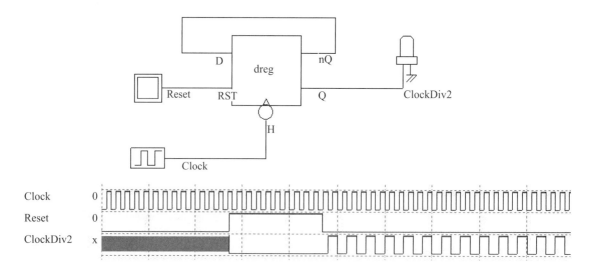

Fig. 8.41   Logic simulation of the divider-by-two (ClockDiv2.SCH)

8.5.1   Maximum Operating Frequency

The most important parameter to be characterized in the clock divider is the maximum frequency f_{max} up to which the cell divides properly. Let us extract this frequency f_{max} by using the compiled version of the clock divider.

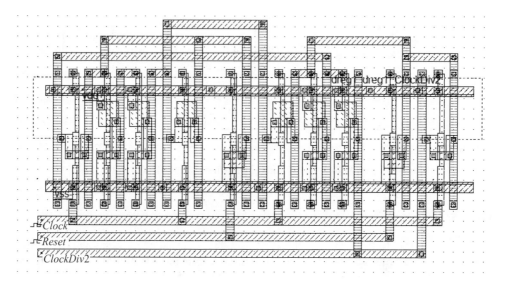

Fig. 8.42   The compiled layout of the clock divider (ClockDiv2.MSK)

Firstly, the default clock assigned to the signal *Reset* should be changed into a pulse property, to avoid a cyclic reset of the divider. Secondly, the input *Clock* is reprogrammed as a piece-wise-linear where the input frequency starts from 1 GHz and rises up to 10 GHz.

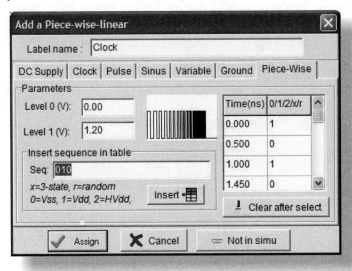

Fig. 8.43 The piece-wise-linear description of the clock with increased frequency (ClockDiv2.MSK)

The recommended simulation mode is **Frequency vs. time**. The frequency variation of the output node *ClockDiv2* appears in the upper part of Figure 8.44. It can be seen that the divider circuit works correctly up to 2.5 GHz, that is, an input frequency of 5.0 GHz.

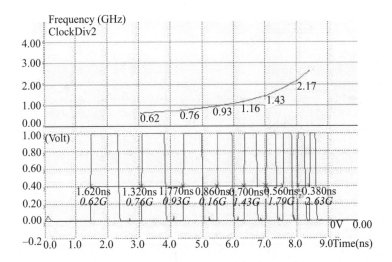

Fig. 8.44 Analog simulation of the divider-by-two for the evaluation of the maximum operating frequency (ClockDiv2.MSK)

8.5.2 Asynchronous Counter 0..15

The *Dreg* cell connected as a one-stage counter cell may be cascaded to create a larger counter circuit. The clocking of each stage is simply carried out by the previous counter stage output, to form an asynchronous counter circuit. The 4-stage binary counter displays numbers from 0 to 15, using a chain of four *Dreg* cells, as illustrated in Figure 8.45. Note the *Reset* signal common to all stages.

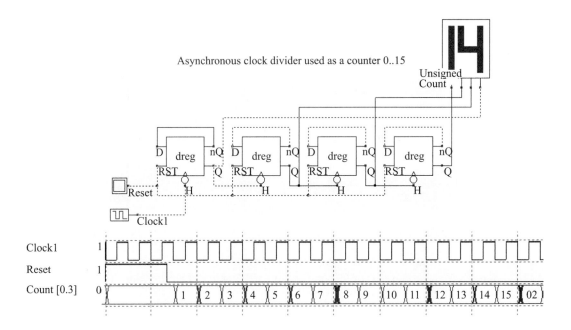

Fig. 8.45 The asynchronous counter circuit principles and logic simulation (CountA16.SCH)

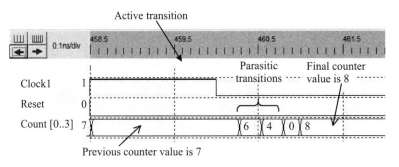

Fig. 8.46 Intermediate counter values during asynchronous switching (CountA16.SCH)

The asynchronous counter has several intermediate values during switching due to cascaded delays in the chain. Before the last stage is correctly settled, several parasitic values appear in the chronograms. In Figure 8.47, a divide-by-13 counter is proposed. It uses an AND gate connected to the *Reset* control, to provoke a general reset of the *Dreg* cells once the desired number has been attained.

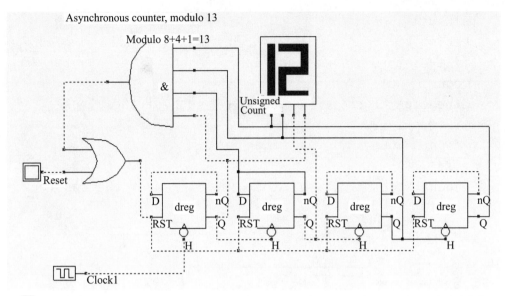

Fig. 8.47   Counter 0..12 using an evaluation circuit for a forced Reset (CountA13.SCH)

An undesired *Reset* may be provoked by the AND gate in the instability time interval during which the counter chain changes its value. This is why asynchronous counters are dangerous to use, and synchronous counters are preferred.

8.6   Synchronous Counters

The main difference between asynchronous and synchronous counters is the clock connection. In the case of a synchronous counter, the signal *clock* is shared by all stages of the counter. The synchronous counter shown in Figure 8.48 [Weste] uses one adder and one *Dreg* for each stage. The counter performs up and down counting. The *Reset* signal is shared by all *Dreg* cells.

In the simulation, we see that the counter increments the output when Up/Down is at zero. Otherwise, the output is decremented.

8.6.1   Counter Model

The behavioural model of a counter is given in Table 8.2. The model description uses the Verilog format, which is very similar to the internal format used to describe models in DSCH.

Table 8.2   The behavioural model of the counter

```
module countUp(clock,reset,count);
  input clock,reset;
  output [3:0] count;
  reg [3:0] count;    // declares a register type for Count
```

```
always
begin
if (reset==0)   // asynchronous Reset of the counter
 count = 0;
@(negedge clock);  // At a fall edge of the clock
if (count==9)
 count = 0;         // After 9, 0
else
 count = count+1;  // otherwise increment the counter
end
endmodule
```

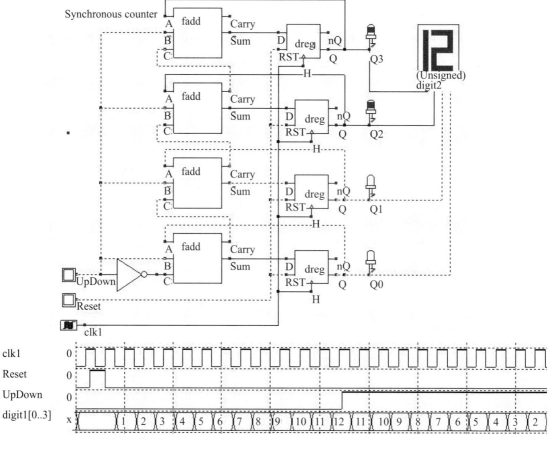

Fig. 8.48 Design of a synchronous Up/Down counter using a full adder and a Dreg (CounterUpDown.SCH)

The *count* value is declared as a register, which is equivalent to a variable in a programming language. The count register is assigned immediate values such as 0, or assigned an incremented value, which requires an adder circuit. Notice that Verilog understands the addition "+", the subtraction "−", and several other operators. The value of *count* remains unchanged until a new assignment is executed. The register *count* is changed on a negative edge of the input clock.

8.7   Shift Registers

Shift registers are used in many integrated circuits to convert serial logic data into parallel data. In many cases, serial data links are preferred for communication between blocs, as it reduces the number of pins and the size of buses with a positive impact on the interconnection cost. However, the serial link is slower than the parallel interface. Edge-sensitive *Dreg* cells are very useful for converting serial data stream into parallel data stream. In Verilog, the shift operation symbol is "<<" for shift left and ">>" for shift right.

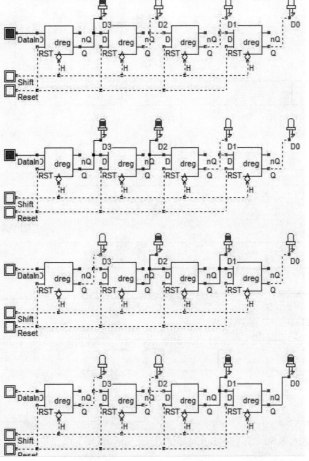

Fig. 8.49   Using Dreg cells to build shift registers (ShiftReg4.SCH)

In the circuit shown in Figure 8.49, four Dreg cells are chained by connecting the *Q* output to the *D* input of the next stage. The serial data enters by the input *DataIn*. At each fall edge of *Clock*, the register copies the input onto the output. In the upper configuration, a '1' has propagated to the first stage. At the next active edge of the clock, it has propagated to the second stage, while a new one passes through the first stage. Then, a zero is presented on *DataIn*. At the end, the four serial information (1,1,0,0) appears in (D0, D1, D2, D3).

8.7.1 Programmable Delay Line

Shift registers may also be used to construct a programmable delay line. The principles of this circuit are to pass the input signal directly, or to delay the information for one period of clock through a *Dreg* cell. The decision circuit is a multiplexor, controlled by the user. The simulation confirms that the input data string *DataIn* is delayed over one, two, three or four clock periods, depending on the number of active *AddDt*.

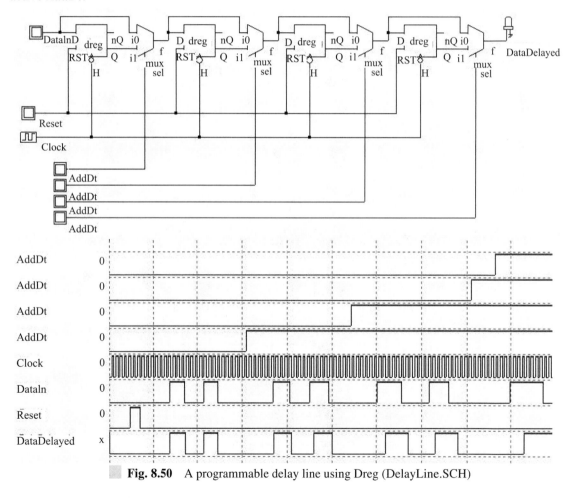

Fig. 8.50   A programmable delay line using Dreg (DelayLine.SCH)

8.8 A 24-hour Clock

In this chapter we describe a circuit to generate a clock that counts hours and minutes. The basic components are described in Figure 8.51. The main clock is divided by 60 to create minutes, which is again divided by 60 to create hours. Several counters are needed for this circuit: a counter up to 10 and 6 for minutes, and a counter up to 24 for hours.

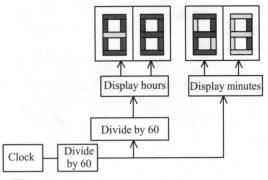

Fig. 8.51 Principles of a 24-hours clock circuit

8.8.1 Basic Cell

A possible implementation is based on synchronous counters. In that case, the basic element, called *T latch*, has two ways of resetting the output *Q*. The synchronous reset (*~Sclear*) is effective on a fall edge of the clock, while the asynchronous *Reset* assigns a zero to *Q* independently of the clock. The synchronous counter makes an extensive use of the synchronous *Reset*, while the asynchronous *Reset* is kept for initializing the whole hardware, and for starting with a determined state.

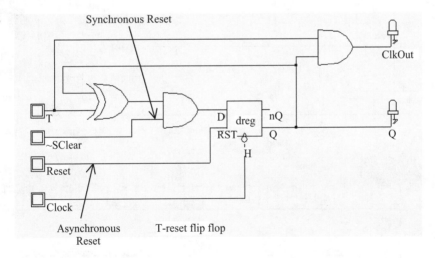

Fig. 8.52 The T-flip flop used for synchronous counters (ClkBascT.SCH)

The T-flip flop architecture is derived from the synchronous counter of Figure 8.48. The full adder circuit is replaced by a half adder as the decrement function is useless. The XOR and AND cells of the half adder circuit may be identified in the T-flip flop design.

8.8.2 Synchronous Counter

Consider the counter design shown in Figure 8.53. Four T-flip flop circuits have been cascaded to create a synchronous counter up to 15. The problem is to build an interrupting circuit which resets the registers once the number 9 is attained. This is done with a NAND gate situated on the right lower side of the schematic diagram. The NAND output is connected to the synchronous *Clear* of the latches. When asserted, it should be held during one complete period of the clock before the circuit is reset. The circuit starts to count if *Enable* is active (high level) and if the asynchronous Reset is inactive (low level). The synchronous clear *~Clear* should also be high. In that condition, the circuits count from 0 to 9 and are then reset. Notice in the chronograms of Figure 8.54 the aspect of *Sup9*, which corresponds to a clean pulse during one period of the clock. This signal will be used to control the next stages of the counters.

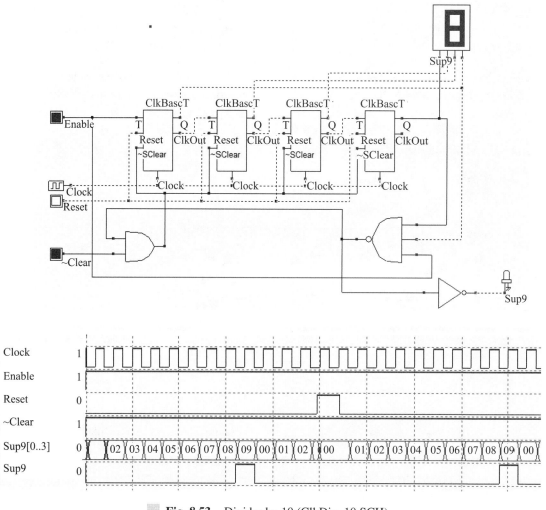

Fig. 8.53   Divider by 10 (ClkDiv_10.SCH)

The synchronous counter shown in Figure 8.54 is used to count from 0 to 5. The structure is very similar to the counter from 0 to 9, except that only three T-flip flop circuits are needed. We combine the two counters to create the minute counter from 0 to 59. The most important connection is between the output *Sup9* and the Enable input of *Clk_Div6*. When *Sup9* is asserted, the counter from 0 to 5 is activated, and the next clock edge will increase the slave counter. Otherwise, only the master counter from 0 to 9 is working.

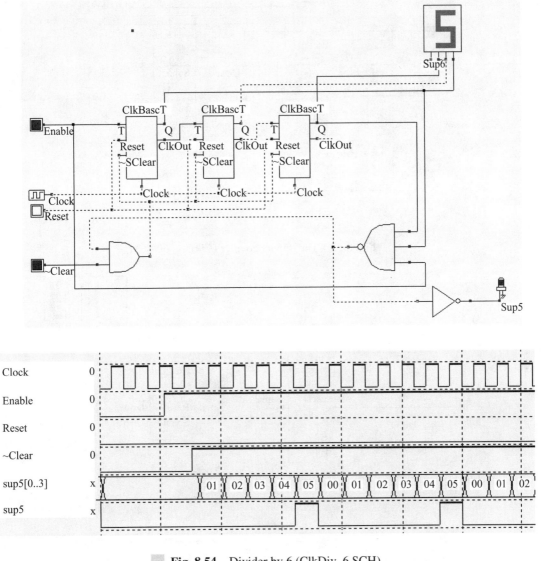

Fig. 8.54 Divider by 6 (ClkDiv_6.SCH)

The circuit for counting hours re-uses the circuits *ClkDiv_10* and needs a new circuit *ClkDiv_3*. The signal *Equ23* is generated when the number 3 appears in the master counter [0..9] and the number 2 appears in the slave counter [0..2] (Figure 8.56).

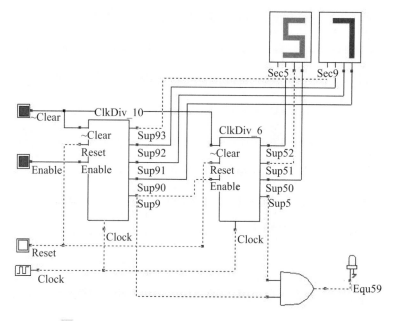

Fig. 8.55 The minute counter (ClkDiv_60.SCH)

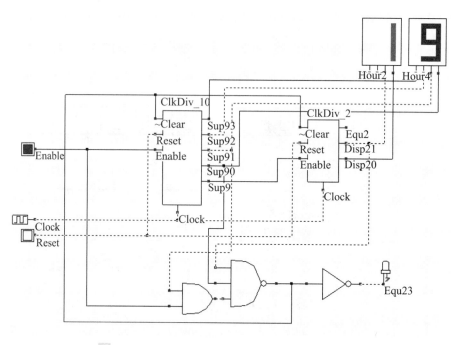

Fig. 8.56 The hour counter (ClkDiv_24.SCH)

The final schematic diagram of the clock is shown in Figure 8.57. The circuit works properly up to a maximum frequency. If we increase the frequency of the main clock, the clock resets erratically. This is due to the cumulated delay, which is the sum of all delay stages. Also notice the buffer on the clock signal, used to delay the fall edge of the clock on the *ClkDiv_24* circuit, as the enable signal is issued from the *ClkDiv_60* circuit through *Equ59* with a significant delay.

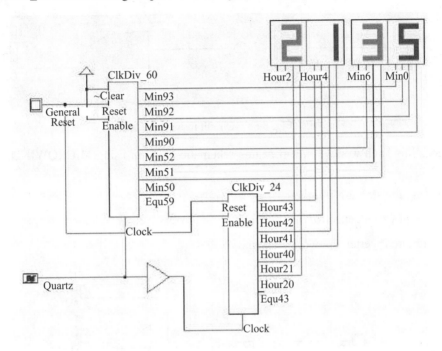

Fig. 8.57 The complete schematic diagram of the 24H00 clock counter (Clk24H00.SCH)

8.9 Conclusion

In this chapter, we have introduced the elementary latch based on two inverters, and the RS latch with its NAND and NOR implementation. The D latch has also been detailed, and the timing performances have been analyzed at the layout level. The edge-trigged register was described from a behavioural and structural point of view. Several versions of this register have been reviewed, and some main applications have been illustrated: the clock divider, the asynchronous counter, the shift register and the programmable delay line. At the end of this chapter, synchronous counter circuits have been presented, with an application to a complete 24-hour clock.

References

John P. Uyemura "Introd to VLSI Circuits and Systems", Wiley, 2002.

EXERCISES

1. Compare the current consumption and switching performances of two versions of D latch: the complex gate implementation and the AND/NOR implementation.

2. Using DSCH, realize the following schematic diagram, and simulate its behaviour with a clock at the input *In*. What is the functionality of that circuit?

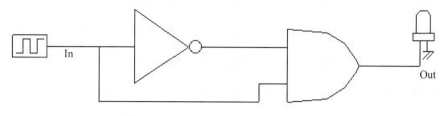

Fig. 8.58   Circuit to be analyzed

3. How does this latch work? Implement this schematic diagram using MICROWIND. What is the effect of:

- Degrading the strength of inverter *Inv1*?

- Adding a physical capacitor on node *n1*?

- Adding more inverters between *Inv1* and *And1*?

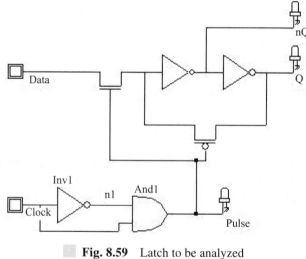

Fig. 8.59   Latch to be analyzed

Analog Cells

This chapter deals with analog basic cells, from the simple resistor and capacitor to the operational amplifier.

9.1 Resistor

An area-efficient resistor available in CMOS process consists of a strip of polysilicon [Hasting]. The resistance between $s1$ and $s2$ is usually counted in a very convenient unit called "ohm per square", noted $\Omega/\square$. The default value polysilicon resistance per square is 10 Ω, which is quite small, but rises to 200 Ω if the salicide material is removed (Figure 9.1).

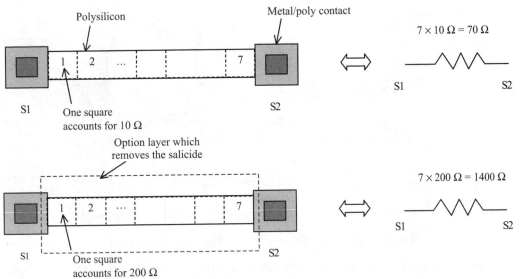

Fig. 9.1   The polysilicon resistance with unsalicide option

In the cross-section shown in Figure 9.2, the salicide material deposited on the upper interface between polysilicon and the oxide creates a metal path for current that reduces the resistance dramatically.

Notice the shallow trench isolation and surrounding oxide that isolate the resistor from the substrate and other conductors, enabling very high voltage biasing (up to 100 V). However, the oxide is a poor thermal conductor which limits the power dissipation of the polysilicon resistor.

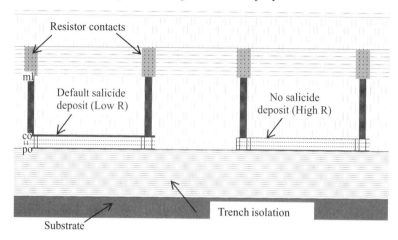

Fig. 9.2   Removing the salicide material to increase the sheet resistance (ResPoly.MSK)

The salicide is part of the default process, and is present at the surface of all polysilicon areas. However, it can be removed thanks to an option layer programmed by a double click in the option layer box, and a tick at "Remove Salicide". In the example shown in Figure 9.3, the default resistance is 33 Ω, and the unsalicide resistance rises to 330 Ω.

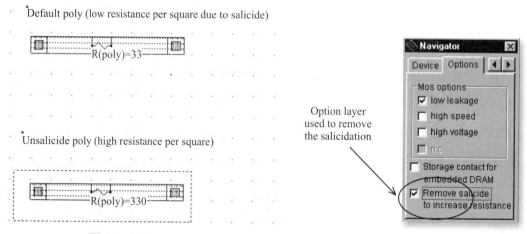

Fig. 9.3   Removing the salicide material thanks to an option layer

Other resistors consist of N+ or P+ diffusions. An interesting feature of the diffusion resistor is the ability to combine a significant resistance value and a diode effect. As illustrated in Figure 9.4, when *V1* goes below 0 V, the P-substrate/N+ diode is turned on and a path is created to the ground. Remember that the P-substrate is usually considered as a common ground reference. The diffusion resistor is also used in input/output protection devices.

The command **Help** → **Design Rules** gives access to the square resistance and unsalicide square resistance of all materials, as shown in Figure 9.4.

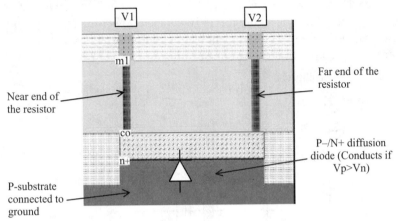

Fig. 9.4   The diffusion resistor combines a resistance effect and a diode (ResDiff.MSK)

Design rules for CMOS 0.12μm - 6 Metal

Design rules and electrical parameters

Layer	Width	Spacing	Surface	Surf capa	Lin capa	Ctk capa	Res	Unsalicid	Thickn	Height	Permitt
	lambda	lambda	lambda2	af/μm2	af/μm	af/μm	ohm	ohm	μm	μm	
via3	2	4	0				2.00/via		0.50	3.30	4.00
metal3	3	4	16	160.00		30.00	0.05/sq	1.00/sq	0.40	2.90	3.10
via2	2	4	0				2.00/via		0.50	2.40	4.00
metal2	3	4	16	180.00		30.00	0.05/sq	1.00/sq	0.40	2.00	3.10
via	2	4	0				2.00/via		0.50	1.50	4.00
metal	3	4	16	200.00		30.00	0.05/sq	1.00/sq	0.40	1.10	3.10
poly	2	3	16				4.00/sq	40.00/sq	0.20		
poly2	2	2	8				4.00/sq	40.00/sq	0.20		
contact	2	4	0				20.00/via		1.10	0.00	4.00
diffn	4	4	16	350.00	100.00		25.00/sq	250.00/sq	0.40	0.00	4.00
diffp	4	4	16	300.00	100.00		30.00/sq	300.00/sq	0.40	0.00	4.00
nwell	10	11	144	250.00			120.00/sq		1.00	0.00	4.00

Default resistance per square

Resistance if unsalicide

✓ OK   Techno: CMOS 0.12μm - 6 Metal loaded from file "default.rul"

Fig. 9.5   Resistance parameters in CMOS 0.12 μm

9.1.1   Process Variations

The process variations have a strong impact on the physical value of the resistor. Most processes specify square resistance within +/–25 per cent [Hastings]. This means that the value of the resistor is linked to a statistical distribution, usually between a minimum-maxium range, and not an exact value. In reality, the spread of resistance is usually less than 10 per cent within a single integrated circuit. However, two different integrated circuits may produce a significantly different resistance distribution. In Figure 9.6, the average resistance $R1$ measured on a test chip n°1 is 5 per cent higher than the target resistance R_{typ}, and the average resistance $R2$ on a test chip n°2 is 10 per cent lower than the typical value R_{typ}. In his process specifications, the integrated circuit manufacturer only warranties that the measured resistance will not be larger than +/–25 per cent of the typical value R_{typ}.

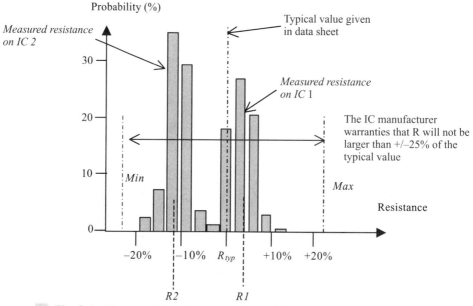

Fig. 9.6   The spread in resistance value and the typ/min/max specifications

The resistor value varies because of lithography and process variations. In the case of the poly resistance, the width, height and doping may vary (Figure 9.7 left). Polysilicon resistors are rarely designed with the minimum 2 lambda width, but rather 4 or 6 lambda, so that the impact of the width variations is smaller. But the equivalent resistance is smaller, meaning less silicon efficiency. As illustrated in Figure 9.7, a variation ΔW of 0.2 λ on both edges results in a 20 per cent variation of the resistance on a 2 λ width resistor, but only a 10 per cent variation for a larger resistor designed with a width of 4 λ.

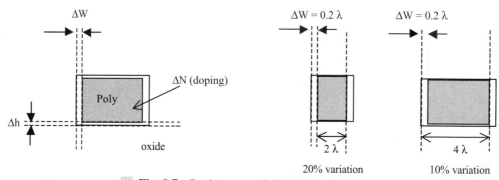

Fig. 9.7   Resistance variations with the process

9.1.2 Resistor Design

There exist efficient techniques to reduce the resistance variation within the same chip. In Figure 9.8, the resistor design on the left upper part is not regular, uses various polysilicon widths, and sometimes uses too narrow conductors. Although the design rules are not violated, the process variations will enlarge the spread of resistance values. In order to minimize the effects of process variations, resistors should:

- Always be laid out with an identical width.

- Use at least twice the minimum design rules.

- Use the same orientation.

- Use dummy resistance. These boxes of poly have no active role. Their role is to have a regular width variation on the active part.

Dog-bone resistors (Figure 9.8) may not pack as densely as serpentine resistors as two metal layers are used and induce supplementary design rules [Baker][Hastings] but are said to be less sensitive to process variations.

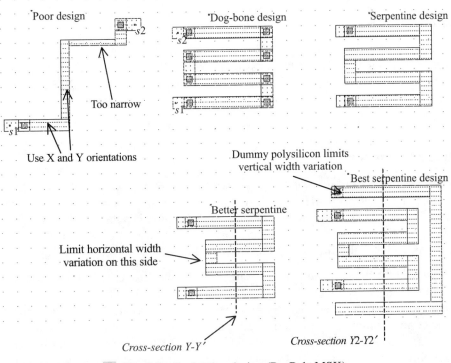

Fig. 9.8 Resistance design (ResPoly.MSK)

The use of dummy devices is an efficient technique to avoid irregularities in resistor elements. In the case of a polysilicon serpentine without dummy devices, the cross-section Y-Y' in Figure 9.9 exhibits important irregularities between bones. This results in an important process dependence and resistance variation. Now, if dummy components are inserted at the bottom and top of the design (Cross-section Y2-Y2′), the irregularities impact the dummy bones, but not the active parts of the resistor, which is much less impacted by the process variations. Typically, the resistance variation between two identical designs within the same chip is around 5 per cent without dummy devices, and less than 1 per cent with dummy devices.

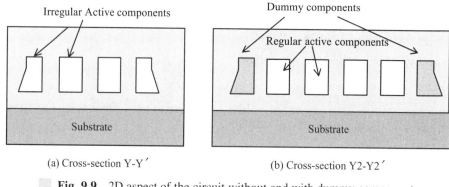

(a) Cross-section Y-Y′   (b) Cross-section Y2-Y2′

Fig. 9.9   2D aspect of the circuit without and with dummy components

9.2 Capacitor

Capacitors are used in analog design to build filters, compensation, decoupling, etc. Ideally, the value of the capacitor should not depend on the bias conditions, so that the filtering effect would be situated at constant frequencies. We describe in this paragraph the diode capacitor, the MOS capacitor, the poly-poly2 and inter-metal capacitor. Some of these capacitors exist between an active node and a fixed voltage.

9.2.1 Diode Capacitor

Diodes in reverse mode exhibit a capacitor behaviour. However, the capacitance value is strongly dependent on the bias conditions [Hastings]. A simple N+ diffusion on a P-substrate is an NP diode, which may be considered as a capacitor as long as the N+ region is polarized at a voltage higher than the P-substrate voltage, which is usually the case as the substrate is grounded (0 V). In 0.12 μm, the capacitance is around 300 aF/μm2 (1 atto-Farad is equal to 10^{-18} Farad).

Let us briefly recall the diode operation. A strong current flows from the P to the N region when the voltage difference is significantly higher that the threshold voltage V_T. For example, the P-/N+ diode is in such conditions when V_P is significantly higher that $V_N + V_T$, where V_T is approximately 0.3 V in 0.12 μm. The current is limited by the serial resistance of the diode, which is of the order of 100 ohm to 1000 ohm, depending on the surface of the diode (Figure 9.10). A very small current flows between the P and the N region as long as $V_P < V_N + V_T$. In such conditions, the diode may be considered as a capacitance. The order of the parasitic current is between the nano-ampere and the pico-ampere (10^{-9} to 10^{-12} A). As the substrate is grounded, V_N is always higher than V_P meaning that the N+/P- combination is equivalent to a capacitor.

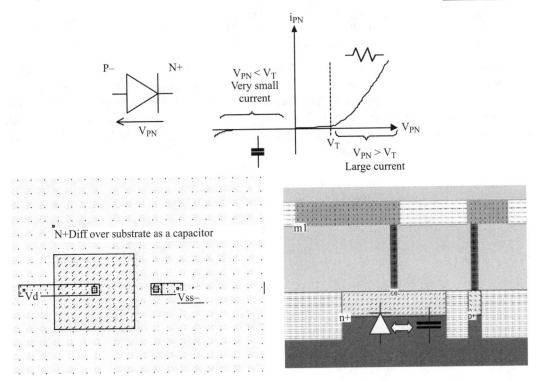

Fig. 9.10   The diffusion over substrate as a non-linear capacitor (Capa.MSK)

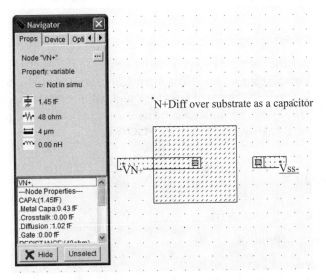

Fig. 9.11   Extraction of the diffusion capacitance (Capa.MSK)

The capacitance may be extracted by a double click in the N+ diffusion area, or by the icon **View Electrical Node**. The N+diffusion has an equivalent parasitic capacitance of around 1fF, as extracted in Figure 9.11. Notice that the value of this capacitance is an average value, computed for V_N equal to

VDD/2. This capacitor suffers from two main drawbacks. Firstly, the capacitance is small as compared to the silicon area. Secondly, the capacitance depends significantly on the value of V_N, with a non-linear law.

The typical variation of the capacitance with the diffusion voltage V_N is given in Figure 9.12. The capacitance per μm^2 provided in the electrical rules is a rude approximation of the capacitance variation. A large voltage difference between V_N and the substrate result in a thick zone with empty charges, which corresponds to a thick insulator, and consequently to a small capacitance. When V_N is lowered, the zone with empty charges is reduced, and the capacitance increases. If V_N goes lower than the substrate voltage, the diode starts to conduct.

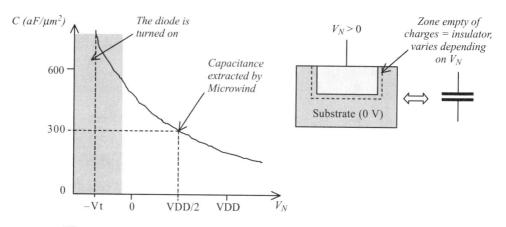

Fig. 9.12   The diffusion capacitance varies with the polarization voltage

9.2.2   MOS Capacitor

The MOS transistor is often the simplest choice to build a capacitor. In 0.12 μm, the gate oxide has an equivalent thickness of 2 nm (20 angstrom, also written 20 Å), which leads to a capacitance evaluated by Equation 9.1. There is also a thicker oxide used for high voltage MOS devices, called dual-oxide (5 nm or 50 Å in CMOS 0.12 μm), for which Equation 9.2 is applicable. The parameter in the design rules used to configure the gate oxide capacitor is CPOOxide.

$$C_{thinox} = \frac{\varepsilon_0 \varepsilon_r}{e} = \frac{8,85e^{-12} \times 3.9}{2.0 \times 10^{-9}} = 17e^{-3} \text{ F/m}^2 = 17e^{-15} \text{ F/}\mu m^2 = 17f \text{ F/}\mu m^2 \qquad \text{(Equation 9.1)}$$

where

e = gate oxide thickness (m)

ε_0 = vacuum permittivity (F/m)

ε_r = relative permittivity (no unit)

$$C_{dualox} = \frac{\varepsilon_0 \varepsilon_r}{e2} = \frac{8,85e^{-12} \times 3.9}{5.0 \times 10^{-9}} = 6,8 f \text{F/}\mu m^2 \qquad \text{(Equation 9.2)}$$

where

e_2 = dual-oxide thickness (m)

The design of a gate capacitor using a large MOS device is shown in Figure 9.13, with a capacitance of around 300 fF. In analog design, the gate capacitor is often surrounded by a guard ring. It is usually very difficult to integrate capacitors of more than a few hundred pico-farads. If nano-farad capacitors are required, these components are too large to be integrated on-chip, and must be placed off-chip.

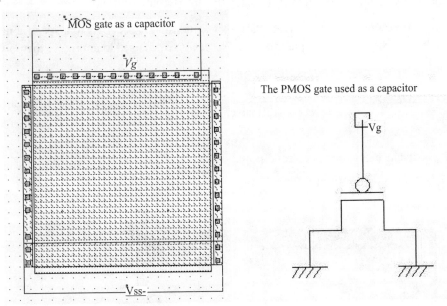

Fig. 9.13   Generating an efficient capacitor based on an MOS device with a very large length and width (CapaPoly.MSK)

Using the ultra-thin gate increases the capacitance, however, the risk of oxide damage due to over-stress is significantly increased. The dielectric strength of the SiO_2 oxide is the critical field above which the electric field damages the dielectric material. The dielectric strength E_{crit} ranges from 4 to 10 MV/cm, depending on the fabrication technique [Hasting]. The critical voltage V_{crit} above which the oxide is damaged is approximated by Equation 9.3. Be careful with the unusual units: the oxide thickness is directly expressed in nanometers and the dielectric strength in MV/cm.

$$V_{crit} = 0,1 \cdot e \cdot E_{crit}$$ (Equation 9.3)

where

e = equivalent thickness (nm)

E_{crit} = dielectric strength (MV/cm)

A SiO_2 gate dielectric of 2 nm leads to a critical voltage ranging between 0.8 V (dry oxide growth) and 2 V (high quality gate oxide deposit). Long term reliability requires the supply voltage to keep below half of this critical voltage. This is why VDD is around 1 V, and a voltage stress significantly above VDD may lead to the gate oxide destruction. The gate oxide used in high voltage MOS devices is fitted with the user requirements: to handle 3.3 V, a reliable value for the gate oxide is 7 nm. To handle 5 V, the minimum gate oxide rises to 10 nm.

9.2.3 Poly-Poly2 Capacitor

Most deep-submicron CMOS processes incorporate a second polysilicon layer (poly2) to build floating gate devices for EEPROM. An oxide thickness of around 20 nm is placed between the poly and poly2 materials, which induces a plate capacitor around 1,7 fF/μm^2 (Equation 9.3).

In MICROWIND, the command **Edit → Generate → Capacitor** gives access to a specific menu for generating capacitor (Figure 9.14). The parameter in the design rules used to configure the poly-poly2 capacitor is CP2PO.

$$C_{PolyPoly2} = \frac{\varepsilon_0 \varepsilon_r}{e_{pp}} = \frac{8,85e^{-12} \times 3.9}{20 \times 10^{-9}} = 1700\ aF/\mu m^2 \qquad \text{(Equation 9.4)}$$

where

e_{pp} = distance between poly1 and poly2 (m)

ε_0 = vacuum permittivity (F/m)

ε_r = relative permittivity (no unit)

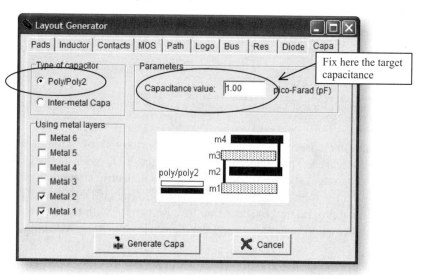

Fig. 9.14 The generator menu handles the design of poly/poly2 capacitor and inter-metal capacitors

The poly/poly2 capacitor simply consists of a sheet of polysilicon and a sheet of poly2, separated by a specific dielectric oxide which is 20 nm in the case of the default CMOS 0.12 μm process. The contacts are placed on both sides of the capacitor, between poly and metal on the left, between poly2 and metal on the right (Figure 9.14). The gate oxide is not used here because of its low breakdown voltage. A dual-oxide (5 nm) would also suffer from voltage over-stress that may occur in many analog designs, such as power amplifiers in radio frequency (see Book Part II). Moreover, a thick oxide suffers from less process variations, which ensures a better control of the final capacitance.

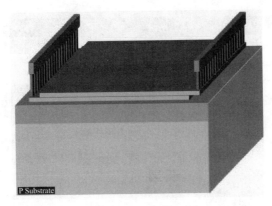

Fig. 9.15 The 3D aspect of the poly-poly2 capacitor (CapaPolyPoly2.MSK)

9.2.4 High Precision Poly-Poly2 Capacitor

As process variations mainly affect the peripheral aspect of the layers, square geometry performs better than rectangular geometry. The optimum dimensions lie between $10 \times 10\ \mu m$ and $50 \times 50\ \mu m$ [Hastings]. If larger sizes are used, gradient effects affect the quality of the oxide which is no longer uniform in the whole dielectric surface. Consequently, the capacitor should be split into $50 \times 50\ \mu m$ units. Also, no device or diffusion region should be designed next to the capacitor. It is highly recommended to shield the capacitor area by using a guard ring of contacts which limit the substrate noise that may couple to the lower capacitor plate.

Also, the high impedance node should be connected to the upper plate of the metal, which is more isolated from the substrate and lateral noise. Finally, dummy capacitors can also be placed on the layout, for high precision matching. An example of high precision capacitor using

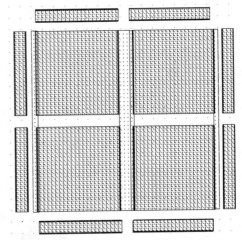

Fig. 9.16 The layout of a 16 pF poly-poly2 capacitor (CapaPoly15pF.MSK) using four units and dummy capacitor

4 units of 4 pF each is shown in Figure 9.16. Dummy capacitors do not have the same size as capacitor units for the reason relative to silicon area saving. However, the spacing between the dummy capacitor and the active capacitor is preserved. The dummy devices serve as electrostatic shielding.

9.2.5 Inter-metal Capacitor

The multiplication of metal layers creates lateral and vertical capacitance effects of rising importance. Although the inter-metal oxide is 10 to 50 times thicker than the ultra-thin gate oxide, the spared silicon area in upper metal layers may be used for small size capacitance, which might be attractive for compensation or local decoupling.

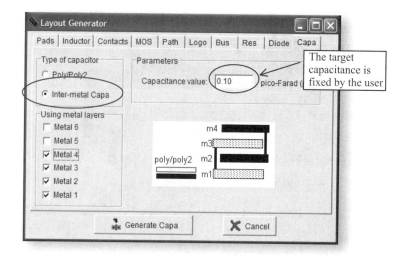

Fig. 9.17   Menu for generating an inter-metal capacitor

The menu for generating the inter-metal capacitor is the same as for the poly/poly2 capacitor, except that the type of capacitor is changed (Figure 9.17). Depending on the desired capacitor value, MICROWIND computes the size of the square structure, made of metal plates, that reaches the capacitance value. In Figure 9.17, a sandwich of metal 1, metal 2, metal 3 and metal 4 is selected, for a target value of 100 fF. The comparative aspect of the poly/poly2 capacitor and inter-metal capacitor is given in Figure 9.18 for an identical capacitance value of 100 fF. We confirm the poor efficiency of inter-metal capacitor due to the thick oxide, in comparison with the area-effective poly/poly2 structure.

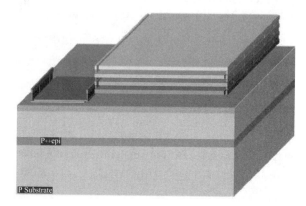

Fig. 9.18   Generating an inter-metal capacitor (CapaPolyMetalComp.MSK)

9.2.6 Capacitor Cell

Notice that a capacitor cell also exists in most logic cell libraries. This cell is inserted regularly in the design to add voluntary capacitance between the power rails VDD and VSS. This capacitor acts as a noise decoupling and reduces the external parasitic noise provoked by logic gate switching. The main drawback of this gate oxide capacitor is its low breakdown voltage, and the non-negligible possibility of gate-oxide defect, which may result in a permanent conductive path between the supply rails. An implementation of the capacitor cell in a silicon area identical to the basic inverter is proposed in Figure 9.20 left. Using a larger cell area, the gate area can be enlarged, which raises the equivalent decoupling capacitance very rapidly.

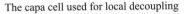

The capa cell used for local decoupling

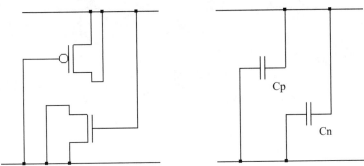

Fig. 9.19   Principles and equivalent diagram of the capacitor cell, inserted in the logic circuit core for improved noise decoupling (CapaCell.MSK)

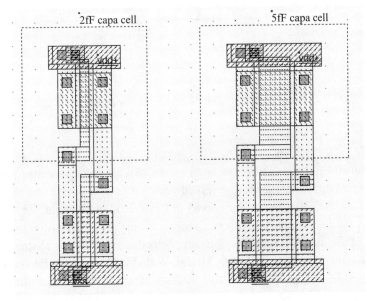

Fig. 9.20   Two implementations of the capacitor cell (2 fF, 5 fF) (CapaCell.MSK)

9.3   The MOS Device for Analog Design

The MOS has been used in previous chapters mainly as a switch. The most important parameters are the maximum available current *Ion* current and the parasitic leakage current *Ioff*. The *Ion* current corresponds to a maximum *Vgs* and maximum *Vds* (upper right point in Figure 9.21). In the case of analog design, the MOS does not operate only in cut-off or saturation regime. It also operates in the so-called quadratic zone, that appears in the static characteristics shown in Figure 9.21. In that case, *Vds* is rarely very large, as well as *Vgs*. The MOS device operates in an intermediate regime which is attractive for most analog applications.

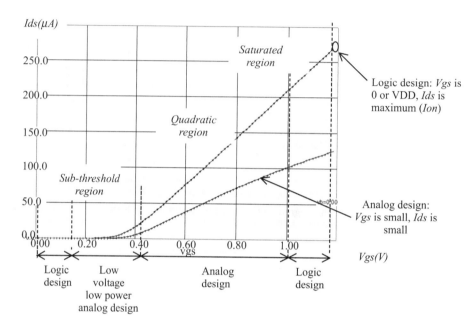

Fig. 9.21 In analog design, the MOS device also operates in the quadratic zone

9.3.1 Using Bsim4

The MOS model "level 3" is reasonably accurate in the case of logic circuits. When dealing with analog circuits, significant mismatches are observed between MOS Level 3 and the advanced model BSIM4. This is because the equations of the model in Level 3 do not account for many second order parameters, that have a very important impact when running in small voltages and currents. We recommend the use of the model BSIM4 in all analog cells described in this chapter. Select BSIM4 in the menu of the command **Simulate → Using Model → BSIM4**. An alternative consists in selecting the model BSIM4 in the list proposed in the simulation parameters (**Simulate → Simulation parameters**).

You may automatically select BSIM4 by adding a text in the layout that starts with "BSIM4". In the layout example shown in Figure 9.22, we added a text "BSIM4", which forces the simulator to use the BSIM4 model instead of the default MOS Level 3. This text appears for example at the left lower part of the layout of the transmission gate, described in Figure 9.22.

9.3.2 Analog Switch

The analog switch is able to transfer or interrupt an analog signal. It can be constructed using the pass gate described in Chapter 2 which used one n-channel MOS and one p-channel MOS in parallel. The transmission gate lets an analog signal flow if *Enable* = 1 and *~Enable* = 0. In that case both the n-channel and p-channel devices are on.

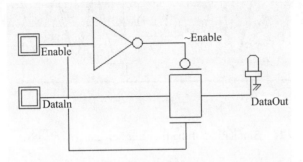

Fig. 9.22 Schematic diagram of the analog switch (TGATE.Sch)

The layout of the analog switch is shown in Figure 9.23. The inverter is situated on the left, the transmission gate on the right. A sinusoidal wave with a frequency of 2 GHz is assigned to *DataIn*. The sinusoidal property may be found in the palette of Microwind2, near the clock and pulse properties. With a zero on *Enable* (And a 1 on *~Enable*), the switch is off, and no signal is transferred (Figure 9.24). When *Enable* is asserted, the sinusoidal wave appears nearly identical to the output.

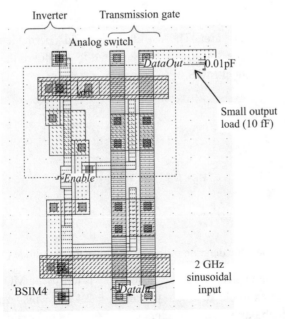

Fig. 9.23 Simulation of the analog switch (AnalogSwitch.MSK)

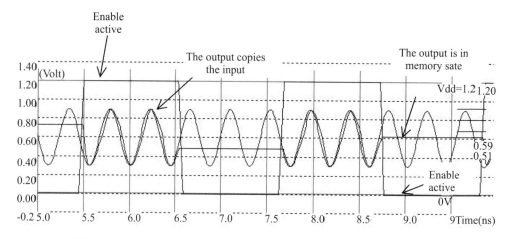

Fig. 9.24 Simulation of the transmission gate (AnalogSwitch.MSK)

As the output is loaded with a 10 fF virtual capacitance, the output is slightly distorted. This is due to the non-linear resistance variation of the transmission gate with the input voltage. Consequently, the transmission gate reacts faster for a low voltage or a high voltage, and slower for an intermediate voltage where n-channel and p-channel devices have less current capabilities, even if both conduct at the same time.

9.4 Diode-connected MOS

The schematic diagram of the diode-connected MOS is proposed in Figure 9.25. This circuit features a high resistance within a small silicon area. The key idea is to build a permanent connection between the drain and the gate. Most of the time, the source is connected to ground in the case of n-channel MOS, and to VDD in the case of p-channel MOS.

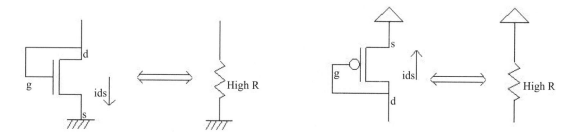

Fig. 9.25 Schematic diagram of the MOS connected as a diode (MosRes.SCH)

In order to create the diode-connected MOS, the easiest way is to use the MOS generator. Enter a large length and a small width, for example W = 0.24 μm and L = 2.4 μm. This sizing corresponds to a long channel, featuring a very high equivalent resistance.

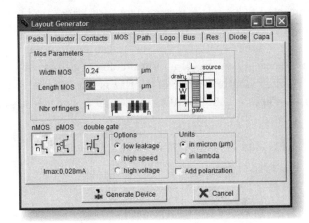

Fig. 9.26 Using the MOS generator to create a n-channel MOS with a large length and small width.

Add a poly/metal contact and connect the gate to one diffusion. Add a clock on that node. Add a VSS property to the other diffusion. The layout result is shown in Figure 9.27.

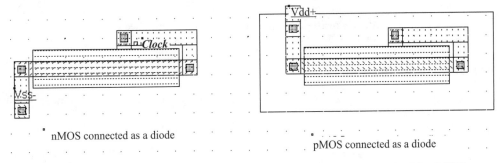

nMOS connected as a diode   pMOS connected as a diode

Fig. 9.27 Schematic diagram of the MOS connected as a diode (ResMos.MSK)

Now, click **Simulation on Layout**. In a small window, the MOS characteristics are drawn, with the functional point drawn as a colour dot (Figure 9.28). It can be seen that the *I/V* characteristics correspond to a diode. The resistance is the invert value of the slope in the *Id/Vd* characteristics. For *Vds* larger than 0.6 V, the resistance is almost constant. As the current *Ids* increases to 10 µA in 0.4 V, the resistance can be estimated to be around 40 KΩ. A more precise evaluation is performed by MICROWIND if you draw the slope manually. At the bottom of the screen, the equivalent resistance appears, together with the voltage and current.

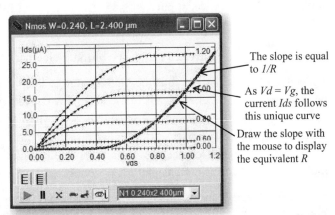

The slope is equal to *1/R*

As *Vd = Vg*, the current *Ids* follows this unique curve

Draw the slope with the mouse to display the equivalent *R*

Fig. 9.28 Using the simulation on layout to follow the characteristics of the diode-connected MOS (ResMos.MSK)

In summary, the MOS connected as a diode is a capacitance for *Vgs<Vt*, a high resistance when *Vgs* is higher than the threshold voltage *Vt*. The resistance obtained using such a circuit can easily reach 100 KΩ in a very small silicon area. The same resistance can be drawn in poly but would require a much larger area. The resistance per square of an unsalicide polysilicon serpentine is approximately 40 ohm. In Figure 9.29, a polysilicon resistance of 30 KΩ is drawn close to the MOS device with a 30 KΩ resistance. The advantage of using a MOS resistance rather than a polysilicon resistance is obvious in terms of the silicon area.

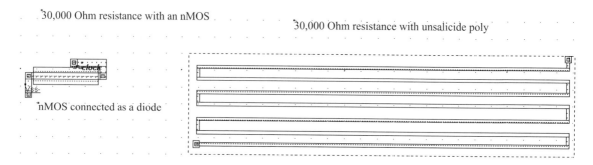

Fig. 9.29   An MOS device resistance compared to the same resistance in poly (ResMos.MSK)

9.5   Voltage Reference

The voltage reference is usually derived from a voltage divider made from resistance. The output voltage *Vref* is defined by Equation 9.5.

$$V_{\text{ref}} = \frac{R_N}{R_N + R_P} V_{DD}$$ (Equation 9.5)

with

V_{DD} = power supply voltage (1.2 V in 0.12 μm)

R_N = equivalent resistance of the n-channel MOS (ohm)

R_P = equivalent resistance of the p-channel MOS (ohm)

The value of the resistance must be high enough to keep the short-circuit current low, to avoid wasted power consumption. A key idea is to use MOS devices rather than polysilicon or diffusion resistance to keep the silicon area very small. Notice that two nMOS or two pMOS properly connected feature the same function. pMOS devices offer higher resistance due to lower mobility, as compared to n-channel MOS. Four voltage reference designs are shown in Figure 9.30. The most common design uses one p-channel MOS and one n-channel MOS connected as diodes.

The alternative solutions consist in using two n-channel MOS devices only (left lower part of Figure 9.30), or their opposites built from p-channel devices only. It is not only one reference voltage that may be created, but three, as shown in the right part of Figure 9.30, which use four n-channel MOS devices connected as diodes.

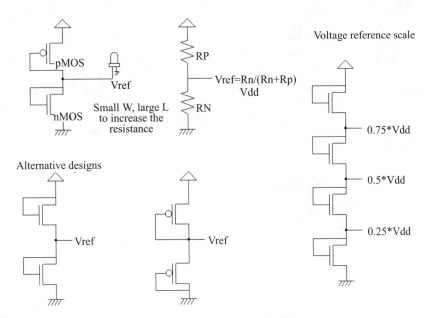

Fig. 9.30 Voltage reference using p-MOS and n-MOS devices as large resistance

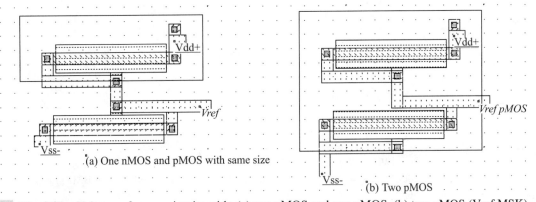

Fig. 9.31 Voltage reference circuits with: (a) one nMOS and one pMOS, (b) two pMOS (Vref.MSK)

In the layout of Figure 9.31, the pMOS and nMOS have the same size. Due to lower pMOS mobility, the resulting *Vref* is a little lower than VDD/2. Using BSIM4 instead of model 3, we see that the voltage reference obtained with two identical pMOS devices is not VDD/2 either, as shown in the simulation of Figure 9.32. This is due to the non-symmetrical polarization of the pMOS regarding the substrate voltage *Vbs* which has a significant impact on the current (Figure 9.33). Consequently, a good VDD/2 voltage reference requires a precise adjustment of MOS sizing, a good confidence in the accuracy of the model, and several iterations of design/simulation until the target reference voltage is reached.

The value of the voltage reference *Vref* versus the size of the n-channel and p-channel MOS is quite difficult to calculate as the resistance of the channel is highly non-linear. In [Baker], the formulation is deduced from the equations of model 1:

$$V_{ref} = \frac{V_{DD} - V_{tp} + \sqrt{\frac{\beta_N}{\beta_p}}V_{tn}}{\sqrt{\frac{\beta_N}{\beta_p}} + 1}$$ (Equation 9.6)

with

$$\beta_N = \mu_N \frac{W_N}{L_N} \quad \text{and} \quad \beta_p = \mu_p \frac{W_p}{L_p}$$

where

μ_N = mobility of electrons (600 cm²/V.s)

μ_p = mobility of holes (250 cm²/V.s)

W_n = nMOS width (μm)

L_n = nMOS length (μm)

W_p = pMOS width (μm)

L_p = pMOS length (μm)

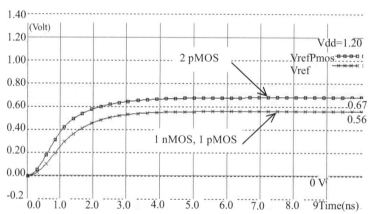

Fig. 9.32   Simulation of the two voltage reference circuits (Vref.MSK) using BSIM4 model

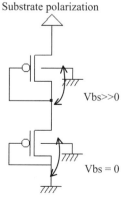

Fig. 9.33   The polarization of the two pMOS is not identical due to the substrate effect (Vref.SCH)

9.5.1 Multiple Voltage Reference

Not more than three MOS devices can be connected in series to produce intermediate voltage references. The limiting factor is the threshold voltage. When trying to simulate a series of more than 3 MOS connected as diodes, the operating regime stays in sub-threshold mode, which is attractive for very low power consumption, but introduces important set-up delays and creates very weak voltage references. When more than three reference voltages are needed, the network is built by using resistance, as shown in Figure 9.34. The static power consumption is 100 µW, which is due to the dc current flowing through the resistors between VDD and VSS. Larger resistance would decrease this static power, down to the specified user requirement.

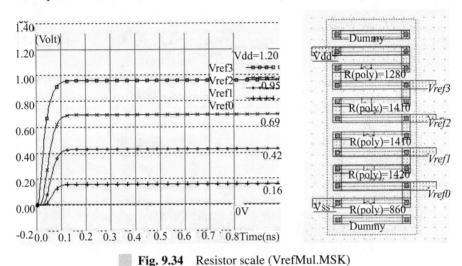

Fig. 9.34 Resistor scale (VrefMul.MSK)

9.5.2 Shielding

The voltage reference *Vref* created by our circuits is very weak, in the sense that the current which flows in the MOS branch is small. In other words, the *Vref* signal is highly resistive, which is also referred to as "high impedance". This means that a parasitic signal that couples with the *Vref* connection may induce some noise, for example by proximity effect and capacitance coupling *Cx*, as shown in Figure 9.35.

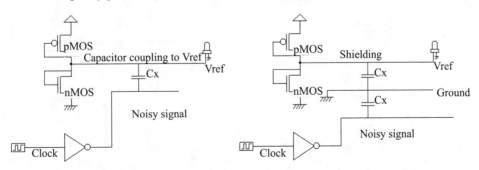

Fig. 9.35 Without shielding, the Vref voltage may be altered by noisy signals routed close to its interconnect (VrefNoise.SCH)

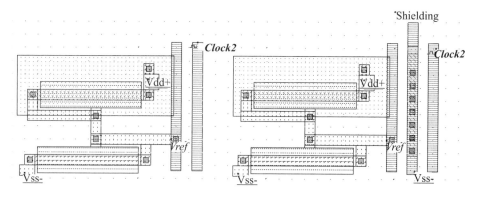

Fig. 9.36   Unshielded voltage reference (left) and shielded voltage reference (right) (VrefNoise.MSK)

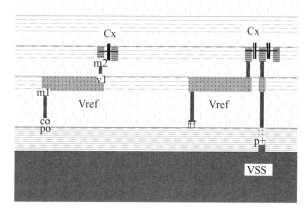

Fig. 9.37   2D cross-section of the unshielded (left) and shielded voltage reference (VrefNoise.MSK)

Adding a metal shielding acts as a noise barrier and protects *Vref* from the clock coupling (Figure 9.36). The 2D cross-section shows the two structures: on the left side, the output signal *Vref* has a parasitic coupling capacitance *Cx* connected to the noise signal. When the noisy signal switches, the voltage reference is altered. On the right structure, a barrier made of a P-type diffusion, metal and metal 2 create an electrostatic screen. The coupling still exists, but *Cx* is now connected to a cold signal, meaning that the ground capacitance is increased, and the noise is eliminated.

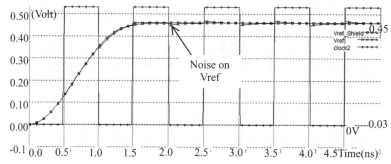

Fig. 9.38   Simulation of the unshielded and shielded Vref showing a small crosstalk noise (VrefNoise.MSK)

The effect of the shielding is demonstrated in the simulation shown in Figure 9.38. In MICROWIND, the crosstalk effect due to lateral coupling is extracted but not simulated by default. The crosstalk simulation is activated by the command **Simulate → With crosstalk**. Although the coupling capacitor still exists, the shielding is connected to the ground and prevents the clock from interfering with the *Vref_shield* signal which remains flat.

9.5.3   Influence of Temperature

You may change the temperature (**Simulate → Simulate Options**) and see how the voltage reference is altered by temperature. The circuit based on one nMOS and one pMOS, already presented in Figure 9.30, is not greatly influenced by temperature. The temperature coefficient (TC), to a first order, is almost equal to zero. A similar result is obtained using model 3 or BSIM4. This means that a stable on-chip voltage reference around VDD/2 is quite simple to achieve.

The design of a reference voltage VDD/3 leads to unbalanced resistivity, which has a direct impact on the temperature coefficient. This time, the voltage is no longer a reference voltage as it only coincides with VDD/3 at a room temperature of 25°C. At high temperature, the voltage reference is much too low. The simulation results are summarized in Figure 9.39. The parametric analysis has been used to compute the voltage *Vref* iteratively at the end of a 5 ns simulation, with increased temperature from – 40 to 120°C.

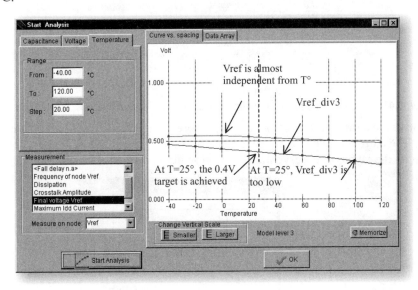

Fig. 9.39   Simulation of the influence of temperature on the reference voltage (Vref.MSK)

9.6   Current Mirror

The current mirror is one of the most useful basic blocs in analog design. It is primarily used to copy currents. When a current flows through an MOS device *N1*, an almost identical current flows through the device *N2*, as soon as *N1* and *N2* are connected as current mirrors. In its most simple configuration, the current mirror consists of two MOS devices connected as shown in Figure 9.40. A current *I1* flowing through the device *Master* is copied onto the MOS device *Slave*. If the size of *Master* and *Slave* are identical, in most operating conditions, the currents *I2* and *I1* are identical. The remarkable phenomenon is that the current is almost independent from the load, represented in Figure 9.40 by a resistor R_{load}.

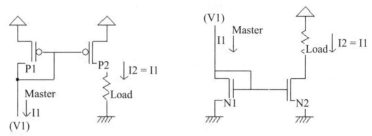

Fig. 9.40   Current mirror principles in nMOS and pMOS version (CurrentMirror.SCH)

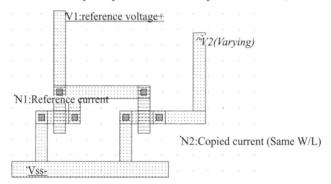

Fig. 9.41   Layout of an n-channel current mirror with identical size (CurrentMirror.MSK)

The illustration of the current mirror behaviour is proposed in the case of two identical n-channel MOS (Figure 9.41). The current of the master *N1* is fixed by *V1*, which is around 0.6 V in this case. We use the simulation on the layout to observe the current flowing in *N1* and *N2*.

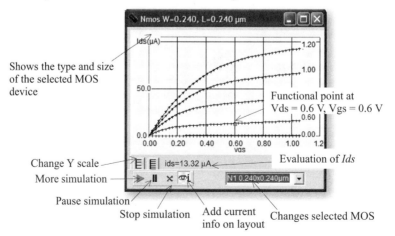

Fig. 9.42   The nMOS N1 has a fixed current of around 12 µA flowing between drain and source (CurrentMirror.MSK)

Concerning *N1*, the gate and drain voltage is fixed to 0.6 V, which corresponds to a constant current of around 13 µA, as shown in Figure 9.42. The voltage *V2* (Figure 9.43) varies thanks to a clock. We observe that the current *I2* is almost equal to 13 µA, independently of *V2*, except when *V2* is lower than 0.2 V. More precisely, the variation of *I2* is between 12 and 16 µA when *V2* varies from 0.2 to 1.2 V.

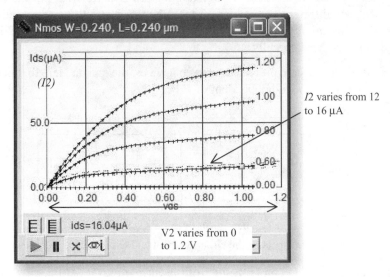

Fig. 9.43 Illustration of the nMOS current mirror principles (CurrentMirror.MSK)

9.6.1 Improving the Current Mirror

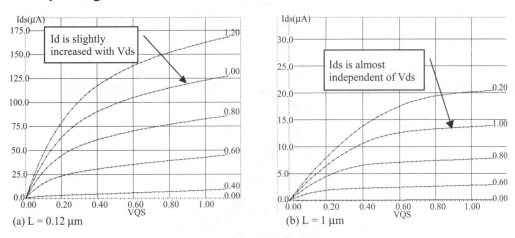

Fig. 9.44 Long channel MOS are preferred for high performance current mirrors

As the basic principle of the current copy is the assumption that *Id* is independent of *Vds*, the long channel MOS (Figure 9.44 right) is a better candidate than the short channel MOS (Figure 9.44 left). Although the short channel MOS works faster, the long channel MOS is preferred for its higher precision when copying currents.

9.6.2 MOS Matching

A set of design techniques can improve the current mirror behaviour, which are described as follows.

- All MOS devices should have the same orientation. During fabrication, the chemical process has proven to be slightly different depending on the orientation, resulting in variations of effective channel length. This mismatch alters the current duplication if one nMOS is implemented horizontally, and the other vertically.

- Long channel MOS devices are preferred. In such devices, the channel length modulation is small, and consequently *Ids* is almost independent of *Vds*.

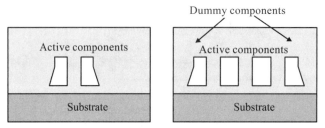

Fig. 9.45 2D aspect of the circuit without and with dummy components

- Dummy gates should be added on both sides of the current mirror. Although some silicon area is lost, due to the addition of inactive components, the patterning of active gates leads to very regular structures, ensuring a high quality matching (see Figure 9.45).

- MOS devices should be in parallel. If possible, portions of the two MOS devices should be interleaved to reduce the impact of an always-possible gradient of resistance, or capacitance with the location within the substrate.

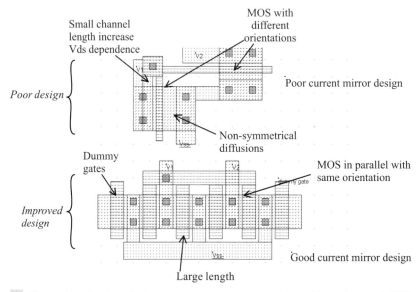

Fig. 9.46 Design of high performance current mirrors (MirrorMatch.MSK)

A synthesis of these recommendations is proposed in the two designs of Figure 9.46. The current mirror situated at the top of the figure cumulates design weaknesses: short channel length, non-symmetrical drain and source design, and different orientations for the devices. The design at the bottom realizes the same current copying function, and complies with the most important rules for a good current copying: dummy components, same orientation, symmetrical design, and MOS devices with large length.

9.6.3 Current Multiplier

If the ratio W/L of the *Slave* (Transistor *P2*) is ten times the ratio of the *Master*, the current *I2* on the right branch is ten times the current *I1* on the left branch. This is illustrated by the schematic diagram on the left of Figure 9.47. In the case of the pMOS current mirror, if the ratio W/L of the *Slave* (*P2*) is five times the ratio of the *Master*, the current *I2* is five times the current *I1*.

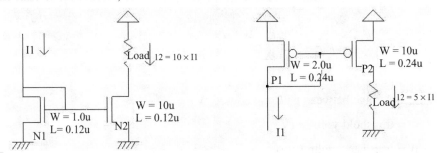

Fig. 9.47   Multiplying currents by changing the size of the MOS (MirrorMatch.MSK)

9.7   The MOS Transconductance

In its most simple form, the MOS can be represented as a current generator controlled by the voltage, or Voltage-Controlled Current Source (VCCS). The schematic diagram of the VCCS is given in Figure 9.6. We add to V_{GS} a small sinusoidal input v_{gs} which provokes a small variation of current i_{ds} to the static current I_{DS}. For small variations of v_{gs}, the link between the variation of current i_{ds} and the variation of voltage v_{gs} can be approximated by (Equation 9.6).

$$i_{ds} = g_m v_{gs}$$ (Equation 9.6)

The transconductance g_m has the dimension of the ratio of current to voltage, that is the invert of the resistance, and its definition is given in Equation 9.7.

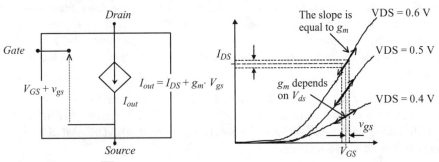

Fig. 9.48   The MOS transconductance g_m.

$$g_m = \frac{\partial I_{DS}}{\partial V_{GS}}$$ (Equation 9.7)

The derivation of the transconductance g_m from the MOS equations in level 1 leads to a quite simple expression reported in Equations 9.8 (linear region) and 9.9 (saturation region).

Linear region
$$g_m = \frac{\partial\left(\beta(V_{gs} - vt)\cdot V_{ds} - \frac{(V_{ds})^2}{2}\right)}{\partial V_{GS}} = \beta(V_{ds})$$ (Equation 9.8)

Saturation region
$$g_m = \frac{\partial(\beta(V_{gs} - vt)^2)}{\partial V_{GS}} = \frac{\beta(V_{gs} - vt)}{2}$$ (Equation 9.9)

with

$$\beta = UO\frac{\varepsilon_0\varepsilon_r}{TOX}\cdot\frac{W}{L}$$

where

V_{gs} = voltage between gate and source (V)

Vt = threshold voltage (V)

W = transistor width (μm)

L = transistor length (μm)

UO = electron mobility (m/V^{-2})

TOX = gate oxide (m)

For deep-submicron technology, more accurate expressions of the transconductance g_m are proposed such as those proposed in. The most important point to remember is the dependence (linear in a first approximation) of g_m with the width and V_{ds}.

9.8 Single Stage Amplifier

The goal of the amplifier is to multiply by a significant factor the amplitude of a sinusoidal voltage input V_{in}, and deliver the amplified sinusoidal output V_{out} on a load. Such a circuit may be found at the input stage and the output stage of all telecommunication devices such as mobile phones. The input stage amplifier increases the amplitude of the captured signal from around 0.1-1 mV to 10-100 mV for further processing, while the output stage amplifier delivers a high voltage on the antenna to emit a significant power (Figure 9.49).

The single stage amplifier may consist of a MOS device (we choose here an n-channel MOS) and a load. The load can be a resistance or an inductance. In the circuit, we use a resistance made with a p-channel MOS device with gate and drain connected (Figure 9.50). The pMOS which replaces the passive load is called an active resistance.

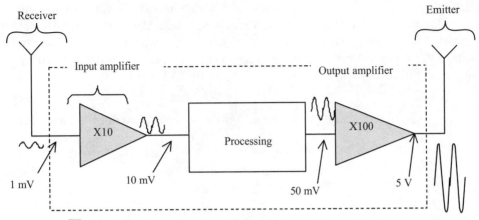

Fig. 9.49 Example of amplifier circuits used in mobile devices

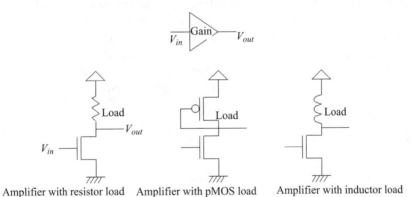

Fig. 9.50 Single stage amplifier design with MOS devices (AmpliSingle.SCH)

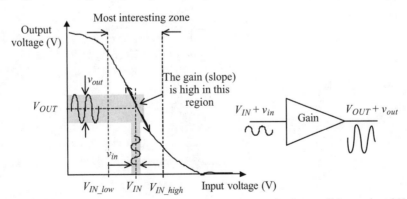

Fig. 9.51 The amplifier has a high gain at a certain input range, where a small input signal V_{in} is amplified to a large signal V_{out}

The single stage amplifier characteristics between V_{in} and V_{out} have a general shape as shown in Figure 9.51. The most interesting zone corresponds to the input voltage range wherein the transfer function has a linear shape, that is, between *VIN_low* and *VIN_high*. Outside this voltage range, the behaviour of the circuit does not correspond anymore to an amplifier. If we add a small sinusoidal input v_{in} to V_{IN}, a small variation of current i_{ds} is added to the static current I_{DS}, which induces a variation v_{out} of the output voltage V_{OUT}. The link between the variation of current i_{ds} and the variation of voltage v_{in} can be approximated by Equation 9.10.

$$i_{ds} = g_m v_{gs} \qquad \text{(Equation 9.10)}$$

Consequently, the gain of the amplifier for small signals can be expressed by Equation 9.11.

$$\text{Gain} = \frac{v_{in}}{v_{out}} = \frac{-i_{ds}\dfrac{1}{g_{mp}}}{i_{ds}\dfrac{1}{g_{mn}}} = -\frac{g_{mn}}{g_{mp}} \qquad \text{(Equation 9.11)}$$

In other words, the gain of the amplifier is high if g_{mp} is low, which is equivalent to a high pMOS pass resistance. The sign minus in Equation 9.11 illustrates the fact that an increase of *Vin* corresponds to a decrease of *Vout*. The diode-connected p-channel MOS creates a high resistance when the channel width is minimum and the channel length is very large. Such a design means a high amplifier gain. In Figure 9.52, an nMOS device with large width and minimum length is connected to a high resistance pMOS load. A 50 mV sinusoidal input (*Vin*) is superimposed to the static offset 0.6 V (V_{IN}). What is expected is a 500 mV sinusoidal wave (*Vout*) with a certain dc offset (V_{OUT}).

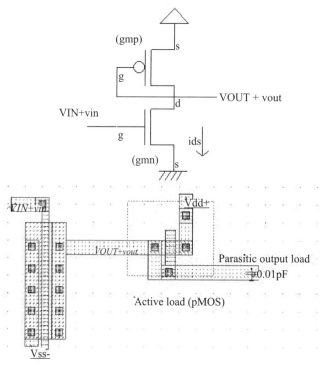

Fig. 9.52   Single stage amplifier layout with a pMOS as a load resistor (AmpliSingle.MSK)

The time-domain simulation of the amplifier with a 1 GHz sinusoidal input exhibits very poor performances. The gain is almost 0 and the output is very low, close to the ground. This is because the offset V_{IN} has been fixed to a default value of VDD/2 (0.6 V) which does not correspond to the region where the circuit provides a high gain. We are probably higher than V_{IN_high}.

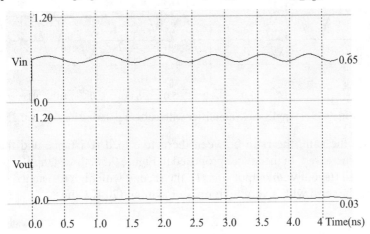

Fig. 9.53 Simulation of the amplifier response to a 50 mV input with a 0.6 V offset (AmpliSingle.MSK)

What we need now is to find the characteristics *Vout/Vin* in order to tune the offset voltage V_{IN}. In the simulation window, click "**Voltage vs voltage**" and **More**, to compute the static response of the amplifier (Figure 9.54). The range of voltage input that exhibits a correct gain appears clearly. For V_{DS} higher than 0.25 V and lower than 0.4 V, the output gain is around 3. Therefore, an optimum offset value is 0.35 V. Change the parameter **Offset** of the input sinusoidal wave to place the input voltage in the correct polarization.

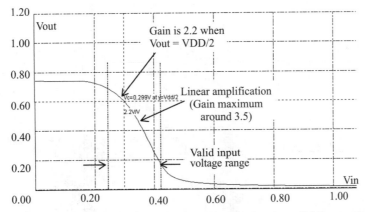

Fig. 9.54 Single stage amplifier static response showing the valid input voltage range

We change the sinusoidal input offset and start again the simulation. A gain of 3.5 is observed when the offset V_{IN} is 0.35 V. In Figure 9.55, the input amplitude is 100 mV peak to peak, while the output amplitude is 350 mV peak-to-peak. These pieces of information appear in the information bar of the main window.

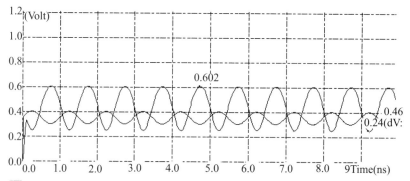

Fig. 9.55 Single stage amplifier with correct polarization V_{IN} = 0.35 V

In order to increase the gain, the ratio between the active load resistance and the n-channel MOS resistance should be increased. In the layout proposed in Figure 9.56, three amplifiers are implemented: one with a pMOS load (layout with output *sinus1*), the second with high resistance pMOS (layout with output *sinus2*), and the third with a very high resistor symbol (20 K Ohm).

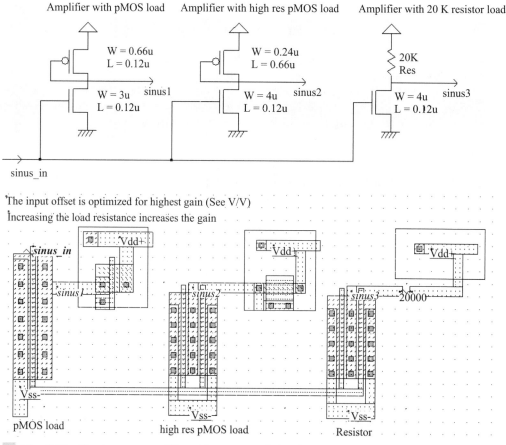

Fig. 9.56 Single stage amplifier with correct polarization V_{IN} = 0.35 V (AmpliSingle2.MSK)

The gain for *sinus2* is increased to 4.5, as observed in the static simulation (Figure 9.57), with sharper characteristics, but the input voltage range that features amplification is significantly reduced. A further increase of the pMOS resistance does not increase the gain. The gain saturates to around 8. If we replace the pMOS device by a resistor symbol, we also observe that the gain is limited to around 9.0.

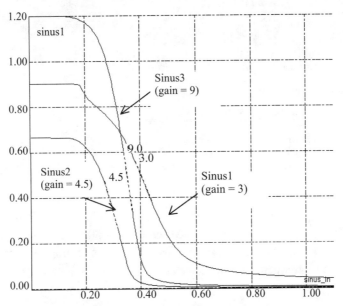

Fig. 9.57   The active load sizing acts on the gain, but reduces the input voltage range for amplification (AmpliSingle2.MSK)

9.8.1   Transit and Cutt-off Frequency

The transit frequency f_t is a parameter well representative of the "speed" of the MOS device. It corresponds to the frequency at which the current *ids* starts being lower than the current flowing through the gate *igs*. The *igs* current is due to the charge and discharge of the capacitor Cgs. The *ids* current is the main amplifier current that flows between the drain and the source. Thus, f_t is the frequency for which the current gain of the MOS device is unity. Based on the equations of level 1, an analytical approximation of f_t is reported in equation 9.12 [Baker].

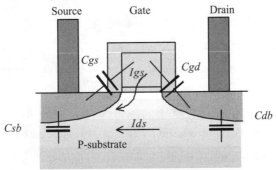

Fig. 9.58   Cross-scetion of the MOS device showing the main parasitic capacitors

$$f_t = \frac{\varepsilon_0 \varepsilon r \cdot \mu_n W}{2\pi \cdot TOX \cdot LC_{gs}} (V_{gs} - V_t)$$

(Equation 9.12)

with

$$C_{gs} \approx \frac{2}{3}(WLC_{ox})$$

with

$$\mu n = \text{electron mobility (m.V}^{-2})$$

$$W = \text{channel width (μm)}$$

$$L = \text{channel length (μm)}$$

$$C_{gs} = \text{gate to source capacitance (F)}$$

$$V_{gs} = \text{gate to source voltage (V)}$$

$$V_t = \text{threshold voltage (V)}$$

Replacing the value of C_{gs} in the expression of ft, the transit frequency becomes independent of the channel width (Equation 9.13). For an n-channel MOS device in 0.12 μm, the measured value of f_t is around 50 GHz, for V_{gs} around Vdd/2. The transit frequency f_t is an important metric of technology performances for very high frequency circuit design.

$$f_t = \frac{\varepsilon_0 \varepsilon r \cdot \mu_n}{\frac{4}{3}\pi^2 \cdot TOX \cdot L^2 \cdot C_{ox}} (V_{gs} - V_t)$$

(Equation 9.13)

A similar parameter, the cut-off frequency, is the frequency at which the gain starts decreasing. We consider G0 as the gain for low frequency input signals. The cut-off frequency is usually defined as the frequency when the gain is decreased by 1.4, according to Equation 9.14.

$$f_{cut\text{-}off} = f\left(Gain = \frac{G0}{\sqrt{2}}\right)$$

(Equation 9.14)

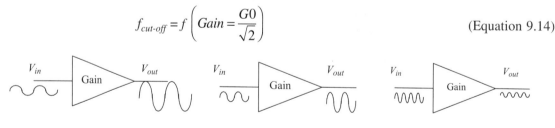

Low frequency: high gain G0   *Cut-off: gain divided by 1.4*   *Higher than cut-off: gain tends to 0*

Fig. 9.59   The cut-off frequency corresponds to the input frequency where the gain starts to decrease

MICROWIND does not perform frequency analysis. However, we can create a sinusoidal wave with an increasing frequency, thanks to the parameter **Increase f** of the sinus property assigned to *Vin* which is fixed in Figure 9.60 to 0.2. At each period, the frequency is increased by 20 per cent.

The best simulation mode is **Frequency vs. time**. The upper screen displays the input frequency while the lower screen shows the amplitude of the input and output signals (Figure 9.61). There is no precise tool to locate the cut-off frequency. However, we observe a significant decrease of the gain from 2 GHz onwards, when the output is loaded with a 10 fF parasitic capacitance. The parasitic capacitance of the output node has a direct impact on the cut-off frequency.

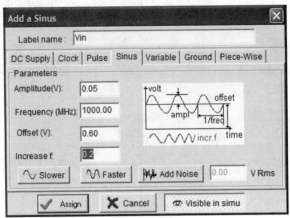

Fig. 9.60   Time-domain simulation with a sinusoidal wave with frequency sweep (AmpliSingle.MSK)

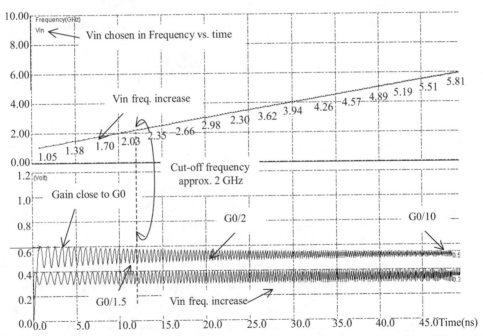

Fig. 9.61   Extracting the cut-off frequency from sweep sinusoidal input response (AmpliSingle.MSK)

9.8.2   The Inverter as an Amplifier?

Could the logic CMOS inverter act as an amplifier? Theoretically yes, as the static characteristics of the CMOS inverter are very much like the static response of the basic amplifier described earlier. Using the mode **Voltage vs. Voltage**, we find a gain of 10 for the basic inverter.

In order to operate in the amplifier zone, we should inject a signal around VDD/2, otherwise there is no chance of taking advantage of the high amplification. The commutation point varies according to the ratio between the nMOS and pMOS size, as illustrated in Figure 9.62 wherein the static characteristics of three inverters are compared. The BSIM4 model is mandatory for reliable simulations.

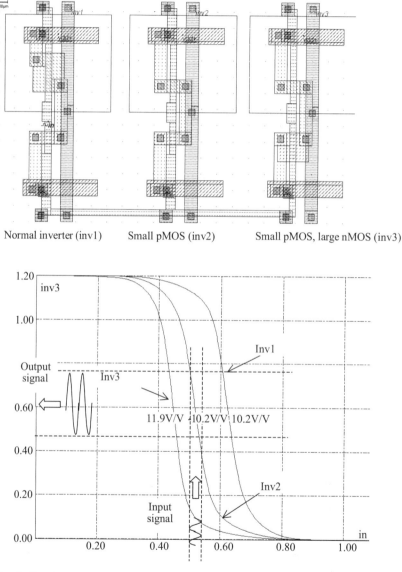

Normal inverter (inv1)   Small pMOS (inv2)   Small pMOS, large nMOS (inv3)

Fig. 9.62   Static characteristics of the CMOS inverter in 0.12 µm (invAmpli.MSK)

In 0.35 µm CMOS technology, the static characteristics of the three inverters exhibit a much higher gain (Figure 9.63). Use the command **File → Select Foundry** and choose **cmos035.RUL** to switch to the 0.35 µm technology. Again the BSIM4 model is forced thanks to the label 'BSIM4' in the layout, for accurate results. The measured gain is around 15. As the process parameters are not well controlled, the commutation point of the inverter may fluctuate over a significant range, depending on the location of the die on the wafer, or even on the die itself. As a consequence, placing the input voltage in the exact region of amplification is difficult and requires a specific control circuitry.

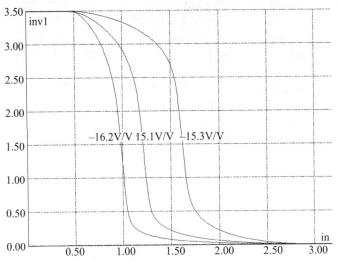

Fig. 9.63   CMOS inverter characteristics in 0.35 μm exhibit higher gain (invAmpli.MSK)

High gain amplifiers are preferably built from two medium gain stages of amplifiers, rather than one very high gain stage (Figure 9.64). The constraints for the input voltage range are easier to handle in the case of two stage amplifiers. Also, voltages higher than the logic voltage supply are used to increase the voltage range of the circuit. In 0.12 μm technology, the majority of analog amplifiers is based on dual-oxide MOS devices, operating at 2.5 V.

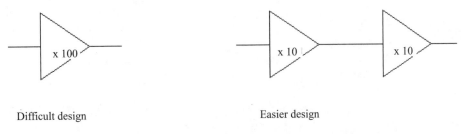

Fig. 9.64   High gain is usually achieved by two stages of amplifiers

9.9   Simple Differential Amplifier

The goal of the differential amplifier is to compare two analog signals, and to amplify their difference. The differential amplifier formulation is reported below (Equation 9.15). Usually, the gain K is high, ranging from 10 to 1000. The consequence is that the differential amplifier output saturates very rapidly, because of the supply voltage limits.

$$Vout = K(Vp - Vm) \qquad \text{(Equation 9.15)}$$

The usual symbol for the differential amplifier is given at the top of Figure 9.65, with its most simple MOS implementation. The differential amplifier principles are illustrated in Figure 9.66. We suppose that both *Vp* and *Vm* have an identical value *Vin*. Consequently, the two branches have an identical current *I1* so that no current flows to charge or discharge the output capacitor, which is connected to the output Vout (left figure). Now, if the gate voltage of the *N1* device is increased to *Vin + dV*, the current through the left branch is increased to *I1'*, greater than *I1*. The current mirror copies this *I1'* current on the right branch, so that *I1'* also flows through *P2*. As the *N2* gate voltage remains at *Vin*, the over-current *I1'-I1* is evacuated to the output stage and charges the capacitor. The output voltage *Vout* rises.

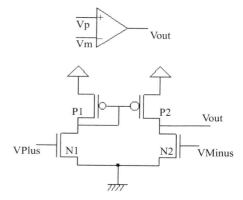

Simplest differential amplifier

Fig. 9.65 Symbol and schematic diagram of the differential amplifier

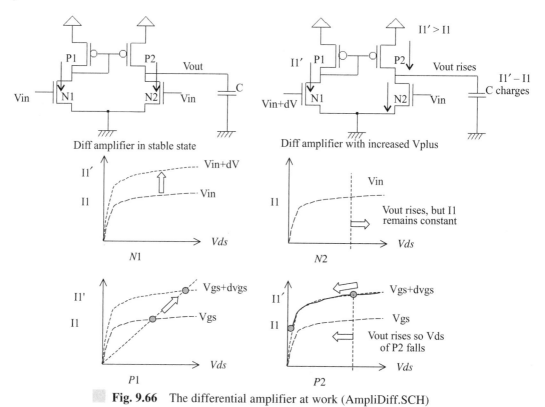

Fig. 9.66 The differential amplifier at work (AmpliDiff.SCH)

The process will end when *Vout* is high enough so that *P2* is no longer a good current mirror, which means that *I1'* is finally decreased to *I1*. The key idea is that a small variation *dV* of the input voltage is transformed into a huge variation of the output voltage *Vout*, which is the definition of a high gain amplifier.

9.9.1 A Poor Design

A direct translation of the differential amplifier into layout is performed as illustrated in Figure 9.67. The differential pair is built from short channel nMOS devices. Their size is kept identical, and drawn with the same orientation, to minimize the offset generated by the transistor mismatch. In the simulation, it can be seen that a 50 mV voltage difference between *Vp* and *Vm* is amplified by the circuit. However, the output is very far from saturation. The gain of the circuit is very small. The main reasons are the use of very short channel MOS devices for which the current mirror performances are quite poor, and the high voltage difference between the drain and source of the differential pair which forces the devices to operate at a low performance regime, with several saturation effects.

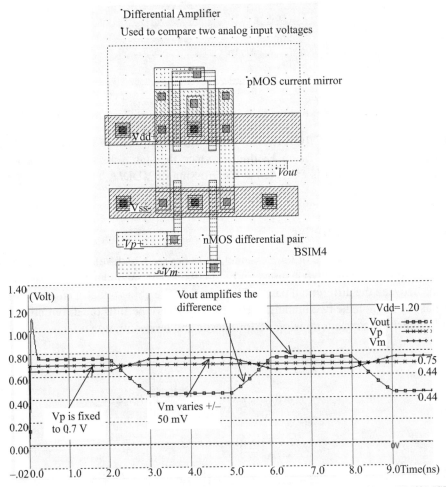

Fig. 9.67 Layout and transient simulation of the differential amplifier (AmpliDiff.MSK)

The mismatch between Level 3 and BSIM4 models is quite important in this particular circuit. This is why it is highly recommended to use BSIM4 to simulate properly the performances of analog circuits. Remember that a convenient way of forcing MICROWIND to use BSIM4 instead of the default Model 3 is to add a text in the layout starting with "BSIM4".

9.9.2 Measure the Gain

The gain of the differential amplifier is the *K* factor appearing in Equation 9.15. This equation is only valid for small differences between *Vp* and *Vm*, otherwise *Vout* saturates near VSS or near VDD.

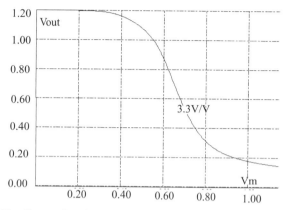

Fig. 9.68 Computing the gain of the differential amplifier (AmpliDiff.MSK)

The gain can be computed by MICROWIND in the Voltage versus Voltage simulation mode, by selecting the item **Slope** appearing in the menu **Evaluate**. Once the static characteristics of the differential amplifier are obtained, the gain is extracted at the crossing of VDD/2. The gain is very low: 3.3 V/V (Figure 9.68).

9.9.3 Improving the Amplifier

The first action consists in the use of long channel MOS device which suffers less channel modulation effects. This is proposed in the layout shown in Figure 9.69. The second action consists in inserting an nMOS device between the differential pair and the ground. The gate voltage *Vbias* controls the amount of current that can flow on the two branches. This pass transistor permits the differential pair to operate at lower *Vds*, which means better analog performances and less saturation effects.

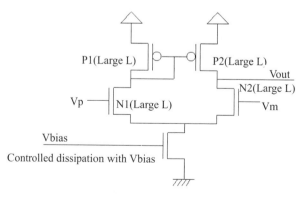

Fig. 9.69 An improved differential amplifier (AmpliDiff.SCH)

9.9.4   Find the Input Range

The best way to measure the input range is to connect the differential amplifier as a follower, that is *Vout* connected to *Vm*, as shown in Figure 9.70. The *Vm* property is simply removed, and a contact poly/metal is added at the appropriate place to build the bridge between *Vout* and *Vm*. The new differential amplifier layout is shown in Figure 9.71. A slow ramp is applied on the input *Vin* and the result is observed on the output. We use again the «Voltage vs Voltage» to draw the static characteristics of the follower. The BSIM4 model is forced for simulation by a label "BSIM4" on the layout.

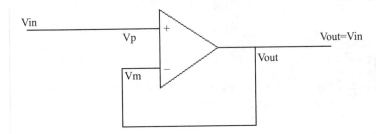

Fig. 9.70   The connection between Vout and Vm creates a follower (AmpliDiff2.MSK)

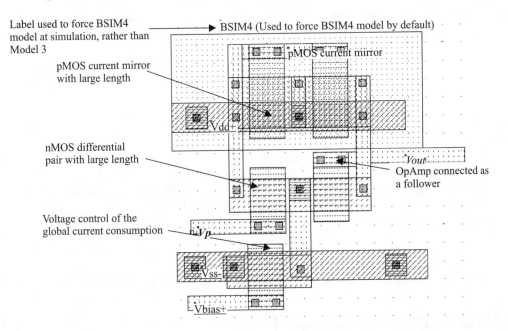

Fig. 9.71   The layout corresponding to the improved differential amplifier (AmpliDiffLargeLength.SCH)

One convenient way of simulating the follower response is to assign *Vp* a clock with a very slow rise and fall time. As can be seen from the resulting simulation shown in Figure 9.72, a low *Vbias* features a larger voltage range, specifically at high voltage values. The follower works properly starting at 0.4 V, independently of the *Vbias* value.

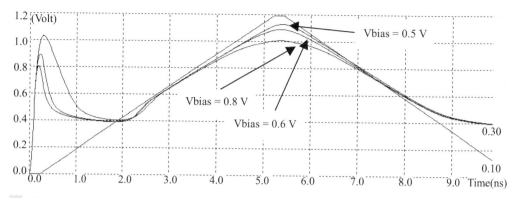

Fig. 9.72   Effect of Vbias on the differential amplifier performance (AmpliDiffVbias.MSK)

A high *Vbias* leads to a slightly faster response, but reduces the input range and consumes more power as the associated nMOS transistor drives an important current. The voltage *Vbias* is often fixed to a value that is a little higher than the threshold voltage *Vtn*. This corresponds to a good compromise between the switching speed and input range.

The differential amplifier may be constructed by using nMOS devices for the current mirror and pMOS devices for the differential pair. This circuit, proposed in Figure 9.73, features a symmetrical behaviour, that is, a good follower performance for low voltages and an intrinsic limitation near VDD-0.4 V.

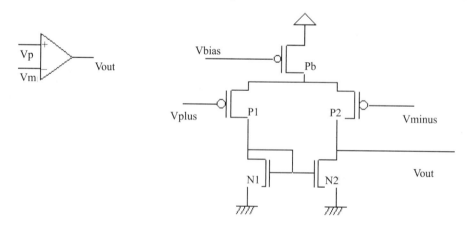

Fig. 9.73   A differential amplifier based on a pMOS differential pair and an nMOS current mirror

9.9.5   Double Differential Amplifier

The double differential amplifier is built by using the two previous differential amplifiers connected to a common output. This circuit is valid because the output stage works alternatively: one for the high voltages, and another for the low voltages. The result, shown in Figure 9.74, is quite correct: the follower copies the input with a reduced error, from 0.1 V to 1.1 V, that is, 100 mV close to the supply voltage. This amplifier is close to a rail-to-rail operational amplifier that may be found in most CMOS analog cell libraries.

Still the layout is incomplete: neither dummy devices nor proper isolation circuits have been placed. The devices could also be arranged in a more compact way, to decrease the silicon area. Finally, to further decrease the channel length modulation effect, MOS devices with larger length could be used. The speed performances could be kept identical with an increased width, at the price of a larger silicon area.

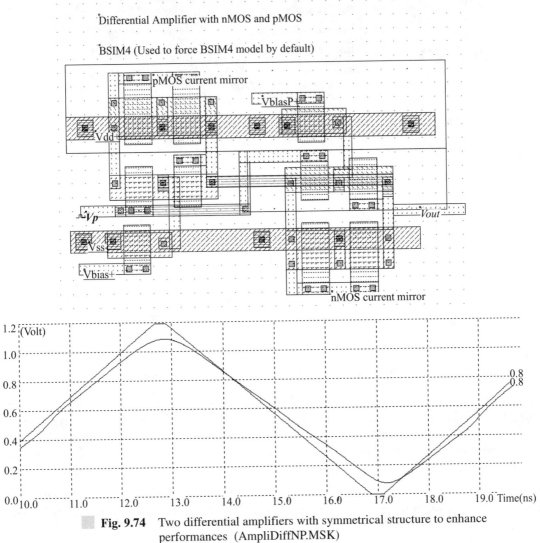

Fig. 9.74　Two differential amplifiers with symmetrical structure to enhance performances (AmpliDiffNP.MSK)

9.9.6　Push-pull Amplifier

The push-pull amplifier is built by using a voltage comparator and a power output stage. Its schematic diagram is shown in Figure 9.75, with some details about the important voltage nodes. The difference between *Vp* and *Vm* is amplified and produces a result, codified *Vout*. Transistors *Nb* and *Pb* are connected as diodes in series to create an appropriate voltage reference V_{bias}, fixed between the nMOS threshold voltage *Vtn* and half of VDD. The differential pair consists of transistors *N1* and *N2*. This time, two stages of current mirrors are used: *P1, P2* and *P3, PO*.

Node	Description	Typical value
V+	Positive analog input	Close to VDD/2
V–	Negative analog input	Close to VDD/2
Vbias	Bias voltage	A little higher than VTN
Vout	Analog output	0..VDD

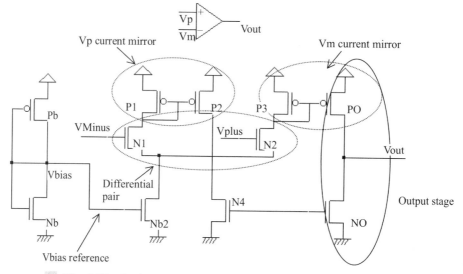

Fig. 9.75 Push-pull operational amplifier (AmpliPushPull.SCH)

The output stage consists of transistors PO and NO. These transistors are designed with large widths in order to lower the output resistance. Such a design is justified when a high current drive is required: high output capacitor, antenna dipole for radio-frequency emission, or more generally a low impedance output. The ability to design the output stage according to the charge is a key advantage of this structure compared to the simple differential pair presented earlier.

The implementation shown in Figure 9.76 uses NO and PO output stage devices with a current drive that is around five times larger than the other devices. In practice, the ratio may rise up to 10-20.

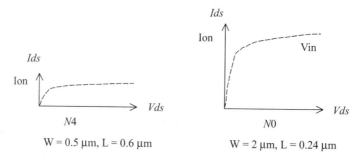

Differential amplifier with push-pull output stage

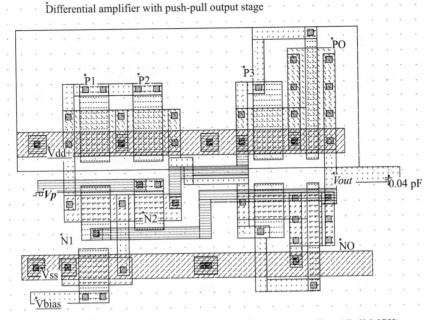

Fig. 9.76   Push-pull operational amplifier (AmpliDiffPushPull.MSK)

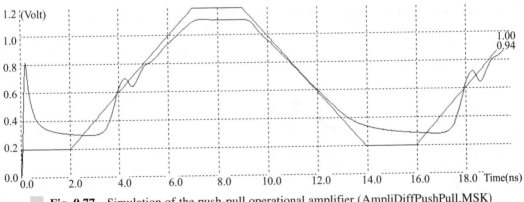

Fig. 9.77   Simulation of the push-pull operational amplifier (AmpliDiffPushPull.MSK)

The transient simulation (Figure 9.77) shows an interesting phenomenon called ringing. The oscillation appearing at time 4.0 ns is typical for a feedback circuit with a large loop delay and a very powerful output stage. Although an extra 10 fF has been added to load the output artificially, its voltage is strongly driven by the powerful devices PO and NO. The oscillation is not dangerous in itself. However, it signifies that the output stage is too strong as compared to its charge. If you use the **Voltage vs. Voltage** simulation mode to get the transfer characteristics Vout/V+, you may see the consequence of the oscillation effect: the simulator hardly converges to a stable result. Increasing the precision does not improves the design significantly (Figure 9.78).

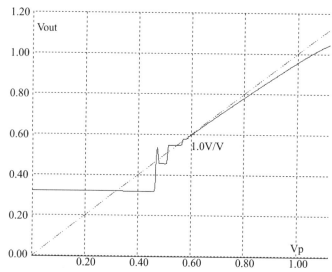

Fig. 9.78 The oscillation of the push-pull operational amplifier is also observed when trying to obtain the transfer characteristics of the circuit (AmpliDiffPushPull.MSK)

9.10 Wide Range Amplifier

Another popular operational amplifier design is shown in Figure 9.79. The amplifier is built by using a classical voltage comparator and a power output stage similar to the push-pull circuit. However, the pMOS device *PO* has a constant gate voltage, thus acting as a current generator, while the nMOS device *NO* is controlled as seen before.

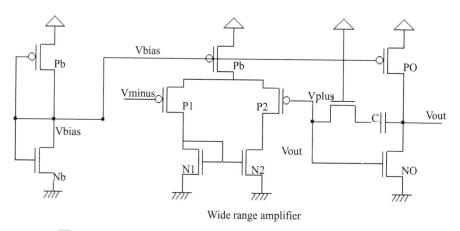

Wide range amplifier

Fig. 9.79 Schematic diagram of the wide range amplifier

The circuit shown in Figure 9.80 has been implemented in a 0.35 μm CMOS process. The corresponding layout is shown in Figure 9.81. Not all design rules for a high quality analog design have been observed at that time:

- The orientation of the upper pMOS is not identical. This may impact the quality of the mirror (Design warning 1).

- The differential pairs use channels with minimum length. This strongly increases the second order effects, the offset and non-linearities (Design warning 2).

- No dummy device has been used to improve the quality of the differential pair response (Design warning 3).

- The arrangement is far from being optimal. A lot of silicon area remains unused (Design warning 4).

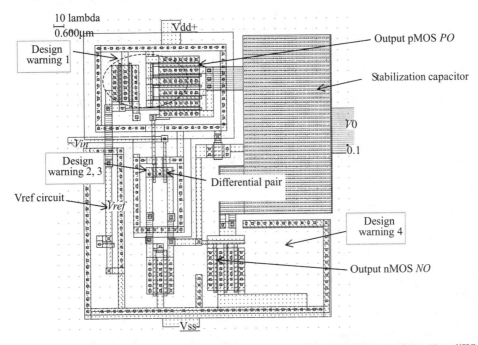

Fig. 9.80 Implementation of the wide range amplifier using a 0.35 μm CMOS technology (AmpliWide.MSK)

In order to minimize the parasitic offset, the critical MOS devices (N1, N2, P1 and P2) should be divided into sub-elements N1a, N1b, etc. and placed in a centroid geometry as illustrated in Figure 9.81. Dummy elements are also added around the amplifier. The effects of process non-uniformity are efficiently compensated and the parasitic electrical effects are consequently reduced.

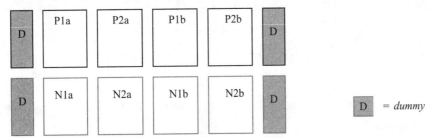

Fig. 9.81 Design efforts to limit the impact of process variations on the sense amplifier

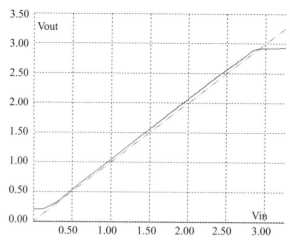

Fig. 9.82 Simulation of the wide range amplifier in 0.35 µm CMOS technology (AmpliWide.MSK)

The simulation of the wide range amplifier is shown in Figure 9.82. It matches very nicely with real case measurements made on a CMOS test chip fabricated in 0.35 µm, which included the circuit without any modification, except, of course, the addition of supply rails and input/output pads.

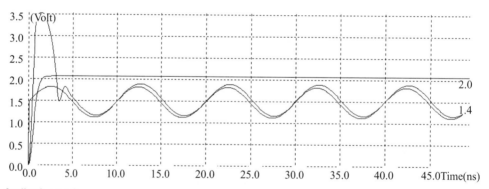

(a) With feedback capacitor

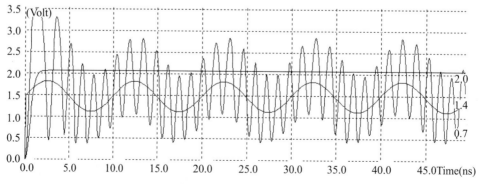

(b) Without feedback capacitor

Fig. 9.83 The stabilization circuit is required to ensure a correct follower response and avoid instability (AmpliWide.MSK)

It is important to point out that the compensation capacitor has a very important role. When the stabilization circuit is active, we notice almost no ringing effect except at the early stages when the circuit is turned on [Figure 9.83(a)]. In contrast, if we delete one connection in the stabilization circuit, an enormous ringing effect is superimposed on the output voltage [Figure 9.82(b)].

9.11   On-chip Voltage Regulator

In deep-submicron technology, the use of very thin gate oxide implies a low supply voltage. This supply decrease is mainly due to the increased risk of damaging the oxide that separates the gate from the drain and source regions, when high voltage differences are present. In contrast, the input/output interface of the integrated circuit must meet standard requirements in terms of voltage, basically 5 V or 3.3 V supply. This means that a specific circuit must be designed to generate a low voltage source internally from an external high voltage supply.

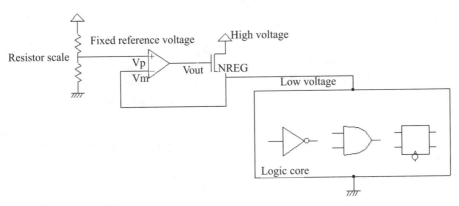

Fig. 9.84   Principles of an on-chip voltage regulator based on an operational amplifier

A circuit that realizes the voltage shift is proposed in Figure 9.84. The basic idea consists in using an operational amplifier to control the gate of an n-channel MOS device. When the logic core switches, the core voltage connected to *Vm* is lowered. Consequently the operational amplifier tends to increase *Vout*, which reduces the *NREG* device resistance, and increases the voltage until the reference voltage is attained. The negative feedback creates a stable feedback loop to recover from the voltage drops during the switching of circuits.

From a design point of view, the NREG device must be designed very large, even when the supplied logic circuit is small (Figure 9.85). If the equivalent width of the NREG is too small, the gate voltage saturates and the feedback is inefficient. In the case of Figure 9.85, the initial design using a width of 100 lambda did not work properly. Using a 200 lambda MOS device, the regulation is effective.

In the example shown in Figure 9.86, the voltage *Vref_Low* is fixed to approximately 1.2 V. The high voltage supply is 3.3 V, which requires high voltage MOS devices for the amplifier and the regulator device *NREG*. When a strong current flows due to concurrent switching in the core, the regulator reacts quite rapidly. In the case of very large logic core, the size of the pass transistor is increased to very impressive values: the combined width may be larger than a millimeter. Such a giant MOS device is obtained by placing MOS gates in parallel. A design example is shown in Figure 9.87, which has a width of 1 mm, with an *Ion* current of around 600 mA.

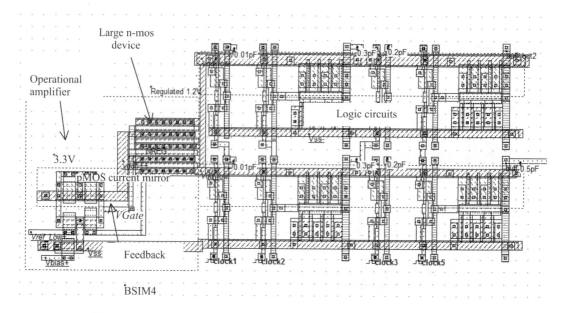

Fig. 9.85 Implementing a small on-chip voltage regulator (Vreg.MSK)

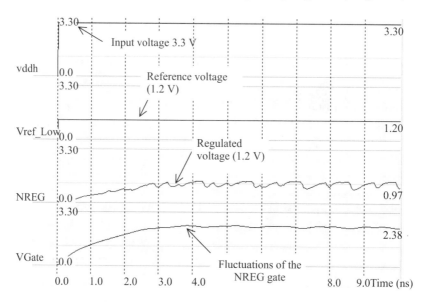

Fig. 9.86 The on-chip voltage regulator at work during multiple transitions of the logic core (Vreg.MSK)

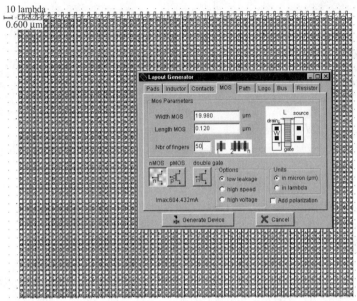

Fig. 9.87   Giant MOS built from MOS devices in parallel, to regulate very high density core circuits (VregBigMos.MSK)

9.12   Noise

The random motion of electrons in conductors create an unwanted signal called noise. The two major sources of noise in CMOS circuits are the resistors and MOS devices. The thermal noise is a very important parasitic effect in the resistor. The parasitic voltage due to thermal noise is shown in Figure 9.88. Without thermal noise in resistors, the voltage should be constant. Taking into account the thermal noise, a small random fluctuation is observed.

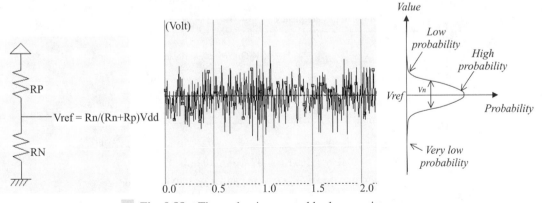

Fig. 9.88   Thermal noise created by large resistances

The voltage oscillates around the desired voltage with a Gaussian distribution. The amplitude of the noise is linearly proportional to temperature and to the value of the resistor. The noise is almost independent of the nature of the resistor material and the frequency.

The signal is not periodic due to its random nature. However, its properties are well predictable when we consider a long time and average parameters. It can be seen in a spectrum analyzer that the energy is homogenously spread over the whole range of measurable frequencies (Typically 1 KHz, 10 GHz). The thermal noise is considered as a "white noise" by analogy to light where all colours added together create a white colour. As the thermal noise is Gaussian distributed, it is called a "Gaussian white noise". Its spectral power density (SPD) can be approximated by the following formulation.

$$SPD \approx 4kTR \tag{Equation 9.16}$$

where

k = Boldzmann constant = 1.38×10^{-23} J/K

T = temperature (°K)

R = resistance value (Ohm)

An example of measured noise power density is given in Figure 9.89. The thermal noise appears as predicted by Equation 9.16, at a level approaching 10^{-17} V^2/Hz. A new noise source is also found at low frequencies. The noise is called $1/f$ noise as it is proportional to $1/f$. We also observe another important noise of around 100 MHz, which corresponds to the FM radio emission. Any discrete device with dimensions larger than some millimeter acts as an antenna and then captures a small portion of radio-frequency signals. The same effect is observed in mobile phone bands (900 MHz, 1800 MHz, 1900 MHz).

The MOS device has significant serial access resistance on the drain, source, gate and channels. Consequently, the MOS device is also an important source of noise. There is no noise model in MICROWIND MOS devices. But the noise analysis may be performed in SPICE simulations, as the noise sources and parameters are extensively described [WinSpice].

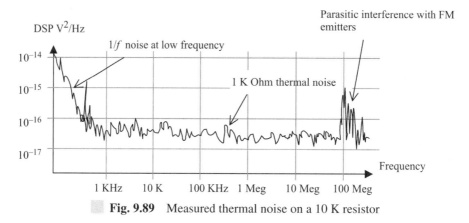

Fig. 9.89 Measured thermal noise on a 10 K resistor

Low noise design refers to specific techniques which try to limit the noise effect and its possible consequences on analog signal quality. The most efficient techniques consists in replacing resistors by inductors, by minimizing the values of resistors when such components are absolutely required, and by using MOS devices with large channel length, which have better noise performances than short channel devices.

9.13 Conclusion

Several aspects of analog design have been described in this chapter. Firstly, we detailed the implementation of resistor and capacitor elements. Secondly, we focused on several analog properties of the MOS device, as an analog switch and a high value resistance. We created voltage references, and analyzed the impact of the device size, temperature and shielding. In the third part, the current mirror was presented and simulated. Some guidelines for device matching were illustrated. The transient and frequency characteristics of the single stage amplifier were also studied, and several designs for the differential amplifier were reviewed, including a rail-to-rail amplifier. The gain and input range were characterized for each design. An example of fabricated wide-range amplifier was proposed. Finally, we detailed an on-chip voltage regulator, with a focus on very large MOS devices.

References

[Baker] R.J. Baker, H.W. Li, D.E. Boyce "CMOS circuit design, layout and simulation", IEEE Press, ISBN 0-7803-34 16-7, 1998.

[Goval] Goval R. "High frequency analog integrated circuit design", Wiley, 1995, ISBN 0-471-53043-3.

[Hastings] Alan Hastings "The art of analog layout", Prentice-Hall, 2001, ISBN 0-13-087061-7.

[WinSpice] can be down loaded from *www.winspice.com*.

EXERCISES

1. Considering the switch N1 off and switch N2 on, what is the value of the inverter output and input? When N1 is turned on and N2 is turned off, what is the value of the output if the input rises from *Vin* to *Vin* + *dV*? We assume an inverter slope with a maximum gain G of -5.

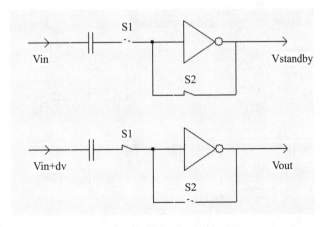

Fig. 9.90   Exercise Q.1

2. The cascode current mirror has several advantages over the simple current mirror. The output impedance is higher, and the current mirroring capabilities are better in terms of accuracy. The schematic diagram of the cascode current mirror is given in Figure 9.91. The disadvantage of the structure is the minimum level of the output voltage which is reduced as compared to the regular current mirror. Design an nMOS or pMOS cascode mirror and compare it to the standard structure.

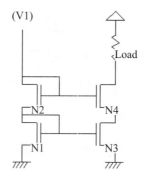

Fig. 9.91   The cascode current mirror

3. Let us consider the amplifier of Figure 9.92.

   - What is the typical value for *Vo*? What is the impact of a change in the voltage *Vo*?

   - If the width of *P2* W_{P2} is $10 \times W_{P1}$, and of *P3* is $50 \times W_{P1}$, what is the value of the current flowing through *P2* and *P3*, compared to *I1* crossing *P1*?

   - Where are the *V–* and *V+* inputs located?

   - What is the role of *P4* and *P5*?

   - What is the equivalent function for *N1* and *N2*?

   - Locate the two-stage output driver.

   - Design the operational amplifier and extract the gain.

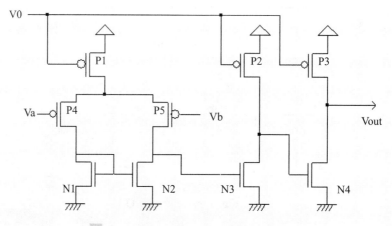

Fig. 9.92   An operational amplifier

10

Conclusion

This book has described several aspects of the CMOS circuit design, using the MICROWIND tool as an illustration. A very important gap exists between the educational PC-based tool and the professional tools used in industry for real-case designs. However we hope that the readers have caught the essential parts of MOS devices, logic circuits, and analog cells, through the illustrations and numerous examples.

The second book called "Advances CMOS design" covers more topics related to:

- Field programmable gate arrays
- Memories
- Radio-frequency analog cell design
- Converters between analog and logic signals
- Input/output interfacing
- Silicon-on-insulator technology.

No book and no teaching can replace the practical experience. Although the simulations should never be trusted, the access to integrated circuit technology tends to be more and more costly, which justifies the relevance of such simple tools.

The authors have dedicated around two years to build the technical contents of this book, and tried their best to improve the MICROWIND and DSCH tools, trying to make attractive and simple something which tends to be more and more complicated. Still, some bugs needs to be corrected, the user's interface should be improved, and important new features should be included. As the tools are in constant evolution thanks to the user's feedback and comments, we encourage the reader to download the updated versions of Microwind and Dsch from their home page.

We hope that the reader will find the contents of this book and the companion tools useful. It is our hope that the reader will design logic and analog circuits by himself, understand by a practical approach the principles of deep-submicron VLSI design, and later contribute to innovative designs to support the electronic systems of the future.

Design Rules

This section gives information about the design rules used by MICROWIND. You will find all the design rule values common to all CMOS processes. All that rules, as well as process parameters and analog simulation parameters are detailed here.

A.1   Lambda Units

The MICROWIND software works is based on a lambda grid, not on a micro grid. Consequently, the same layout may be simulated in any CMOS technology. The value of lambda is half the minimum polysilicon gate length. Table A.1 gives the correspondence between lambda and micron for all CMOS technologies supported by Microwind.

Table A.1   Correspondence between technology and the value of lambda in μm

Technology file available in the CD-Rom	Minimum gate length	Value of lambda
Cmos12.rul	1.2 μm	0.6 μm
Cmos08.rul	0.7 μm	0.35 μm
Cmos06.rul	0.5 μm	0.25 μm
Cmos035.rul	0.4 μm	0.2 μm
Cmos025.rul	0.25 μm	0.125 μm
Cmos018.rul	0.2 μm	0.1 μm
Cmos012.rul	0.12 μm	0.06 μm
Cmos90n.rul	0.1 μm	0.05 μm
Cmos65n.rul	0.07 μm	0.035 μm
Cmos45n.rul	0.05 μm	0.025 μm

A.2 Layout Design Rules

The software can handle various technologies. The process parameters are stored in files with the appendix '.RUL'. The default technology corresponds to a generic 6-metal 0.12 µm CMOS process. The default file is CMOS012.RUL. In order to select a new foundry, click on **File -> Select Foundry** and choose the appropriate technology in the list.

n-Well

r101	Minimum well size	12 λ
r102	Between wells	12 λ
r110	Minimum well area	144 λ²

Diffusion

r201	Minimum N+ and P+ diffusion width	4 λ
r202	Between two P+ and N+ diffusions	4 λ
r203	Extra nwell after P+ diffusion :	6 λ
r204:	Between N+ diffusion and nwell	6 λ
r205	Border of well after N+ polarization	2 λ
r206	Between N+ and P+ polarization	0 λ
r207	Border of Nwell for P+ polarization	6 λ
r210	Minimum diffusion area	24 λ²

Polysilicon

r301	Polysilicon width	2 λ
r302	Polysilicon gate on diffusion	2 λ
r303	Polysilicon gate on diffusion for high voltage MOS	4 λ
r304	Between two poly silicon boxes	3 λ
r305	Polysilicon vs. other diffusion	2 λ
r306	Diffusion after polysilicon	4 λ
r307	Extra gate after polysilicon	3 λ
r310	Minimum surface	8 λ²

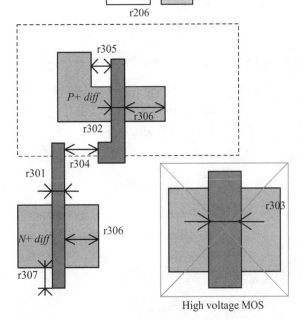

High voltage MOS

A.2.1   Second Polysilicon Design Rules

r311	Polysilicon2 width	2λ
r312	Polysilicon2 gate on diffusion	2λ
r320	Polysilicon2 minimum surface	$8\lambda^2$

MOS Option

rOpt	Border of "option" layer over diff N+ and diff P+	7λ

Contact

r401	Contact width	2λ
r402	Between two contacts	5λ
r403	Extra diffusion over contact	2λ
r404	Extra poly over contact	2λ
r405	Extra metal over contact	2λ
r406	Distance between contact and poly gate	3λ
r407	Extra poly2 over contact	2λ

Metal 1

r501	Metal width	4λ
r502	Between two metals	4λ
r510	Minimum surface	$16\lambda^2$

Via

r601	Via width	2λ
r602	Between two Via	5λ
r603	Between Via and contact	0λ
r604	Extra metal over via	2λ
r605	Extra metal 2 over via:	2λ

Metal 2

r701	Metal width:	$4\,\lambda$
r702	Between two metal 2	$4\,\lambda$
r710	Minimum surface	$16\,\lambda^2$

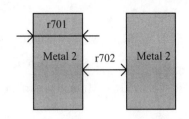

Via 2

r801	Via2 width: $2\,\lambda$
r802	Between two via2: $5\,\lambda$
r804	Extra metal 2 over via2: $2\,\lambda$
r805	Extra metal 3 over via2: $2\,\lambda$

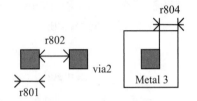

Metal 3

r901	Metal 3 width: $4\,\lambda$
r902	Between two metal 3: $4\,\lambda$
r910	Minimum surface: $32\,\lambda^2$

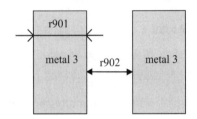

Via 3

ra01	Via3 width: $2\,\lambda$
ra02	Between two via3: $5\,\lambda$
ra04	Extra metal 3 over via3: $2\,\lambda$
ra05	Extra metal 4 over via3: $2\,\lambda$

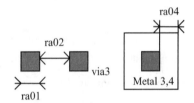

Metal 4

rb01	Metal 4 width: $4\,\lambda$
rb02	Between two metal 4: $4\,\lambda$
rb10	Minimum surface: $32\,\lambda^2$

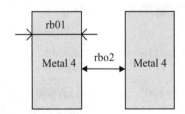

Via 4

rc01	Via4 width: $2\,\lambda$
rc02	Between two via4: $5\,\lambda$
rc04	Extra metal 4 over via2: $3\,\lambda$
rc05	Extra metal 5 over via2: $3\,\lambda$

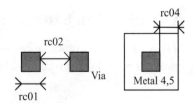

Metal 5

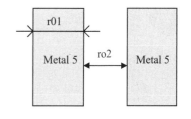

rd01	Metal 5 width:	8 λ
rd02	Between two metal 5:	8 λ
rd10	Minimum surface:	100 λ^2

Via 5

re01	Via5 width:	4 λ
re02	Between two via5:	6 λ
re04	Extra metal 5 over via5:	3 λ
re05	Extra metal 6 over via5:	3 λ

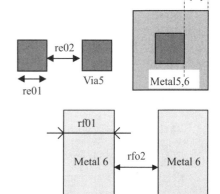

Metal 6

rf01	Metal 6 width:	8 λ
rf02	Between two metal 6:	15 λ
rf10	Minimum surface:	300 λ^2

A.3 Pads

The rules are presented below in μm. In .RUL files, the rules are given in lambda. As the pad size has an almost constant value in μm, each technology gives its own value in λ.

rp01	Pad width:	100 μm
rp02	Between two pads	100 μm
rp03	Opening in passivation v.s via:	5 μm
rp04	Opening in passivation v.s metals:	5 μm
rp05	Between pad and unrelated active area:	20 μm

A.4 Electrical Extraction Principles

MICROWIND includes a built-in extractor from layout to electrical circuit. The MOS devices, capacitance and resistance are worthy of interest. The flow is described in Figure A.1.

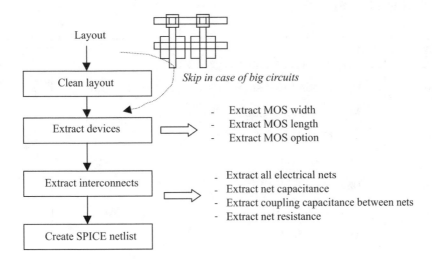

Layout

Clean layout — *Skip in case of big circuits*

Extract devices ⟹
- Extract MOS width
- Extract MOS length
- Extract MOS option

Extract interconnects ⟹
- Extract all electrical nets
- Extract net capacitance
- Extract coupling capacitance between nets
- Extract net resistance

Create SPICE netlist

Fig. A.1 Extraction of the electrical circuit from layout

The first step consists in cleaning the layout. Mainly, redundant boxes are removed, and overlapping boxes are transformed into non-overlapping boxes. In the case of complex circuits, MICROWIND may skip this cleaning step as it requires a significant amount of computational time.

A.5 Node Capacitance Extraction

Each deposited layer is separated from the substrate by a SiO_2 oxide and generated by a parasitic capacitor. The unit is the $aF/\mu m^2$ (atto = 10^{-18}). Basically all layers generate parasitic capacitors. Diffused layers generate junction capacitors (N+/P-, P+/N). The list of capacitance handled by Microwind2 is given below. The name corresponds to the code name used in CMOS012.RUL (CMOS 0.12 μm). Surface capacitance refers to the body. Vertical crosstalk capacitance refers to inter-layer coupling capacitance, while lateral crosstalk capacitance refers to adjacent interconnects.

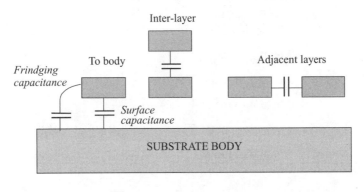

Fig. A.2 Capacitances

SURFACE CAPACITANCE

Name	Description	Linelic(aF/μm)	Surface(aF/μm^2)
CpoOxyde	Polysilicon/thin oxide capacitance	n.c	4600
CpoBody	Polysilicon to substrate capacitance	n.c	80
CMEBody	Metal on thick oxide to substrate	42	28
CM2Body	Metal2 on body	36	13
CM3Body	Metal3 on body	33	10
CM4Body	Metal4 on body	30	6
CM5Body	Metal5 on body	30	5
CM6Body	Metal6 on body	30	4

INTER-LAYER CROSSTALK CAPACITANCE

Name	Description	Value (aF/μm^2)
CM2Me	Metal 2 on metal 1	50
CM3M2	Metal 3 on metal 2	50
CM4M3	Metal 4 on metal 3	50
CM5M4	Metal 5 on metal 4	50
CM6M5	Metal 6 on metal 5	50

LATERAL CROSSTALK CAPACITANCE

Name	Description	Value (aF/μm)
CMeMe	Metal to metal (at 4 λ distance, 4 λ width)	10
CM2M2	Metal 2 to metal 2	10
CM3M3	Metal 3 to metal 3	10
CM4M4	Metal 4 to metal 4	10
CM5M5	Metal 5 to metal 5	10
CM6M6	Metal 6 to metal 6	10

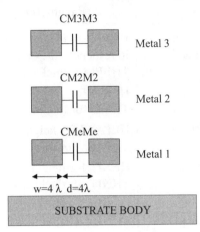

The crosstalk capacitance value per unit length is given in the design rule file for a predefined interconnect width (w = 4 λ) and spacing (d = 4 λ).

In MICROWIND, the computed crosstalk capacitance is not dependent on the interconnect width *w*.

The computed crosstalk capacitance value is proportional to 1/*d* where *d* is the distance between interconnects.

Fig. A.3   Crosstalk capacitance

A.5.1   Parameters for Vertical Aspect of the Technology

The vertical aspect of the layers for a given technology is described in the RUL file after the design rules, using code HE (height) and TH (thickness) for all layers. Figure A.4 illustrates the altitude 0, which corresponds to the channel of the MOS. The height of diffused layers can be negative, for P++ EPI layer, for example.

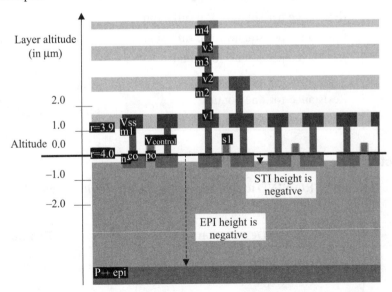

Fig. A.4   Description of the 2D aspect of the CMOS technology

Layer	Description	Parameters
EPI	Buried layer made of P++ used to create a good ground reference underneath the active area.	HEEPI for height (negative in respect to the origin) THEPI for thickness
STI	Shallow trench isolation used to separate the active areas.	HESTI for height THSTI for thickness
PASSIVATION	Upper SiO_2 oxide on the top of the last metal layer	HEPASS for height THPASS for thickness
NITRIDE	Final oxide on the top of the passivation, usually Si_3N_4.	HENIT for height THNIT for thickness
NISO	Buried N- layer to isolate the Pwell underneath the nMOS devices, to enable forward bias and back bias	HENBURRIED for height THNBURRIED for thickness

A.6   Resistance Extraction

Name	Description	Value (Ω)
RePo	Resistance per square for polysilicon	4
RePu	Resistance per square for unsalicide polysilicon	40
ReP2	Resistance per square for polysilicon2	4
ReDn	Resistance per square for n-diffusion	100
ReDp	Resistance per square for p-diffusion	100
ReMe	Resistance per square for metal	0.05
ReM2	Resistance per square for metal 2 (up to 6)	0.05
ReCo	Resistance for one contact	20
ReVi	Resistance for one via (up to via5)	2

A.6.1   Dielectrics

Some options are built in MICROWIND to enable specific features of ultra-deep submicron technology. Details are provided in the table below.

Code	Description	Example Value
HIGHK	Oxide for interconnects (SiO_2)	4.1
GATEK	Gate oxide	4.1
LOWK	Inter-metal oxide	3.0
LK11	Inter-metal1 oxide	3.0
LK22	Inter-metal2 oxide (up to LK66)	3.0

LK21	Metal2-Metal1 oxide	3.0
LK32	Metal3-Metal2 oxide (up to LK65)	3.0
TOX	Normal MOS gate oxide thickness	0.004 μm (40 Å)
HVTOX	High voltage gate oxide thickness	0.007 μm (70 Å)

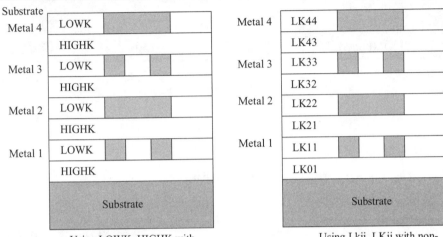

Fig. A.5 Illustration of the use of LOWK, HIGHK dielectric constants (left figure) or detailed permittivity for each layer (right figure)

A.7 Simulation Parameters

The following list of parameters is used in MICROWIND to configure the simulation.

Code	Description	Typical Value
VDD	Supply voltage of the chip	2.0 V
HVDD	High voltage supply	3.3 V
DELTAT	Simulator minimum time step to ensure convergence. You may increase this value to speed up the simulation but instability problems may rise.	0.5e-12 s
TEMPERATURE	Operating temperature of the chip	25 °C

A.7.1 Models Level1 and Level3 for Analog Simulation

Up to four types of MOS devices may be described. In the rule file, the keyword "MOS1", "MOS2", "MOS3" and "MOS4" are used to declare the device names appearing in menus. In 0.12 μm technology, three types of MOS devices are declared as follows. NMOS & PMOS keywords are used to select n-channel MOS or p-channel MOS device parameters.

Parameter	MOS1	MOS2	MOS3
Default name	High Speed	Low Leakage	High voltage
Vt (nmos)	0.3	0.5	0.7
Vt (pmos)	–0.3	–0.5	–0.7
KP (nmos)	300	300	200
KP (pmos)	150	150	100

```
* MOS definition
MOS1 low leakage
MOS2 high speed
MOS3 high voltage
```

Fig. A.6   Description of MOS options in 0.12 µm technology (cmos012.RUL)

The list of parameters for level 1 and level 3 is given below:

Parameter	Keyword	Definition	Typical Value 0.25 µm	
			nMOS	pMOS
VTO	l3vto	Threshold voltage	0.4 V	–0.4 V
U0	l3u0	Low field mobility	0.06 m^2/V.s	0.025 m^2/V.s
PHI	l3phi	Surface potential at strong inversion	0.3 V	0.3 V
LD	l3ld	Lateral diffusion into channel	0.01 µm	0.01 µm
GAMMA	l3gamma	Bulk threshold parameter	0.4 V$^{0.5}$	0.4 V$^{0.5}$
KAPPA	l3kappa	Saturation field factor	0.01 V^{-1}	0.01 V^{-1}
VMAX	l3vmax	Maximum drift velocity	150 Km/s	100 Km/s
THETA	l3theta	Mobility degradation factor	0.3 V^{-1}	0.3 V^{-1}
NSS	l3nss	Sub-threshold factor	0.07 V^{-1}	0.07 V^{-1}
TOX	l3tox	Gate oxide thickness	3 nm	3 nm
CGSO	L3cgs	Gate to source lineic capacitance	100.0 pF/m	100.0 pF/m
CGDO	L3cgd	Gate to drain overlap capacitance	100.0 pF/m	100.0 pF/m
CGBO	L3cb	Gate to bulk overlap capacitance	1e-10 F/m	1e-10 F/m
CJSW	L3cj	Side-wall source and drain capacitance	1e-10 F/m	1e-10 F/m

For MOS2, MOS3 and MOS4, only the threshold voltage, mobility and oxide thicknesses are user-accessible. All other parameters are identical to MOS1.

Parameter	Keyword	Definition	Typical Value 0.25 µm	
			nMOS	pMOS
VTO MOS2	l3v2to	Threshold voltage for MOS2	0.5 V	–0.5 V
VTO MOS 3	l3v3to	Threshold voltage for MOS3	0.7 V	–0.7 V
U0 MOS 2	l3u2	Mobility for MOS2	0.06	0.025
U0 MOS 3	l3u3	Mobility for MOS3	0.06	0.025
TOX MOS 2	l3t2ox	Thin oxide thickness for MOS2	3 nm	3 nm
TOX MOS 3	l3t3ox	Thin oxide thickness for MOS3	7 nm	7 nm

A.7.2 BSIM4 Model for Analog Simulation

The list of parameters for BSIM4 is given below:

Parameter	Keyword	Description	nMOS value in 0.12 µm	pMOS value in 0.12 µm
VTHO	b4vtho	Long channel threshold voltage at Vbs = 0 V	0.3 V	0.3 V
VFB	b4vfb	Flat-band voltage	–0.9	–0.9
K1	b4k1	First-order body bias coefficient	$0.45 \text{ V}^{\frac{1}{2}}$	$0.45 \text{ V}^{\frac{1}{2}}$
K2	b4k2	Second-order body bias coefficient	0.1	0.1
DVT0	b4d0vt	First coefficient of short-channel effect on threshold voltage	2.2	2.2
DVT1	b4d1vt	Second coefficient of short-channel effect on Vth	0.53	0.53
ETA0	b4et	Drain-induced barrier lowering coefficient	0.08	0.08
NFACTOR	B4nf	Sub-threshold turn-on swing factor. Controls the exponential increase of current with *Vgs*	1	1
U0	b4u0	Low-field mobility	0.060 m2/Vs	0.025 m^2/V-s
UA	b4ua	Coefficient of first-order mobility degradation due to vertical field	11.0e-15 m/V	11.0e-15 m/V
UC	b4uc	Coefficient of mobility degradation due to body-bias effect	–0.04650e-15 V-1	–0.04650e-15 V-1
VSAT	b4vsat	Saturation velocity	8.0e4 m/s	8.0e4 m/s
WINT	b4wint	Channel-width offset parameter	0.01e-6 µm	0.01e-6 µm
LINT	b4lint	Channel-length offset parameter	0.01e-6 µm	0.01e-6 µm

PSCBE1	b4pscbe1	First substrate current induced body-effect mobility reduction	4.24e8 V/m	4.24e8 V/m
PSCBE2	b4pscbe2	Second substrate current induced body-effect mobility reduction	4.24e8 V/m	4.24e8 V/m
KT1	b4kt1	Temperature coefficient of the threshold voltage.	– 0.1 V	– 0.1 V
UTE	b4ute	Temperature coefficient for the zero-field mobility U0	– 1.5	– 1.5
VOFF	b4voff	Offset voltage in sub-threshold region	– 0.08 V	– 0.08 V
PCLM	b4pclm	Parameter for channel length modulation	1.2	1.2
TOXE	b4toxe	Gate oxide thickness	3.5e-9 m	3.5e-9 m
NDEP	b4ndep		0.54	0.54
XJ	b4xj	Junction depth	1.5e –7 m	1.5e –7 m

For MOS2, MOS3 and MOS4, only the threshold voltage, mobility and oxide thicknesses are user-accessible. All other parameters are identical to MOS1.

A.8   Technology Files for Dsch

The logic simulator includes a current evaluator. In order to run this evaluation, the following parameters are proposed in a TEC file (example: cmos012.TEC):

Dsch 2.0 - technology file

NAME "CMOS 0.12um"

VERSION 14.12.2001

* Time unit for simulation

TIMEUNIT = 0.01

* Supply voltage

VDD = 1.2

* Typical gate delay in ns

TDelay = 0.02

* Typical wire delay in ns

TWireDelay = 0.07

* Typical current in mA

TCurrent = 0.5

* Default MOS length and width

ML = "0.12u"

MNW = "1.0u"

MPW = "2.0u"

Microwind Program Operation & Commands

B.1   Getting Started

Download the software from *http://www.microwind.net* and follow the installation instructions.

The software runs on Windows 95, 98, NT and XP operating systems.

B.1.1   Command Line Parameters

The command line may include two parameters:

The first parameter is the default mask file loaded at initialization.

The second parameter is the design rule file loaded at initialization.

For example, the command "microwind3 test.MSK cmos018.rul" executes MICROWIND with a default mask file "test.MSK" and the rule file "cmos018.RUL".

B.1.2   Configure the Microwind Icon

You may program the MICROWIND icon by a click with the right button, and then "properties". The default target does not contain any parameter. Just add the default layout file name and the default design rule file. In the example given in Figure B.2, Microwind uses the file "TEST.MSK" and the design rule file "CMOS018.RUL" as initial parameters.

Fig. B.1   Access to Microwind2 icon properties

Fig. B.2   Configuring MICROWIND with a default file name "test.MSK" and default technology "cmos018.RUL"

B.2   List of Commands in Microwind

B.2.1   About Microwind

Information about the software release and contact for support.

B.2.2 Add Text to Layout

Use this icon to fix a text to one box or location in the design. That text illustrates the layout and should be used as much as possible for each significant node such as inputs and outputs. In order to add some text to a particular place, proceed as follows:

- Click on the icon.
- Set the text location with the mouse. A dialog box appears.
- Enter the text in front of **Label name** and press **Assign**. The text is set in the drawing.

A text can be modified as follows: click on the icon, click inside the existing text. The old text appears. Modify it and click on **Assign**. You may add a clock, a pulse, a VDD or VSS voltage source to the text.

B.2.3 Convert Into

Microwind converts the MSK layout into CIF using a specific interface, invoked by (File -> Make CIF file). The CIF file can be exported to VLSI CAD software. The right table of the screen (Figure B.3) gives the correspondence between Microwind layers and CIF layers, the number of boxes in the layout and the corresponding over-etch. The over-etch is used to modify the final size of the CIF boxes in order to fit the exact design rules.

Click on **Convert to CIF** to start conversion. Some parts of the result appear in the left window.

The main unit is 1 nm. You may change it to fit the requirements of the target CAD tool.

For CMOS 0.25 µm rule file (cmos025.RUL), notice the over-etch applied to contact and via. This over-etch is mandatory to obey the final design rules, while keeping the user-friendly and portable lambda-based design.

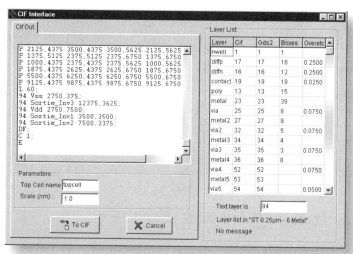

Fig. B.3 The CIF generation screen

Concerning diffusions, notice that the CIF generator produces active areas and implants. MICROWIND uses simple n+diffusion and p+diffusion while most industrial layout design tools use the concept of active area surrounded by implants, either n+ or p+, as illustrated below.

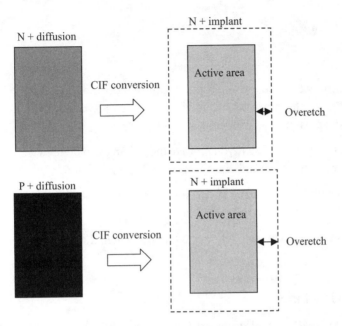

Fig. B.4 The CIF conversion produces active areas and implants, to be compatible with the industrial process

This means that each n+ diffusion box drawn in MICROWIND is converted into two boxes, one linked with "Active Area", with a code name declared in the design rule file, and a second box is linked to "diffn", with a given over-etch. Each p+ diffusion box is converted into two boxes too. On is linked with the same "Active Area", and a second box is linked to "diffp", with a given over-etch.

B.2.4 Colours

- Switch to monochrome: the layout is drawn in black and white. This type of drawing is convenient to build monochrome documentation. Press "Alt"+"Print Screen" to copy the screen to the clipboard. Then, open "Microsoft Word™", click **Edit → Paste**. The screen is inserted into the document.

- White background. The layers appear with a palette of colours on a white background.

B.2.5 Copy

Click on the **Copy** icon. Move the cursor to the design window, and delimit the active area with the mouse. Consequently, all the graphics included in this area are copied. The external shape of the copied

elements appears. Fix those copied elements at the desired location by a click on the mouse.
Click on **Undo** to cancel the copy command.

B.2.6 Cut

Click on the **Cut** icon. Move the cursor to the design window, and delimit the active area with the mouse. Consequently, all the graphics included in this area are erased. Click on **Undo** to fix those elements back into the design.

- A layer is protected from erasing if you remove the tick in the palette twice. In the palette, an empty square to the right of the layer indicates a protected layer.

- A layer is unprotected from erasing if you select it again in the palette. A tick in the square to the right of the layer indicates an unprotected layer.

- One box only can be erased by a click inside that box when the cut command is active. The box is then erased.

B.2.7 Compile One Line

The cell compiler is a specific tool designed for the automatic creation of CMOS cells from logic description. Click on **Compile → Compile One Line**. The menu below appears. The default equation corresponds to a 3-input NOR gate. If needed, one can use the keyboard in order to modify the equation and then click on **Compile**. The gate is compiled and its corresponding layout is generated.

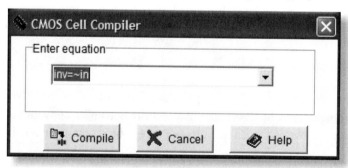

Fig. B.5 The cell compiler window

- The first item of the one-line syntax corresponds to the output name.

- The latter is followed by the sign « – », by the list of input names separated by operators AND '&', OR '|', XOR '^', NOT '~', XNOR '~^'. If need be, parenthesis can be added.

- The input and output names are a maximum of eight character strings.

Table B.1   Examples of logic cell descriptions

Cell	Formula
Inverter	out = ~in
NAND gate	n = ~(a & b)
3 Input OR	s = a | b | c
3 Input NAND	out = ~(a & b & c)
AND-OR Gate	cgate = a & (b | c)
CARRY Cell	cout = (a & b) | (c & (a | b))

The p-channel transistors are located on the top of the n-channel transistor net. If some layout already exists near those icons, the cell origin is moved to the right until enough free space is found. If the NOT operator (Symbol '~') has not been specified after the '=' sign, an inverter is added at the right hand side of the compiled cell. That is why an AND gate is compiled as a NAND gate followed by an inverter.

B.2.8   Compile Verilog File

The cell compiler can handle layout generation from a primitive-based Verilog description text into a layout form automatically. Click on **Compile → Compile Verilog File**. Select a Verilog text file and click on "**Generate**". For instance, the MICROWIND directory contains the ≪ FADD.TXT ≫ file which corresponds to the description of a full adder.

```
// Dsch 2.5d
// 21/04/02 14:00:50
// C:\Dsch2\Book on CMOS\fadd.sch

module fadd(C,B,A,Sum,Carry);
  input C,B,A;
  output Sum,Carry;
  xor #(12) xor2(w4,A,B);
  nand #(10) nand2(w5,A,C);
  nand #(10) nand2(w6,B,C);
  nand #(10) nand2(w7,B,A);
  xor #(12) xor2(Sum,w4,C);
  nand #(10) nand3(Carry,w7,w6,w5);
endmodule

// Simulation parameters
// C CLK 10 10
// B CLK 20 20
// A CLK 30 30
```

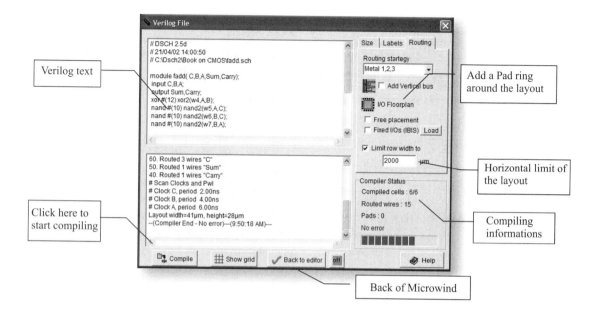

Fig. B.6   The Verilog compiler window

PRIMITIVE	NODES	EXAMPLE
dreg	Inputs : Data,RESET, CLOCK Outputs: Q, nQ	dreg reg1(d,rst,h,q,nq);

Inv, not	Inputs : IN
	Outputs: OUT
		inv inv1(s,e); // both 'inv' and 'not'
		not inv1(s,e); // can be used

and	Inputs : 2 to 4 Outputs: S	and and1(s,a,b,c,d); // limit inputs to 4
nand	Inputs : 2 to 4 Outputs: S	nand nand1(s,a,b,c,d);
or	Inputs : 2 to 4 Outputs: S	or or3(s,a,b,c);
nor	Inputs : 2 to 4 Outputs: S	nor my_nor4(s,a,b,c,d);
xor	Inputs : a,b Outputs: S	xor xor_gate(xor_out,d0,d1);
Nmos	Inputs: gate, source Outputs: drain	nmos nmos1(d,s,g);

Fig. B.7   The Verilog primitives supported by the CMOS compiler

The input/output nodes are routed on the top and the bottom of the active parts, with a regular spacing to ease automatic channel routing between cells. Click **Compile → Show grid** to superimpose the routing grid on the layout. In Figure B.8, the routing of input/output between basic cells is presented. Notice that the routing is performed either at the top or at the bottom of the active parts.

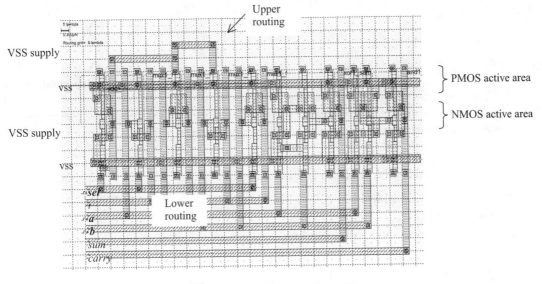

Fig. B.8   The compiler grid

B.2.9   Design Rule Checker

The design rule checker (DRC) scans all the designs and verifies that all the minimum design rules are respected. Click on the icon above or on **Analysis → Design Rule Checker** to run the DRC. The errors are highlighted in the display window, with an appropriate message giving the nature of the error. Details about the position and type of the errors appear on the screen.

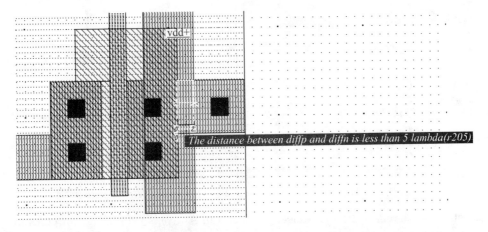

Fig. B.9   Example of design rule error

B.2.10 Draw a Box

The «Draw Box» icon is the default icon. It creates a box in the selected layer. The default layer is polysilicon. If the «Draw Box» icon is not selected, click on it. Then, move the cursor to the display window and fix the first corner of the box with a press of the mouse. Keep the mouse pressed and drag it to the opposite corner of the box. Release the mouse and see how the box is created.

- The active layer is selected in the palette.
- The red colour indicates the active layer.
- The gray key on the right of the layer button specifies that all boxes using the layer can be erased, stretched or copied.
- A click on the gray key turns the key to a red color. A red key protects the layer.

B.2.11 Duplicate XY

The command **Duplicate XY** is very useful for generating an array of identical cells such as RAM cells. Click on **Edit → Duplicate XY**, include the elements to duplicate in an area defined by the mouse, and the screen shown in Figure B.10 appears. In both X and Y, the default multiplication factor is x 2. You may adjust the space between cells. By default, the cells touch each other. Selected boxes appear in the right window, and also in yellow on the main layout window.

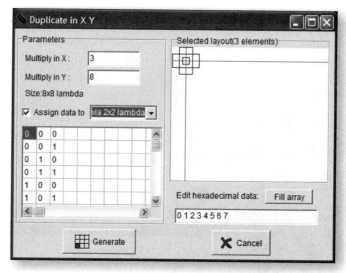

Fig. B.10   The Duplicate X,Y menu, used for a metal1/metal2 crosspoint

The data option (Assign data, Edit Hexadecimal data, Fill Array) is very useful for generating ROM masks or decoder arrays. An example of decoder array is given in Figure B.11. The programming affects the via between metal1 and metal2, according to the list of hexadecimal values given in the editing list.

- Click the desired data on the **Edit hexadecimal data** editing area. The values must be separated by a space.

- Click **Fill array** to convert these data into Boolean values, according to the X,Y array size.

- Select the appropriate box for programming. In this case, the 0/1 create of not a via at each X,Y location.

- Click **Generate**. The following result appears:

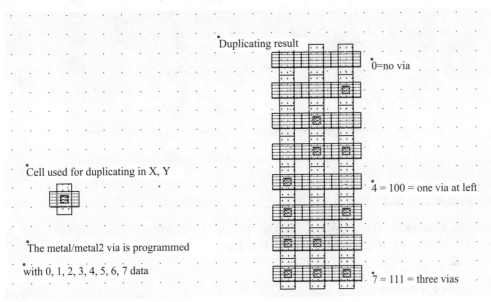

Fig. B.11 Example of duplicating a pattern with programmed via (DuplicateXYExample.MSK)

B.2.12 Flip

↕ Flip horizontal
≚ Flip vertical

In order to apply a rotation or a flip to one part of the design, click on **Edit → Flip and Rotate**, and choose the appropriate Flip command (Horizontal or vertical). Delimit the active area of the boxes in the layout which will be modified.

B.2.13 Help

This icon provides an on-line help for using **Microwind**. It includes a summary of commands, and some details about the design rules, as shown in Figure B.12.

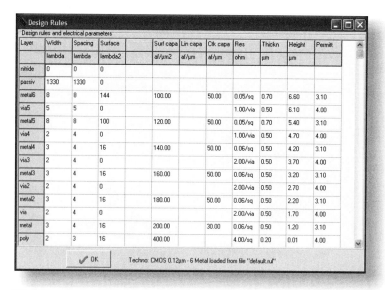

Layer	Width	Spacing	Surface	Surf capa	Lin capa	Ctk capa	Res	Thickn	Height	Permitt
	lambda	lambda	lambda2	af/µm2	af/µm	af/µm	ohm	µm	µm	
nitride	0	0	0							
passiv	1330	1330	0							
metal6	8	8	144	100.00		50.00	0.05/sq	0.70	6.60	3.10
via5	5	5	0				1.00/via	0.50	6.10	4.00
metal5	8	8	100	120.00		50.00	0.05/sq	0.70	5.40	3.10
via4	2	4	0				1.00/via	0.50	4.70	4.00
metal4	3	4	16	140.00		50.00	0.06/sq	0.50	4.20	3.10
via3	2	4	0				2.00/via	0.50	3.70	4.00
metal3	3	4	16	160.00		50.00	0.06/sq	0.50	3.20	3.10
via2	2	4	0				2.00/via	0.50	2.70	4.00
metal2	3	4	16	180.00		50.00	0.06/sq	0.50	2.20	3.10
via	2	4	0				2.00/via	0.50	1.70	4.00
metal	3	4	16	200.00		30.00	0.06/sq	0.50	1.20	3.10
poly	2	3	16	400.00			4.00/sq	0.20	0.01	4.00

✓ OK   Techno: CMOS 0.12µm - 6 Metal loaded from file "default.rul"

Fig. B.12 Design rules and electrical rules proposed in the help menu

B.2.14   Insert Layout

The command **File → Insert Layout** is used to add an MSK file to the existing files. The inserted layout is fixed at the right lower side of the existing layout. The current file name remains unchanged.

B.2.15   Leave Microwind

Click on **File → Leave Microwind** (Or CTRL+Q) in the main menu. If you have made a design or if you have modified some data, you will be asked to save it. After confirmation, you can return to Windows.

B.2.16   Generate

- Box
- Contacts
- nMOS Device
- pMOS Device
- Resistor
- Metal bus
- Metal path
- Inductor
- Diode
- Capa
- Logo
- I/O Pads

The layout generator includes a set of pre-defined layout macros such as box, contacts, nMOS and pMOS devices, resistor, metal bus, metal path, inductor, diode, capacitor, logo and I/O pads. Those cells are built according to design rules, and user-size parameters.

B.2.17   Generate Box

This is the same as the **Draw Box** command described on page 386.

B.2.18   Generate Contacts

This macro generates contacts such as polysilicon/metal, n-diffusion/metal, p-diffusion/metal and metal/metal2 metal2/metal3, (etc.), or stacked contacts can be obtained here. You may also click the above icon in the palette. Multiple contacts can be generated when entering number of contacts in X and Y greater than one.

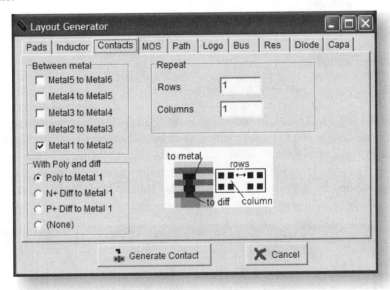

Fig. B.13   The contact menu

B.2.19   Generate nMOS, pMOS Devices

This macro generates either an n-channel or a p-channel transistor. The double gate MOS is also available in some CMOS technologies, for building the EEPROM memory. The parameters of the cell are the channel length (default value is given by the design rules), its width, and the number of gates. Once those parameters are defined, the device outline appears. Click on the mouse to place it in the appropriate place.

Width and length,
given in μm or λ

Number of fingers,
1 by default

NMOS, pMOS or
double gate MOS

MOS options, depending on the
selected technology

Enter W, L in lambda or in μm

Add polarization on the
left side of the MOS

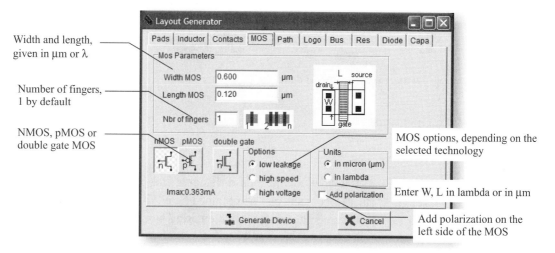

Fig. B.14 The MOS generator menu

B.2.20 Generate Resistor

This command generates a resistor in n-well, polysilicon or poly2, N+ or P+ diffusion. The default aspect of the resistor is a Z with three bars (parameter n in the menu). A virtual resistor symbol may be inserted in the resistor layout, to ensure the handling of the resistance effect during simulation. By default, an option layer configured to remove salicidation is added to the resistance layout. By default, all polysilicon and diffusions have a salicide surface metallization to decrease by a factor of around 10, the sheet resistance. The unsalicide option is recommended for a high resistance value in a small area. Finally, contacts are added by default to the near and far ends of the resistance to facilitate further interconnection.

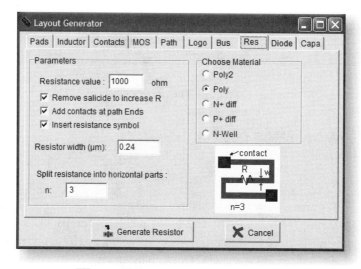

Fig. B.15 The resistor generator menu

B.2.21 Generate Inductor

This command generates a coil made from use-defined metal layers. This item is used for very high frequency oscillators. This inductor is viewed as an inductance thanks to a virtual inductor symbol inserted in the layout. An evaluation of the inductance is proposed at a click of the button **Update L,Q**. The inductor may be a stack of several layers, thanks to the layer menu.

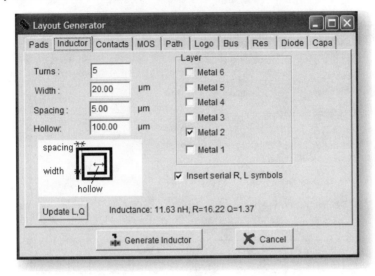

Fig. B.16 The inductor generator menu

B.2.22 Generate Bus

This command generates a set of parallel lines with user-defined layer, width and spacing. This command is useful to build coupled interconnects, or bus path used in the final routing of a chip.

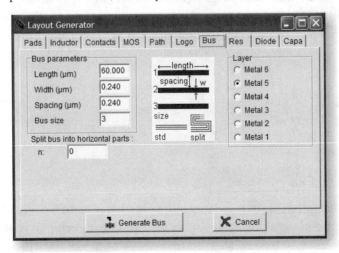

Fig. B.17 The bus generator menu

B.2.23 Generate Path

This command generates a path of interconnects by using one single layer. Both the path width can be changed and the alignment to the routing grid can be changed. A set of contacts can also be placed at both ends of the path. This command is very useful for VDD and VSS supply drawing and single layer interconnects.

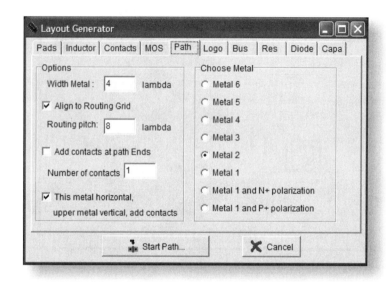

Fig. B.18 The path configuration menu

B.2.24 Generate I/O Pads

It is possible to add various items such as a single pad, (usually 80×80 µm), or even a set of pads all around and to the layout by using the VDD and VSS power rings. In the last case (adding more than one pad), give the number of pads on each side of the chip and if need be, modify the width of the VDD and VSS tracks, as well as the number of VDD/VSS pad pairs.

B.2.25 Generate Diode

You may also generate a polarization seal around the contact to create a pad diode protection, for example.

B.2.26 Make Spice File

Click on **File → Make Spice File** to translate your design into a SPICE compatible description. The circuit extractor included in the software generates the equivalent circuit diagram of the layout and a spice compatible netlist ready to be simulated. You may select the model you will be using for simulation. The choice lies between model 1, model 3 and BSIM4.

- The SPICE description includes the list of n-channel and p-channel transistors and their associated width and length extracted from the layout.
- The text file also details the node names, parasitic capacitances, and device models.
- The SPICE file name corresponds to the current file name with the appendix .CIR

B.2.27 Measure distance

The ruler gives the horizontal and vertical measurements (dx and dy) between two points, directly on the screen in lambda and in micron. The algebraic distance (d) is also given in μm. The ruler is simply erased by the command **View** → **Refresh the screen** or by a press of <ESC>.

B.2.28 MOS Characteristics

Click on the icon. The *Id/Vd* curve of the default MOS (W = 20 μm, L = L minimum) appears.

- The effects of the changing of the model parameters can be seen directly on the screen by a click on the little arrows (up/down), which change the parameter values.
- Click "*Id* vs. *Vg*" to highlight the threshold voltage.
- Click "*Id*(log) vs. *Vg*" to see the sub-threshold behaviour.
- Add measurements by selecting a «.MES» file.
- Skip from nMOS to pMOS device by a click on the corresponding button.
- Select the size of the device in the lower list menu.

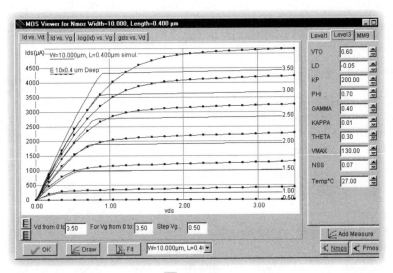

Fig. B.19

Three models can be used:

- MOS Model 1 (Berkeley Spice level 1) for long channel devices: This model is obsolete but remains interesting for comparison with advanced models.
- MOS Model 3 (Simplified version of Berkeley MOS level 3): This is still in use for first order estimation of the circuit performances.
- BSIM4 (Simplified version of Berkeley MOS BSIM4): The state-of the art model for deep-submicron device modelling.

B.2.29   MOS List

Click **Edit → MOS List** to get the list of n-channel and p-channel MOS devices currently edited in the layout. The MOS list is displayed in the navigator window. Click the desired MOS in the list to zoom at the corresponding location in the layout.

B.2.30   Move, Stretch a Box

To move one box, click on the above icon. Using the mouse, create an area that includes the box. Then, drag the mouse to the new location and release the mouse. As a result, the box has been moved the new place. Repeat the same in order to move a set of boxes.

- To protect a layer from moving, click in the rectangle the palette that is situated on the right side of the layer. This will remove the tick.
- To stretch a box, click on one side of the box that you want to stretch. The box outline appears. Drag the mouse to the new location and release the button. The box is stretched.

Tip: To catch the desired border of the box, draw a line perpendicular to the border, entering the box.

B.2.31   Move Step-by- step

To move one box lambda per lambda, click **Edit → Move Step by Step**. Using the mouse, create an area that includes the boxes. The selection appears in yellow. Then, click the arrow until the selection has been moved the new place. The step value (in lambda) is fixed in the edit line.

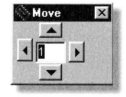

B.2.32   New

Click on **File → New** in order to restart the software with an empty screen. The current design should be saved before asserting this command, as all the graphic information will be physically removed from the computer memory. No **Undo** is available to disable the **New** command.

B.2.33 Open

Click on the above icon. In the list, double-click on the file to load. «.MSK» is the default extension that corresponds to the layout files. The CIF files «.CIF» can also be loaded. The appropriate conversion program transforms the input CIF into MSK format.

B.2.34 Palette

The palette is located on the right side of the screen. A little tick indicates the current layer. The selected layer by default is a polysilicon (PO). The list of layers is given below.

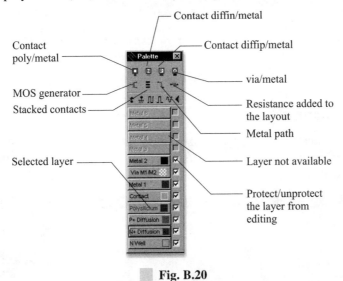

Fig. B.20

- If you remove the tick on the right side of the layer, the layer is switched to protected mode. The Cut, Stretch and Copy commands no longer affect that layer.

- Use **View → Protect all** to protect all layers. The ticks are erased.

- Use **View → Unprotect all** to remove the protection. All layers can be edited.

B.2.35 Process Section in 2D

Click on the above icon to access process simulation. A mouse-operated line is given and embodies the cross-section. The screen shown in Figure B.21 appears. The arrows can be used to move the cross-section to the right or to the left in the X axis, and forward and backward in the Y axis. Zooms in and out are available. Remove the layer names by removing the tick in front of "Layer infos".

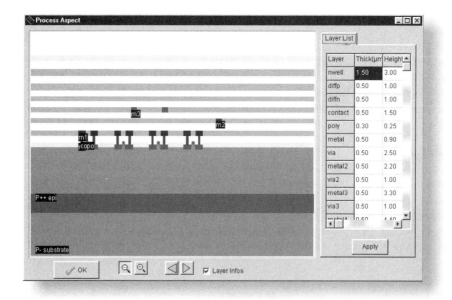

Fig. B.21 Process section

B.2.36 Process Steps in 3D

Click **Simulation → Process Steps in 3D**. Click **Next step** to watch how the layout currently edited on the screen will be fabricated using the selected technology. Use the arrow to shift the displayed portion. Zooms in and out are available.

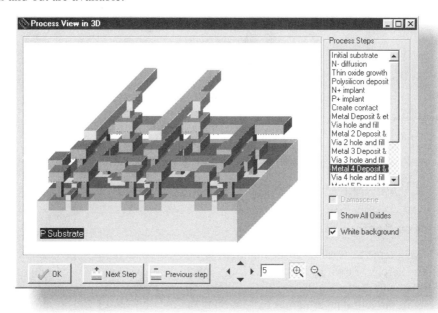

Fig. B.22 Process steps

B.2.37 Protect All

Click on **View** → **Protect All** to protect all layers for editing purpose. All ticks in the palette are removed.

B.2.38 Print Layout

Click on **File** → **Print Layout** to transfer the graphical contents of the screen to the printer. Alternatively, you can make a copy of the window into the clipboard in order to import the screen into your favorite text editor by pressing <Alt>+<Print Screen>. In the text editor or in the graphic editor, simply click on **Edit** → **Paste**. We recommend that you switch to monochrome mode first by invoking the function **File** → **Colors** → **Switch to Monochrom**. In that case the layout will be drawn in a white background color using gray levels and patterns.

B.2.39 Rotate

To apply a rotation to one part of the design, click on **Edit** → **Rotate**. Delimit the active area of the boxes in the layout so that it can be modified using the mouse.

B.2.40 Save

Click on **File** → **Save** to save the layout with its current name. The default name is «EXAMPLE.MSK »

B.2.41 Save As

A new window appears, into which you are to enter the design name. Use the keyboard and type the desired file name. Press **Save**. Your design is now registered within the .MSK appendix.

B.2.42 Search Text

The most convenient way to find a text in the layout is to invoke **Edit** → **Search text**. The list of text labels appears in the navigator menu. If you click on the desired text, the screen is redrawn so that the text label is at the center of the window, with two lines drawing a cross at the text location. Its properties appear in the navigator menu.

- Click on **Hide** to close the navigator window.

- Click on **Extract** to add the electrical properties of the selected text if the layout has not been previously extracted.

- In the case of a very long text list, select the first letter of the text at hand, press that letter on the keyboard. This will automatically effect an alphabetic research and the selector will move to the first label starting with the selected letter.

B.2.43   Select Foundry

Click on **File → Select Foundry**. The list of available processes appears. The default design rule file is written in bold characters. Various technologies are available from 1.2 down to 0.12 μm. Click on the rule file name and the software re-configures itself in order to adapt to the new process

B.2.44   Simulation Parameters

- The default extraction includes the removal of redundant boxes (Purge) and the removal of overlaps (Merge). The fast extraction does not handle the Purge or Merge operations.

- The MOS level can be chosen between level 1, 3 and 9. See Chapter 2 for more details about those models.

- Other options concern the computation of lateral capacitance and vertical crosstalk capacitance.

B.2.45   Simulation on Layout

The simulation is performed directly on the layout with a palette of colours. The most interesting layout files to be simulated in this mode are analog blocks such as the DAC.

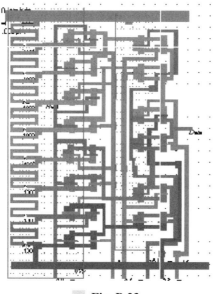

Fig. B.23

During switching, the MOS devices of the inverter behave alternately as closed and opened switches, as illustrated earlier. In a first order approximation, the equivalent model of the switch is a resistance. In the illustration in Figure B.24, we plot the operating point of the n-channel MOS and p-channel MOS during switching. What we see is a complex trajectory of the operating point in the Id/Vd characteristics, for both the nMOS and the pMOS devices. Therefore, the resistance is not a simple function, but rather a varying resistance between several values.

The Figure has been obtained by using a specific simulation mode called **Simulation on Layout** in the simulation menu. The I/V characteristics of the selected devices are updated during simulation to track the functional point. During the analog simulation, the node voltage is superimposed on the layout and appears with a palette of colors:

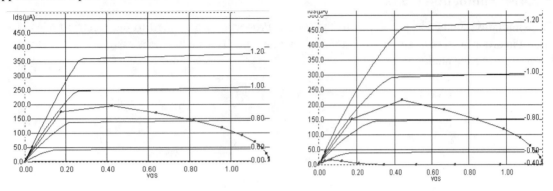

Fig. B.24   The nMOS and pMOS devices during switching (Inv.MSK)

B.2.46   Start Simulation

Both the above icon or the command **Simulate** → **Start Simulation** give access to the automatic extraction and analog simulation of the layout.

- Click on **Voltage vs. Time** to obtain the transient analysis of all visible signals. The delay between the selected start node and selected stop node is computed at VDD/2. You can change the selected start node in the node list, in the right upper menu of the window. You can do the same for the selected stop node.

- Click on **Voltage and Currents** so as to make all voltage curves appear in the lower window, and the VDD, the VSS and the desired MOS currents appear in the upper window. In that mode, the dissipated power within the simulation is also displayed.

- Click on **Voltage vs. Voltage** to obtain transfer characteristics between the X-axis selected node and the Y-axis selected node. Initially the start node is the first clock or pulse of the node list, and the stop node is the first varying node. This mode is useful for the computing of the Inverter characteristics (commutation point), the dc response of the operational amplifier, or for the Schmitt trigger to see the hysteresis phenomenon. The first simulation computes the value of the stop node for start node varying from 0 to VDD. The second click on **Simulate** computes the same for start node varying from VDD to 0. This feature is interesting for circuit with memory effects (Schmidt trigger). Note that the curves may not be exactly the same. You may increase the precision by reducing the computational step "Precision", accessible in the menu, expressed in mV.

Note: You can modify the minimum simulation step Δt, but it may be dangerous. If you increase Δt the simulation speed improves but the numerical error may lead to unstable simulations. If you decrease Δt, the simulation speed is decreased too but the numerical precision is improved. The risk of computing divergence is reduced.

B.2.47 Simulation Icons

The simulation icons add properties to the nodes. Properties are applied to the electric nodes of the circuit in order to serve as simulation guides. You must specify which node is assigned to which voltage before starting the analog simulation.

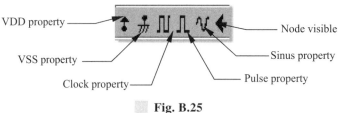

VDD property — Node visible

VSS property — Sinus property

Clock property — Pulse property

Fig. B.25

VDD and VSS: The node is pushed to the power supply voltage with icon VDD, and pulled to the ground 0 V with icon VSS.

CLOCK: When a node becomes a clock, the parameters of the latter are divided as follows: rise time, level one, fall time, and level zero. All values are expressed in nano seconds (ns). If you ask for a second clock, the period will be multiplied by two.

- You may alter level 0 and level 1 by entering a new value with the keyboard.
- To generate a clock starting from VDD instead of VSS, click on Invert L/H.
- Use Period/2 to multiply the clock period by two.
- Use Period/2 to divide the clock period by two.

PULSE: The pulse switches from "Level 0" (0 by default) to "Level 1" (VDD by default" depending on the user-defined time table.

- Enter the string "0101100" and press "Insert". The time-table is updated
- Click "Erase": all lines situated after the selected element of the time-table are erased.

SINUS: The sinusoidal waveform parameters are the amplitude, the offset, frequency and phase. **VISIBLE NODE:** Click on the "eye" and click on the existing text in the layout to make the chronograms of the node appear. Initially, all nodes are invisible, but the clocks and impulse nodes are subsequently made visible.

B.2.48 Statistics

The command **File → Statistics** provides some information about the current technology, the percentage of memory used by the layout and the size of the layout plus its detailed contents. If the layout has previously been extracted or if you click «extract now», the number of devices and nodes will be updated.

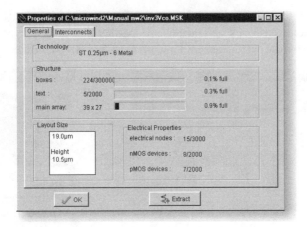

Fig. B.26

B.2.49 Undo

The **Undo** command (**Edit → Undo**) is useful for not taking into account the last editing command. It is possible to undo the commands Cut, Paste, Copy, Move, Stretch, Edit and Compile.

B.2.50 Unprotect All

Click on **View → Unprotect All** to select all layers for editing purpose. All ticks in the palette are asserted.

B.2.51 Unselect All

Click on **View → Unselect All** (or <ESC>) to unselect the layout. This command is useful for drawing the layout back into its default colours after commands such as **View Interconnect** or **View Node** which highlight one single node.

B.2.52 View All

Click **View → View All** to fit the screen with all the graphical elements currently on display.

B.2.53 View Node

Click on the icon above or on **View →View Node**. Then, click in the desired box in the layout. After an extraction procedure has been carried out, you will see that all the boxes are connected to that node. In the case of a large layout, the command may take time. The associated parasitic capacitance, the list of text labels added to the selected boxes, and the node properties are also displayed in a separate navigator window. Click **Unselect**, **Hide**, <Escape> or **View → Unselect All** to unselect the layout.

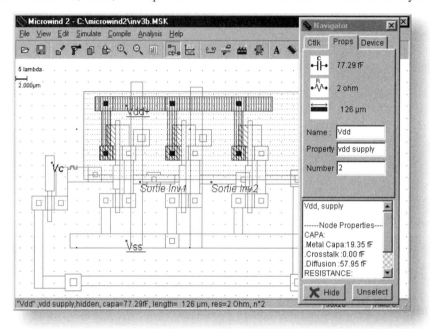

Fig. B.27

B.2.54   View Interconnect

The command **View → View Interconnect** performs an electrical extraction of the metal and polysilicon boxes connected to the desired point. As compared to **View Node**, this command works faster but does not consider diffused layers that can extend the node interconnect network. The command gives the list of connected text labels. Click on <Escape> or on **View → Unselect All** to unselect the layout.

B.2.55   Zoom In and Out

The above icons perform Zoom In and Zoom Out. When zooming in, the area determined by the mouse will be enlarged to fit the display window. When zooming out, the area determined by the mouse will contain the display window.

- If you click once, a zoom is performed at the desired location.

- Press Ctrl+A for **View All**, and Ctrl+o for zoom out.

Dsch Logic Editor Operation and Commands

C.1   Getting Started

Download the software from *http://www.microwind.net* and follow the installation instructions.

The software runs on Windows 98, NT and XP operating systems.

C.2   Commands

C.2.1   About Dsch

Information about the software release and support.

C.2.2   Connect

Use the "Connect" icon to create the electrical contact between crossing interconnects.

C.2.3 Cut

 (CTRL+X)

Click on the Cut icon. Move the cursor to the design window, and delimit the active area with the mouse. Consequently, all the graphics included in this area are erased. Click on **Undo** to fix those elements back into the design. One symbol <u>only</u> can be erased by a click inside its shape when the cut command is active. The symbol is then erased. One single interconnect can be erased by a click on its wire when the cut command is active.

C.2.4 Check Floating Lines

The command **Simulate → Check Floating Lines** may be found in the Simulation menu. The schematic diagram is scanned in order to detect interconnects with a wrong connection to the symbol or other interconnects, as in the example given in Figure C.1.

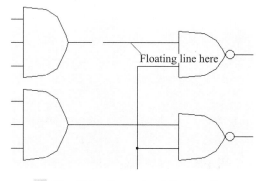

Fig. C.1 Example of floating line

C.2.5 Copy

 (CTRL+C)

Click on the Copy icon. Move the cursor to the design window, and delimit the active area with the mouse. Consequently, all the graphics included in this area are copied. The external shape of the copied elements appears. Fix those copied elements at the desired location by a click on the mouse.

Click on Undo to cancel the copy command.

C.2.6 Critical Path Details

The critical path is the series of logic gates between the output and input with the longest propagation delay. The command **View → Critical Path Details** gives the list of symbols and cumulated delays which build the critical path. The graph of the critical path may be highlighted by using the command **Simulate → Find Critical Path**.

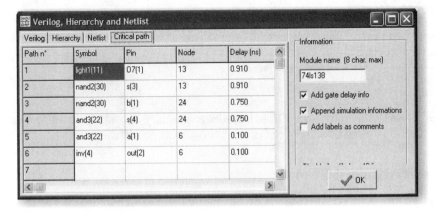

Fig. C.2   Critical path details

C.2.7   Design Hierarchy

The design hierarchy command gives an interesting insight in the hierarchical structure of symbols, together with the list of input and output symbols.

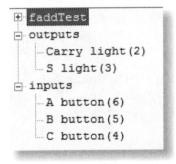

Fig. C.3   An example of design hierarchy

C.2.8   Electrical Net

Click on the icon above or on **View → Electrical Net**. Then, click in the desired interconnect or pin in the schematic diagram. After an extraction procedure has been carried out, you will see that all the wires connected to that node. Click <Escape> or **View → Unselect All** to unselect the diagram.

C.2.9   Find Critical Path

The critical path is the series of logic gates between the output and input with the longest propagation delay. The command **Simulate → Find Critical Path** shows the graph of the critical path. Invoke the command **View → Critical Path Details** to extract the list of symbols and cumulated delays which build the critical path.

C.2.10   Flip Vertical/Horizontal

In order to apply an horizontal or vertical flip to one part of the design, click on **Edit → Flip**. Then, delimit the area inside which the elements will be changed.

C.2.11   Help

This provides an on-line help for using Dsch. Includes a summary of commands, some details about the design rules, and some precision about the current version of the software.

C.2.12   Insert Another Schema

The command "**File → Insert an other Schema**" is used to add an SCH file to the existing files. Its contents are fixed at the right lower side of the existing schematic diagram. The current file name remains unchanged.

C.2.13   Insert User Symbol

The command **Insert → User Symbol** is used to add a user-defined symbol to the existing schematic diagram. The user symbol is created using the command **File → Schema To new Symbol**. The inserted symbol can be fixed at the desired location.

C.2.14   Leave Dsch

Click on "**File → Leave Dsch**" (Or CTRL+Q) in the main menu. If you have made a design or if you have modified some data, you will be asked to save it. After confirmation, you can return to Windows.

C.2.15   Line

(Or right click with the mouse)

The "Line" icon is the default icon. It creates an interconnection between two points in the schematic diagram. If the "Line" icon is not selected, click on it. Then, move the cursor to the display window and fix the start point of the interconnect with a press of the mouse. Keep the mouse pressed and drag it to the interconnect end. Release the mouse and see how the line is created.

C.2.16   List of Symbols

The command gives the complete netlist corresponding to the schematic diagram. The internal structure of hierarchical symbols also appears. The symbol name, list of pins, related node numbers and modelm number are listed.

C.2.17   Make Verilog File

DSCH converts the schematic diagram into Verilog by using a specific interface, invoked by **File → Make Verilog file**. The Verilog text can be exported to VLSI CAD software. The right table of the screen (Figure C.4) gives the list of options: module name, gate delay information, list of labels, and general information about the size of the design. The conversion of the schematic diagram into a Verilog description is useful for compiling the schematic diagram into layout using MICROWIND. The Verilog description is a text with a predefined syntax. Basically, the text includes a description of the module (name, input, output), the internal wires, and the list of primitives. An example of Verilog file generated by DSCH is given in Figure C.4.

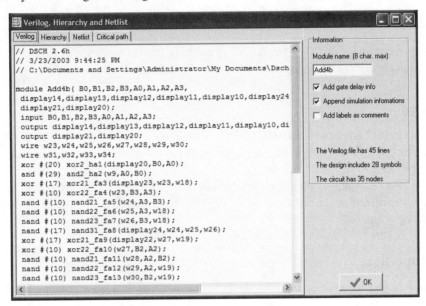

Fig. C.4 Conversion into Verilog

C.2.18 Monochrom/Colour (File Menu, F5)

Switch to monochrome mode: the layout is drawn in black and white. This type of drawing is convenient to build monochrome documentation. Press "Alt"+"Print Screen" to copy the active window to the clipboard. Then, open "Word", click **Edit → Paste**. The screen is inserted into the document.

C.2.19 Move

In order to move one graphical element, click on the "Move" icon or "**Edit → Move**". Then using the mouse, draw an area that includes the elements. Then, drag the mouse to the new location and release the mouse. As a result, the elements are moved the new place. One single line can be moved or stretched (depending where you click) by a direct click on the line. One single text can be moved by a direct click on the text location.

C.2.20 New

Click on **File → New** in order to restart the software with an empty screen. The current design should be saved before asserting this command, as all the graphic information will be physically removed from the computer memory. No Undo is available to disable the New command.

C.2.21 Open

Click on the above icon. In the list, double-click on the file to load. ".SCH" is the default extension that corresponds to the schematic diagrams.

C.2.22 Paste

Invoke the Paste command **Edit → Paste**. All previously copied elements are pasted at the desired location. Deleted elements can be replaced that way. Click on **Undo** to cancel the *paste* command.

C.2.23 Print

Click on **File → Print Layout** to transfer the graphical contents of the screen to the printer. Alternatively, you can make a copy of the window into the clipboard in order to import the screen into your favorite text editor by pressing <Alt>+<Print Screen>. In the text editor or in the graphic editor, simply click on **Edit → Paste**. We recommend that you switch to monochrome mode first by invoking the function **File → Monochrome/color**. In that case the layout will be drawn in a white background color using gray levels and patterns.

C.2.24 Properties

The command **File → Properties** provides some information about the current technology, the percentage of memory used by the schematic diagram and its detailed contents. In the Misc. part, details on the time unit, voltage supply, typical delay and typical wire delay are provided, which configure the delay estimation and current estimation during logic simulation.

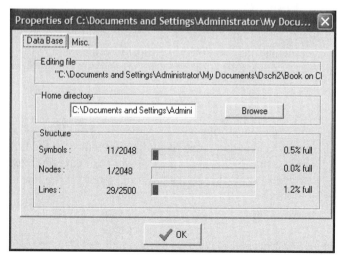

Fig. C.5 File properties, including statistics about the number of symbols, nodes and lines

C.2.25 Rotate

In order to apply a rotation to one part of the design, click on **Edit → Rotate**. Select one of the following proposed actions:

- Rotate right or 90°
- Rotate left or –90°

Then, delimit the area inside which the elements will be rotated.

C.2.26 Save, Save As

Click on **File → Save** to save the schematic diagram with its current name. The default name is "EXAMPLE.SCH". In the case of "Save As…", a new window appears, into which you are to enter the design name. Use the keyboard and type the desired file name. Press **Save**. Your design is now registered within the .**SCH** appendix.

C.2.27 Select Foundry

Click on **File → Select Foundry**. The list of available processes appears. The initial design rule file is "default.tec". Various technologies are available from 1.2 μm down to 90 nm. Click on the rule file name and the software reconfigures itself in order to adapt to the new process.

C.2.28 Simulation Options

The simulation parameters are: the simulation step (10 ps by default), the gate delay, wire delay, supply voltage, and elementary gate current. This parameters are loaded from .TEC files at initialization or with the command **File → Select Foundry**.

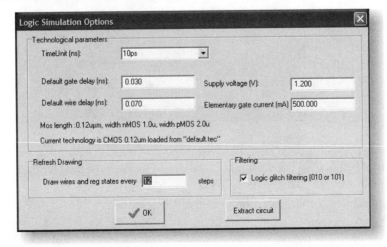

Fig. C.6 Logic simulation parameters

C.2.29 Start Simulation

The command **Start Simulation** launches the electrical net extraction and the logic simulation. The simulation speed may be controlled by the cursor "Fast-Slow". The simulation may be paused, run step-by-step and stopped. By default, the logic state of all interconnects is made visible. You may also see each pin state by a tic in front of "Show pin state".

Fig. C.7 The logic simulation control window

C.2.30 Schema to New Symbol

This command is very important to create user-defined symbols in order to build hierarchical designs. As an example, the full adder diagram based on primitives can be translated into a single symbol which includes the structure, input and outputs, as shown below.

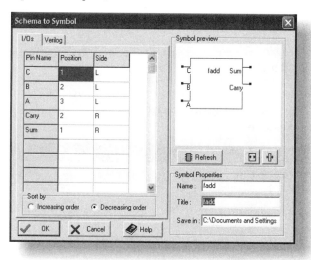

Fig. C.8 Logic simulation parameters

I/Os: The list of I/Os is based on active symbols (buttons, clocks, leds, keyboards, etc.). The position and side in the symbol may be changed in the table.

Verilog: The structural description based on primitives is described in verilog format and included to the symbol description.

Refresh: Update the layout of the user symbol.

Sizing: Act on the icons to change the shape of the user symbol.

Symbol Properties: These properties may be changed by the user.

C.2.31 Symbol Library

The symbol library contains basic logic and electrical symbols, sources, displays and switches. The aspect of the logic library is shown in Figure C.9. Most standard logic symbols (Inverter, Buffer, NAND, AND, NOR, OR, XOR) and D-latches are part of the "Basic" symbol menu. The analog components such as resistor, inductor, capacitor, and operational amplifiers are listed in the "Advanced" menu. Notice several input/output symbols, as well as a variety of switches for programmable arrays. Some more symbols may be found in the IEEE directory, accessible through the command **Insert → User Symbol**.

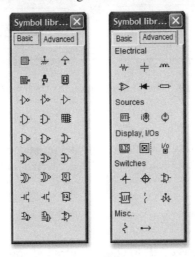

Fig. C.9 The symbol library

C.2.32 Text

Use this icon to fix a text to one box or location in the design. That text illustrates the layout and should be used as much as possible for each significant node such as inputs and outputs. To add some text to a particular place, proceed as follows:

- Click on the icon.
- Set the text location with the mouse. A dialog box appears.
- Enter the text in front of "Text:" and press "Ok". The text is set in the drawing.

A text can be modified as follows: click on the icon, click inside the existing text. The old text appears. Modify it and click on "Ok". Text is added for information only. It has no impact on simulation.

C.2.33 Timing Diagram

The timing diagram gives the time-domain aspect of all input and output nodes. An example of timing diagram is shown in Figure C.10. You may zoom on a specific time window, add the evaluation of the consumed current, and get the exact value of each input/output at a desired location.

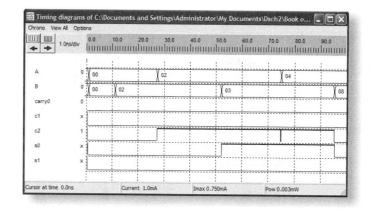

Fig. C.10   The timing diagrams of a logic simulation

C.2.34   Undo

The Undo command (**Edit → Undo**) is useful to not take into account the last editing command. It is possible to undo the commands Cut, Paste, Copy, Move, Stretch, and Edit.

C.2.35   Unselect All

(Escape Key)

Use the command (**View → Unselect All**) to cancel undesired commands, or to redraw the complete schematic diagram.

C.2.36   View All

Click **View → View All** to fit the screen with all the graphical elements currently on display.

C.2.37   View Same

Draw again the schematic diagram without changing the scale. This is used to refresh the screen.

C.2.38   Zoom In and Out

The above icons perform the Zoom In and Zoom Out functions. When zooming in, the area determined by the mouse will be enlarged to fit the display window. When zooming out, the area determined by the mouse will contain the display window.

- If you click once, a zoom is performed at the desired location.
- Press Ctrl + A for «View All», and Ctrl + O for zoom out.

Quick Reference Sheet

D.1 Microwind Menus

D.1.1 File Menu

Reset the program and
start with a clean screen

Read a layout
data file

Insert a layout in the
current layout

Translates the layout
into CIF, SPICE

Save the current layout
into the current filename

Switch to
monochrome/Colour mode

Print the layout

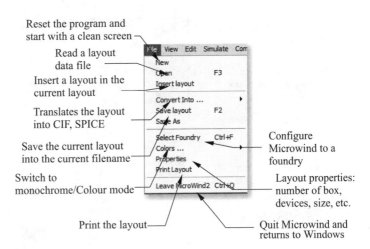

Configure
Microwind to a
foundry

Layout properties:
number of box,
devices, size, etc.

Quit Microwind and
returns to Windows

D.1.2 View Menu

Unselect all layers
and redraw the layout

Fit the window with
all the edited layout

Zoom In, Zoom out
the layout window

Give the list of nMOS
and pMOS devices

Show the palette of layers,
the layout macro and the
simulation properties

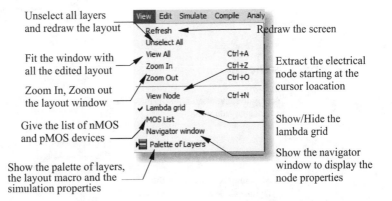

Redraw the screen

Extract the electrical
node starting at the
cursor loacation

Show/Hide the
lambda grid

Show the navigator
window to display the
node properties

D.1.3 Edit Menu

Cancel last editing command

Cut elements included in an area

Duplicate elements included in an area

Flip or rotate elements included in an area

Generate MOS, contacts, pads, diodes, resistors. capacitors, etc.

Connect layers at a desired location

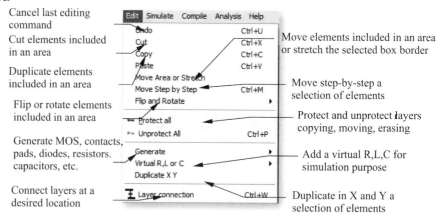

Move elements included in an area or stretch the selected box border

Move step-by-step a selection of elements

Protect and unprotect layers copying, moving, erasing

Add a virtual R,L,C for simulation purpose

Duplicate in X and Y a selection of elements

D.1.4 Simulate Menu

Run the simulation and choose the appropriate mode V(t), I(t), V/V, F(t), etc.

Simulate directly on the layout, with a palette of colors representing voltage

Include crosstalk effects in simulation

View the process steps of the layout fabrication in 3D

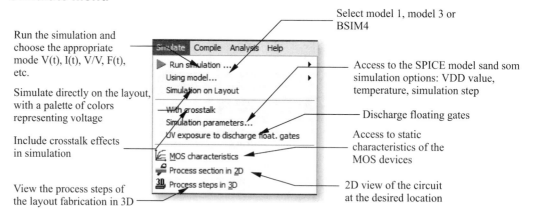

Select model 1, model 3 or BSIM4

Access to the SPICE model sand som simulation options: VDD value, temperature, simulation step

Discharge floating gates

Access to static characteristics of the MOS devices

2D view of the circuit at the desired location

D.1.5 Compile Menu

Compile one single line (on-line)

Compile a Verilog file generated by DSCH

D.1.6 Analysis Menu

Verify the layout and highlight the design rule violations

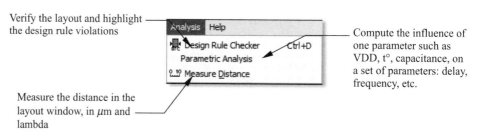

Compute the influence of one parameter such as VDD, t°, capacitance, on a set of parameters: delay, frequency, etc.

Measure the distance in the layout window, in μm and lambda

D.1.7 Palette

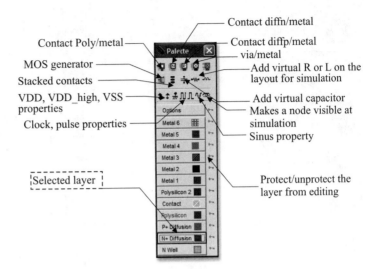

Contact diffn/metal

Contact Poly/metal

Contact diffp/metal
via/metal

MOS generator

Add virtual R or L on the layout for simulation

Stacked contacts

VDD, VDD_high, VSS properties

Add virtual capacitor

Makes a node visible at simulation

Clock, pulse properties

Sinus property

Selected layer

Protect/unprotect the layer from editing

D.1.8 Navigator Window

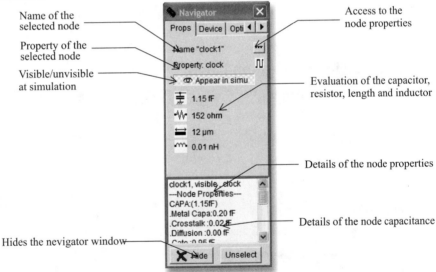

Name of the selected node

Access to the node properties

Property of the selected node

Visible/unvisible at simulation

Evaluation of the capacitor, resistor, length and inductor

Details of the node properties

Details of the node capacitance

Hides the nevigator window

D.1.9 List of Icons

	Open a layout file (MSK format).		Extract and simulate the circuit.
	Save the layout file in MSK format.		Measure the distance in lambda and micron between two points.
	Draw a box using the selected layer of the palette.		2D vertical aspect of the device.

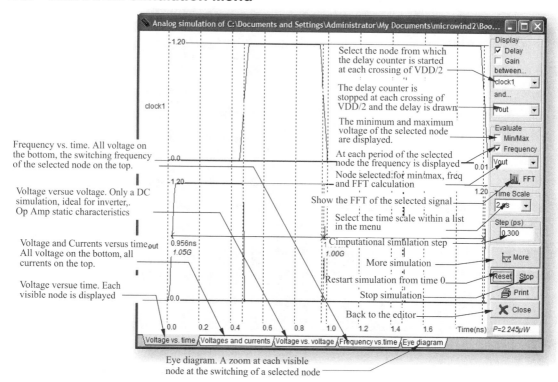	Delete boxes or text.		Step by step fabrication of the layout in 3D.
	Copy boxes or text.		Design rule checking of the circuit. Errors are notified in the layout.
	Stretch or move elements.		Add a text to the layout. The text may include simulation properties.
	Zoom In.		Connect the lower to the upper layers at the desired location using appropriate contacts.
	Zoom Out.		Static MOS characteristics.
	View all the drawing.		View the palette.
	Extract and view the electrical node pointed by the cursor.		Move the layout up, left, right, down.

D.2   Microwind Simulation Menu

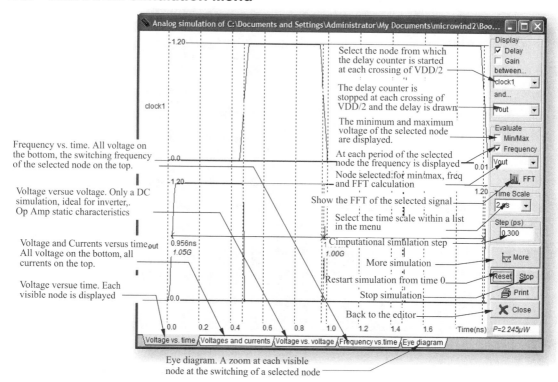

Frequency vs. time. All voltage on the bottom, the switching frequency of the selected node on the top.

Voltage versue voltage. Only a DC simulation, ideal for inverter,. Op Amp static characteristics

Voltage and Currents versus time. All voltage on the bottom, all currents on the top.

Voltage versue time. Each visible node is displayed

Select the node from which the delay counter is started at each crossing of VDD/2

The delay counter is stopped at each crossing of VDD/2 and the delay is drawn

The minimum and maximum voltage of the selected node are displayed.

At each period of the selected node the frequency is displayed

Node selected for min/max, freq and FFT calculation

Show the FFT of the selected signal

Select the time scale within a list in the menu

Cimputational simulation step

More simulation

Restart simulation from time 0

Stop simulation

Back to the editor

Eye diagram. A zoom at each visible node at the switching of a selected node

D.3   Dsch Menus

D.3.1   File Menu

Reset the program and
start with a clean screen

Read a
schematic file

Save the current schematic
diagram into the current
file name

Transform this diagram
into a user symbol

Switch to
monochrom/color mode

Print the schematic diagram

Configure Dsch2
to a given foundry

Design properties:
number of symbols,
nodes, etc.

Quit Dsch2 and
returns to Windows

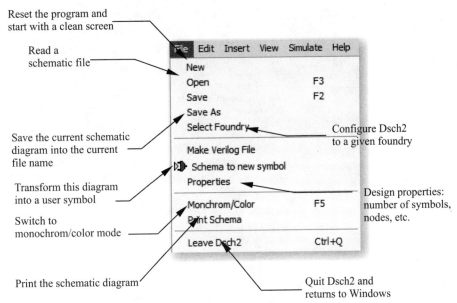

D.3.2   Edit Menu

Cancel last editing command

Cut elements included
in an area

Duplicate elements
included in an area

Flip or rotate elements
included in an area

Create a line

Add text in the
schematic diagram

Move elements included in an area

Add a connection between lines

Search a pin name in this diagram

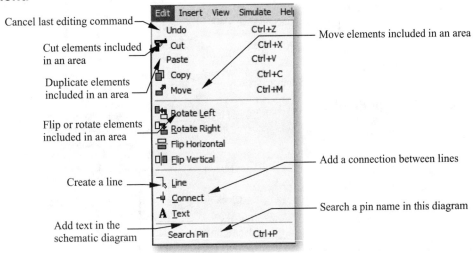

D.3.3   Insert Menu

Insert a user symbol or a
library symbol not accessible
from the symbol palette

Insert an other
schematic diagram

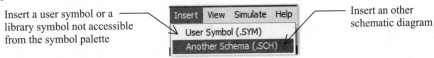

D.3.4   View Menu

Redraw all the schematic diagram

Zoom In, Zoom out the window

Extract the electrical nodes

Show the timing diagrams

Show the palette of symbols

Redraw the screen

Give the list of symbols

Describes the design structure

Show details about the critical path

Unselect all the design

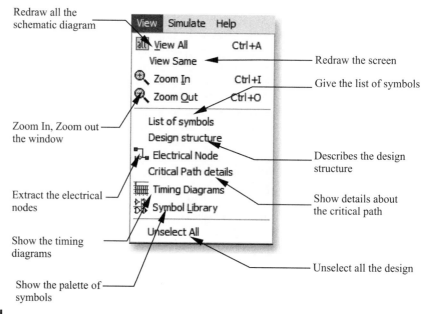

D.3.5   Simulate Menu

Extract the electrical circuit

Detect unconnected lines

Simulate options

Show the critical path (Longest switching path)

Start/stop logic simulation

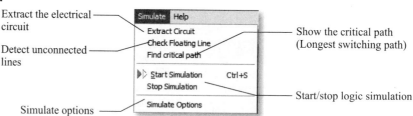

D.3.6   Symbol Palette

Basic logic symbol library

Button

Clock, led

Inv, Inv 3state, buffer

AND gates

NAND gates

NOR gates

XOR gates

nMOS and pMOS

Complex gates

Advanced logic symbol library

VSS, VDD supply

Hexadecimal display

Hexadecimal keyboard

OR gates

Full D latch

Latch

Multiplexor

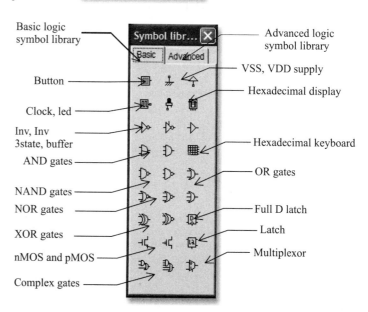

D.4 List of Files

FILE	DESCRIPTION
MICROWIND2.EXE	Microwind2 executable file
DSCH2.EXE	Dsch2 executable file
MICROWIND2.HTMLDSCH2.HTML	Help manuals for Microwind2 and Dsch2
*.RUL	TECHNOLOGY FILES. The MICROWIND2 program reads the rule file to update the simulator parameters, the design rules and parasitic capacitor values. A detailed description of the .RUL file is given at the end of Appendix A.
*.MSK	LAYOUT FILES. The Microwind2 software creates data files with the appendix .MSK. Those files are simple text files containing the list of boxes and layers, and the list of text declarations.
*.CIR	The command File -> Make SPICE File generates a SPICE compatible text description.
*.MES	MOS I/V Measurements
*.TXT	Verilog text inputs
*.TEC	TECHNOLOGY FILES. The Dsch2 program reads the rule file to update the simulator parameters. A detailed description of the .TEC file is given at the end of Appendix A.
*.SCH	Schematic diagram created by Dsch2
*.SYM	Symbols generated and used by Dsch2

D.5 List of Measurement Files

In the package, a selection of measurements are proposed, for comparison between real case measurements and models. Four chips have been fabricated and measured in our laboratory:

Chip «a» (0.35 µm ST)

Chip «b» (0.25 µm ST)

Chip «c» (0.18 µm ST)

Chip "d" (0.8 µm ATMEL-ES2)

Measurement files	Description
0.35 mm CMOS	
Na10 × 0, 4.mes	Nmos W=10 µm, L=0.4 µm (0.35 effective)
Na10 × 10.mes	Nmos W=10 µm, L= 10 µm

Na10 × 2.mes	Nmos W=10 μm, L= 2 μm
Na1 × 0,4. mes	Nmos W=1 μm, L= 0.4 μm
Na80 × 0, 4.mes	Nmos W=80 μm, L= 0.4 μm
Pa10 × 0, 4.mes	Pmos Pmos W=10 μm, L= 0.4 μm
Pa10 × 10.mes	Pmos W=10 μm, L= 10 μm
Pa10 × 2.mes	Pmos W=10 μm, L= 2 μm
Pa1 × 0,4.mes	Pmos W=1 μm, L= 0.4 μm
Pa80 × 0,4.mes	W=80 μm, L= 0.4 μm
0.25 μm CMOS	
Nb10 × 0,25.mes	Nmos W=10 μm, L=0.25 μm
Nb10 × 10.mes	Nmos W=10 μm, L=10 μm
0.18 μm CMOS	
Nc10 × 10.mes	Nmos W=10 μm, L=10 μm
NcHS4 × 0.2.mes	Nmos W=4 μm, L=0.2 μm high speed option
NcHV4 × 0.2.mes	Nmos W=4 μm, L=0.2 μm high voltage option
Nc4 × 0.2.mes	Nmos W=4 μm, L=0.2 μm normal (low leakage)
0.8 μm CMOS	
Nd20 × 20.mes	Nmos W=20 μm, L=20 μm
Nd20 × 0,8.mes	Nmos W=20 μm, L=0.8 μm

D.5.1   Measurement File Example

Measure v3.0 - 4 May 00

LL 10 × 10 um cmos018

nMOS 10.0 10.0

IDVd	5	0.0	2.0	0.5	

41 0

0.0	0.0	0.0	0.0	0.0	0.0
5.0000E-02	2.0226E-12	6.7235E-07	6.4727E-06	1.1584E-05	1.5635E-05
1.0000E-01	2.3586E-12	7.7469E-07	1.2143E-05	2.2459E-05	3.0636E-05
1.5000E-01	2.4540E-12	7.8616E-07	1.7012E-05	3.2623E-05	4.5003E-05
2.0000E-01	2.5151E-12	7.8892E-07	2.1089E-05	4.2079E-05	5.8736E-05
2.5000E-01	2.5716E-12	7.9037E-07	2.4386E-05	5.0826E-05	7.1835E-05

3.0000E-01	2.6275E-12	7.9158E-07	2.6929E-05	5.8867E-05	8.4301E-05
3.5000E-01	2.6834E-12	7.9273E-07	2.8768E-05	6.6202E-05	9.6133E-05
4.0000E-01	2.7393E-12	7.9387E-07	2.9992E-05	7.2834E-05	1.0733E-04

...

2.0000E+00	4.5477E-12	8.3166E-07	3.2021E-05	1.0562E-04	2.0746E-04
IdVg	4	0.0	−1.5	−0.5	0.05
21	0				
0.0	2.0226E-12	5.7123E-13	1.0630E-12	1.5644E-12	
1.0000E-01	3.4083E-11	7.6465E-13	1.0653E-12	1.5644E-12	
2.0000E-01	5.7669E-10	4.6877E-12	1.1181E-12	1.5655E-12	
3.0000E-01	8.9864E-09	8.3526E-11	2.2992E-12	1.5922E-12	

...

1.9000E+00	1.4895E-05	1.3132E-05	1.1692E-05	1.0455E-05	
2.0000E+00	1.5635E-05	1.3899E-05	1.2479E-05	1.1259E-05	

Glossary

TERMS	EXPLANATION
die	Piece of silicon that includes the active devices and the input/output interfaces. Usually 350 to 500µm thick, with an area from 2x2 to 25x25mm.
MOS	metal oxide semiconductor
substrate	The silicon on which active devices are implanted. Usually lightly doped with P impurities.
semiconductor	Depending on biasing conditions, the silicon conducting properties may vary from a conductor to a insulator.
intrinsic carriers	Charges included in the pure silicon crystal.
CMOS technology	Complementary Metal-Oxide-semiconductor, where complementary refers to the use of n-channel and p-channel MOS devices, Metal to a metal gate, oxide to the oxide between the gate and the channel, and semiconductor to the structure of the channel.
nMOS	n-channel MOS device
pMOS device	p-channel MOS device
wafer	The initial substrate used to fabricate integrated circuits. Usually around 500µm thick, its diameter ranges from 4 inches to 12 inches.
hot electrons	Electrons accelerated in the MOS channel may acquire sufficient "temperature" to provoke parasitic effects such as hole/electron generation following the impact on the silicon lattice.

LDD	Lateral drain diffusion. Light doping on the channel borders to prevent the hot electron effect.
parasitic leakage effect	Electrons may cross the channel from the source to the drain even with a zero bias on the gate, that creates a parasitic leakage between drain and source.
transmission gate	Enables or disables the link between two nodes, without any loss of voltage.
process	The technology steps required to complete the fabrication of the integrated circuit.
optical masks	Films used to pattern the layout of the IC on the silicon.
STI	Shallow trench isolation. Deep trenches of silicon insulator to isolate active devices.
passivation	Thick oxide at the surface of the integrated circuit to prevent external contamination and protect active devices.
short channel effect	Accounts for a set of parasitic effects observed for very narrow MOS devices.
DIBL	drain induced barrier lowering
ZTC	Zero temperature coefficient
latchup	Destructive short circuit effect created by NPNP stack of layers that creates a direct path from VDD to VSS, under wrong polarization conditions.
fanout	Number of gate inputs connected to an output node.
commutation point	Voltage value for a gate input at which the output state changes.
3-state	High impedance state. Corresponds to a floating node, meaning that no active device ties the node to any defined voltage value.
low K	Dielectric material with low permittivity. Mainly used between interconnects to reduce coupling effects.
AOI	AND/OR inverted logic
IP	intellectual property blocks
PWL	Piece-wise-linear signal. Used to describe complex signals based on tabulated current or voltage versus time.
salicide	Metal deposit at the surface of a doped silicon area to further decrease its resistance. MOS drain and source, as well as poly gate usually use salicide.
SPD	spectral power density

Index

Software Download Information

The lite versions of Microwind and Dsch can be downloaded from *www.microwind.net* and includes a subset of available commands. The complete version is available through ni2designs, India (*www.ni2designs.com*). Connect to the web page *www.microwind.net* for the latest information about how to download and install the lite versions of Microwind and Dsch.

Once installed, two directories are created, one for Microwind, one for Dsch. In each directory, a sub-directory called html contains help files. Other sub-directories include example files (*.MSK), design rules (*.RUL) and system files. the executable file is microwind3.exe.

In Dsch, other sub-directories include example files (*.SCH and *.SYM), design rules (*.TEC) and system files. The executable file is dsch3.exe.

Authors' Profile

ETIENNE SICARD was born in Paris, in June 1961. He received a B.S degree in 1984 and a Ph.D. in Electrical Engineering in 1987, both from the University of Toulouse. He was granted a scholarship by the Japanese Ministry of Education and stayed on for 18 months at the University of Osaka, Japan. Previously a professor of electronics in the department of physics at the University of Balearic Islands, Spain. Sicard is currently professor at the INSA Electronic Engineering School of Toulouse. His research interests include several aspects of integrated circuit design including noise tolerance and electromagnetic compatibility of integrated circuits. Sicard is the author of several educational softwares in the field of microelectronics and sound processing.

SONIA DELMAS BENDHIA was born in Toulouse, in April 1972. She received an engineering diploma in 1995, and a Ph.D. in Electronic Design from the National Institute of Applied Sciences, Toulouse, in 1998. Sonia Bendhia is currently a senior lecturer at the INSA Electronic Engineering School of Toulouse, Department of Electrical and Computer Engineering. Her research interests include signal integrity in deep sub-micron CMOS Ics, analog design and electromagnetic compatibility of systems. Sonia is the author of technical papers concerning signal integrity and electromagnetic compatibility.